T0309752

Data Mining Techniques in Grid Computing Environments

Data Mining Techniques in Grid Computing Environments

Editor

Werner Dubitzky

University of Ulster, UK

This edition first published 2008

Wiley-Blackwell is an imprint of John Wiley & Sons, formed by the merger of Wiley's global Scientific, Technical and Medical business with Blackwell Publishing.

Registered office: John Wiley & Sons Ltd, The Atrium, Southern Gate, Chichester, West Sussex, PO19 8SQ, UK

Other Editorial Offices:
9600 Garsington Road, Oxford, OX4 2DQ, UK
111 River Street, Hoboken, NJ 07030-5774, USA

For details of our global editorial offices, for customer services and for information about how to apply for permission to reuse the copyright material in this book please see our website at www.wiley.com/wiley-blackwell

Library of Congress Cataloguing-in-Publication Data

Dubitzky, Werner, 1958-
Data mining techniques in grid computing environments / Werner Dubitzky.
p. cm.
Includes bibliographical references and index.
ISBN 978-0-470-51258-6 (cloth)
1. Data mining. 2. Computational grids (Computer systems) I. Title.
QA76.9.D343D83 2008
004'.36–dc22
2008031720

ISBN: 978 0 470 51258 6

A catalogue record for this book is available from the British Library.

Set in 10/12 pt Times by Thomson Digital, Noida, India
Printed in Singapore by Markono Print Media Pte Ltd

First printing 2008

Contents

Preface

Modern organizations across many sectors rely increasingly on computerized information processes and infrastructures. This is particularly true for high-tech and knowledge sectors such as finance, communication, engineering, manufacturing, government, education, medicine, science and technology. As the underlying information systems evolve and become progressively more sophisticated, their users and managers are facing an exponentially growing volume of increasingly complex data, information, and knowledge. Exploring, analyzing and interpreting this information is a challenging task. Besides traditional statistics-based methods, *data mining* is quickly becoming a key technology in addressing the data analysis and interpretation tasks.

Data mining can be viewed as the formulation, analysis, and implementation of an induction process (proceeding from specific data to general patterns) that facilitates the nontrivial extraction of implicit, previously unknown, and potentially useful information from data. Data mining ranges from highly theoretical mathematical work in areas such as statistics, machine learning, knowledge representation and algorithms to systems solutions for problems like fraud detection, modeling of cancer and other complex diseases, network intrusion, information retrieval on the Web and monitoring of grid systems. Data mining techniques are increasingly employed in traditional scientific discovery disciplines, such as biological, medical, biomedical, chemical, physical and social sciences, and a variety of other knowledge industries, such as governments, education, high-tech engineering and process automation. Thus, data mining is playing a highly important role in structuring and shaping future knowledge-based industries and businesses. Effective and efficient management and use of stored data, and in particular the computer-assisted transformation of these data into information and knowledge, is considered a key factor for success.

While the need for sophisticated data mining solutions is growing quickly, it has been realized that conventional data and computer systems and infrastructures are often too limited to meet the requirements of modern data mining applications. Very large data volumes require significant processing power and data throughput. Dedicated and specialized hardware and software are usually tied to particular geographic locations or sites and therefore require the data and data mining tools and programs to be translocated in a flexible, seamless and efficient fashion. Commonly, people and organizations working simultaneously on a large-scale problem tend to reside at geographically dispersed sites, necessitating sophisticated distributed data mining tools and infrastructures. The requirements arising from such large-scale, distributed data mining scenarios are extremely demanding and it is unlikely that a single "killer solution" will emerge that satisfies them all. There is a way forward, though. Two recently emerging computer technologies promise to play a major role in the evolution of future, advanced data mining applications: *grid computing* and *Web services*.

Grid refers to persistent computing environments that enable software applications to integrate processors, storage, networks, instruments, applications and other resources that are managed by diverse organizations in dispersed locations. Web services are broadly regarded as self-contained, self-describing, modular applications that can be published, located, and invoked across the Internet. Recent developments are designed to bring about a convergence of grid and Web services technology (e.g. service-oriented architectures, WSRF). Grid computing and Web services and their future incarnations have a great potential for becoming a fundamental pillar of advanced data mining solutions in science and technology. This volume investigates data mining in the context of grid computing and, to some extent, Web services. In particular, this book presents a detailed account of what motivates the grid-enabling of data mining applications and what is required to develop and deploy such applications. By conveying the experience and lessons learned from the synergy of data mining and grid computing, we believe that similar future efforts could benefit in multiple ways, not least by being able to identify and avoid potential pitfalls and caveats involved in developing and deploying data mining solutions for the grid. We further hope that this volume will foster the understanding and use of grid-enabled data mining technology and that it will help standardization efforts in this field.

The approach taken in this book is conceptual and practical in nature. This means that the presented technologies and methods are described in a largely non-mathematical way, emphasizing data mining tasks, user and system requirements, information processing, IT and system architecture elements. In doing so, we avoid requiring the reader to possess detailed knowledge of advanced data mining theory and mathematics. Importantly, the merits and limitations of the presented technologies and methods are discussed on the basis of real-world case studies.

Our goal in developing this book is to address complex issues arising from grid-enabling data mining applications in different domains, by providing what is simultaneously a *design blueprint*, *user guide*, and *research agenda* for current and future developments in the field.

As *design blueprint*, the book is intended for the practicing professional (analyst, researcher, developer, senior executive) tasked with (a) the analysis and interpretation of large volumes of data requiring the sharing of resources, (b) the grid-enabling of existing data mining applications, and (c) the development and deployment of generic and novel enabling technology in the context of grid computing, Web services and data mining.

As a *user guide*, the book seeks to address the requirements of scientists and researchers to gain a basic understanding of existing concepts, methodologies and systems, combining data mining and modern distributed computing technology. To assist such users, the key concepts and assumptions of the various techniques, their conceptual and computational merits and limitations are explained, and guidelines for choosing the most appropriate technologies are provided.

As a *research agenda*, this volume is intended for students, educators, scientists and research managers seeking to understand the state of the art of data mining in grid computing environments and to identify the areas in which gaps in our knowledge demand further research and development. To this end, our aim is to maintain readability and accessibility throughout the chapters, rather than compiling a mere reference manual. Therefore, considerable effort is made to ensure that the presented material is supplemented by rich literature cross-references to more foundational work and ongoing developments.

Clearly, we cannot expect to do full justice to all three goals in a single book. However, we do believe that this book has the potential to go a long way in fostering the understanding, development and deployment of data mining solutions in grid computing and Web services environments. Thus, we hope this volume will contribute to increased communication and collaboration across various data mining and IT disciplines and will help facilitate a consistent approach to data mining in distributed computing environments in the future.

Acknowledgments

We thank the contributing authors for their contributions and for meeting the stringent deadlines and quality requirements. This work was supported by the European Commission FP6 grants No. 004475 (the DataMiningGrid[1] project), No. 033883 (the QosCosGrid[2] project), and No. 033437 (the Chemomentum[3] project).

Werner Dubitzky

Coleraine and Oberstaufen
March 2008

[1] www.DataMiningGrid.org
[2] www.QosCosGrid.com
[3] http://www.chemomentum.org/

List of Contributors

Andy Brass
University of Manchester
School of Computer Science
Manchester
United Kingdom
andy.brass@manchester.ac.uk

Peter Brezany
Institute of Scientific Computing
University of Vienna
Vienna
Austria
brezany@par.univie.ac.at

Eugenio Cesario
ICAR-CNR
Rende (CS)
Italy
cesario@icar.cnr.it

William K. Cheung
Department of Computer Science
Hong Kong Baptist University
Kowloon Tong
Hong Kong
william@comp.hkbu.edu.hk

Neil P. Chue Hong
EPCC
The University of Edinburgh
Edinburgh
United Kingdom
n.chuehong@omii.ac.uk

Antonio Congiusta
DEIS – University of Calabria and
DIIMA – University of Salerno
Rende (CS) and Salerno
Italy
acongiusta@deis.unical.it

Helen Conover
Information Technology & Systems
Center
The University of Alabama
Huntsville, USA
hconover@itsc.uah.edu

Vasa Curcin
Imperial College London
Department of Computing
London
United Kingdom
vc100@doc.ic.ac.uk

Werner Dubitzky
Biomedical Sciences Research Institute
University of Ulster
Coleraine
United Kingdom
w.dubitzky@ulster.ac.uk

Renato A. Ferreira
Universidade Federal de Minas Gerais
Department of Computer Science
Minas Gerais
Brazil
renato@dcc.ufmg.br

Paul Fisher
University of Manchester
School of Computer Science
Manchester
United Kingdom
pfisher@cs.manchester.ac.uk

Moustafa Ghanem
Imperial College London
Department of Computing
London
United Kingdom
mmg@doc.ic.ac.uk

Carole Goble
University of Manchester
School of Computer Science
Manchester
United Kingdom
carole.goble@manchester.ac.uk

Sara Graves
Information Technology & Systems Center
The University of Alabama
Huntsville, USA
sgraves@itsc.uah.edu

Pierre Gueant
Universidad Politécnica de Madrid
Facultad de Informática
Madrid
Spain
pgueant@fi.upm.es

Dorgival O. Guedes
Universidade Federal de Minas Gerais
Department of Computer Science
Minas Gerais
Brazil
dorgival@dcc.ufmg.br

Yike Guo
Imperial College London
Department of Computing
London
United Kingdom
yg@doc.ic.ac.uk

Pilar Herrero
Universidad Politécnica de Madrid
Facultad de Informática
Madrid
Spain
pherrero@fi.upm.es

Ivan Janciak
Institute of Scientific Computing
University of Vienna
Vienna
Austria
janciak@par.univie.ac.at

Ken Keiser
Information Technology & Systems Center
The University of Alabama
Huntsville, USA
kkeiser@itsc.uah.edu

Eric Kihn
22 National Geophysical Data Center
NOAA
Bolder, CO
USA
eric.a.kihn@noaa.gov

Arie Leizarowitz
Technion -- Israel Institute of Technology
Department of Mathematics
Haifa
Israel
la@techunix.technion.ac.il

Hong Lin
Information Technology & Systems Center
The University of Alabama
Huntsville, USA
alin@itsc.uah.edu

Vassily Lyutsarev
Microsoft Research Cambridge
Microsoft Research Ltd.
Cambridge
United Kingdom
vassilyl@microsoft.com

Manil Maskey
Information Technology & Systems Center
The University of Alabama
Huntsville, USA
mmaskey@itsc.uah.edu

Michael May
Fraunhofer IAIS
Sankt Augustin
Germany
michael.may@iais.fraunhofer.de

Dmitry Medvedev
Geophysical Center RAS
Moscow
Russia
dmedv@wdcb.ru

Wagner Meira Jr.
Universidade Federal de Minas Gerais
Department of Computer Science
Minas Gerais
Brazil
meira@dcc.ufmg.br

Pedro de Miguel
Universidad Politécnica de Madrid
Facultad de Informática
Madrid
Spain
pmiguel@fi.upm.es

Dmitry Mishin
Geophysical Center RAS
Moscow
Russia
dimm@wdcb.ru

Jesús Montes
Universidad Politécnica de Madrid
Facultad de Informática
Madrid
Spain
jmontes@fi.upm.es

Noam Palatin
Technion -- Israel Institute of Technology
Department of Mathematics
Haifa
Israel
noampalatin@gmail.com

José M. Peña
Universidad Politécnica de Madrid
Facultad de Informática
Madrid
Spain
jmpena@fi.upm.es

María S. Pérez
Universidad Politécnica de Madrid
Facultad de Informática
Madrid
Spain
mperez@fi.upm.es

Alexey Poyda
Moscow State University
Moscow
Russia
poyda@wdcb.ru

Rahul Ramachandran
Information Technology & Systems Center
The University of Alabama
Huntsville, USA
rramachandran@itsc.uah.edu

Omer F. Rana
School of Computer Science
Cardiff University
Cardiff
United Kingdom
o.f.rana@cs.cardiff.ac.uk

John Rushing
Information Technology & Systems Center
The University of Alabama
Huntsville, USA
jrushing@itsc.uah.edu

Alberto Sánchez
Universidad Politécnica de Madrid
Facultad de Informática
Madrid
Spain
ascampos@fi.upm.es

Assaf Schuster
Technion -- Israel Institute of Technology
Department of Computer Science
Haifa
Israel
assaf@cs.technion.ac.il

Ali Shaikh Ali
School of Computer Science
Cardiff University
Cardiff
United Kingdom
ali.shaikhali@cs.cardiff.ac.uk

Robert Stevens
University of Manchester
School of Computer Science
Manchester
United Kingdom
robert.stevens@manchester.ac.uk

Martin Swain
Biomedical Sciences Research Institute
University of Ulster
Coleraine
United Kingdom
mt.swain@ulster.ac.uk

Domenico Talia
DEIS – University of Calabria
Rende (CS)
Italy
talia@deis.unical.it

A. Min Tjoa
Institute of Software Technology & Interactive Systems
Vienna University of Technology
Vienna, Austria
tjoa@ifs.tuwien.ac.at

Paolo Trunfio
DEIS – University of Calabria
Rende (CS)
Italy
trunfio@deis.unical.it

Julio J. Valdés
Institute for Information Technology
National Research Council
Ottawa
Canada
Julio.Valdes@nrc-cnrc.gc.ca

Dennis Wegener
Fraunhofer IAIS
Sankt Augustin
Germany
dennis.wegener@iais.fraunhofer.de

Patrick Wendel
InforSense Ltd.
London
United Kingdom
patrick@inforsense.com

Ran Wolff
Technion -- Israel Institute of Technology
Department of Computer Science
Haifa
Israel
rwolff@mis.haifa.ac.il

Jun Zhao
University of Manchester
School of Computer Science
Manchester
United Kingdom
jun.zhao@zoo.ox.ac.uk

Mikhail Zhizhin
Geophysical Center RAS
Moscow
Russia
jjn@wdcb.ru

1

Data mining meets grid computing: Time to dance?

Alberto Sánchez, Jesús Montes, Werner Dubitzky, Julio J. Valdés, María S. Pérez and **Pedro de Miguel**

ABSTRACT

A *grand challenge problem* (Wah, 1993) refers to a computing problem that cannot be solved in a reasonable amount of time with conventional computers. While grand challenge problems can be found in many domains, science applications are typically at the forefront of these large-scale computing problems. Fundamental scientific problems currently being explored generate increasingly complex data, require more realistic simulations of the processes under study and demand greater and more intricate visualizations of the results. These problems often require numerous complex calculations and collaboration among people with multiple disciplines and geographic locations. Examples of scientific grand challenge problems include multi-scale environmental modelling and ecosystem simulations, biomedical imaging and biomechanics, nuclear power and weapons simulations, fluid dynamics and fundamental computational science (use of computation to attain scientific knowledge) (Butler, 1999; Gomes and Selman, 2005).

Many grand challenge problems involve the analysis of very large volumes of data. *Data mining* (also known as *knowledge discovery in databases*) (Frawley, Piatetsky-Shapiro and Matheus, 1992) is a well stablished field of computer science concerned with the automated search of large volumes of data for patterns that can be considered knowledge about the data. Data mining is often described as deriving knowledge from the input data. Applying data mining to grand challenge problems brings its own computational challenges. One way to address these computational challenges is *grid computing* (Kesselman and Foster, 1998). 'Grid' refers to persistent computing environments that enable software applications to integrate processors, storage, networks, instruments, applications and other resources that are managed by diverse organizations in widespread locations.

This chapter describes how both paradigms – data mining and grid computing – can benefit from each other: data mining techniques can be efficiently deployed in a grid environment and operational grids can be mined for patterns that may help to optimize the effectiveness and efficiency of the grid computing infrastructure. The chapter will also briefly outline the chapters of this volume.

Data Mining Techniques in Grid Computing Environments Edited by Werner Dubitzky

1.1 Introduction

Recent developments have seen an unprecedented growth of data and information in a wide range of knowledge sectors (Wright, 2007). The term *information explosion* describes the rapidly increasing amount of published information and its effects on society. It has been estimated that the amount of new information produced in the world increases by 30 per cent each year. The Population Reference Bureau[1] estimates that 800 MB of recorded information are produced per person each year (assuming a world population of 6.3 billion). Many organizations, companies and scientific centres produce and store large amounts of complex data and information. Examples include climate and astronomy data, economic and financial transactions and data from many scientific disciplines. To justify their existence and maximize their use, these data need to be stored and analysed. The larger and the more complex these data, the more time consuming and costly is their storage and analysis.

Data mining has been developed to address the information needs in modern knowledge sectors. *Data mining refers to the non-trivial process of identifying valid, novel, potentially useful and understandable patterns in large volumes of data* (Fayyad, Piatetsky-Shapiro and Smyth, 1996; Frawley, Piatetsky-Shapiro and Matheus, 1992). Because of the information explosion phenomenon, data mining has become one of the most important areas of research and development in computer science.

Data mining is a complex process. The main dimensions of complexity include the following (Stankovski *et al.*, 2004).

- *Data mining tasks.* There are many non-trivial tasks involved in the data mining process: these include data pre-processing, rule or model induction, model validation and result presentation.
- *Data volume.* Many modern data mining applications are faced with growing volumes (in bytes) of data to be analysed. Some of the larger data sets comprise millions of entries and require gigabytes or terabytes of storage.
- *Data complexity.* This dimension has two aspects. First, the phenomena analysed in complex application scenarios are captured by increasingly *complex data structures and types*, including natural language text, images, time series, multi-relational and object data types. Second, data are increasingly located in *geographically distributed* data placements and cannot be gathered centrally for technological (e.g. large data volumes), data privacy (Clifton *et al.*, 2002), security, legal or other reasons.

To address the issues outlined above, the data mining process is in need of reformulation. This leads to the concept of *distributed data mining*, and in particular to grid-based data mining or – in analogy to a data grid or a computational grid – to the concept of a *data mining grid*. A data mining grid seeks a trade-off between data centralization and distributed processing of data so as to maximize effectiveness and efficiency of the entire process (Kargupta, Kamath and Chan, 2000; Talia, 2006). A data mining grid should provide a means to exploit available hardware resources (primary/secondary memory, processors) in order to handle the data volumes and processing requirements of modern data mining applications. Furthermore, it should support data placement, scheduling and resource management (Sánchez *et al.*, 2004).

[1] http://www.prb.org

Grid computing has emerged from distributed computing and parallel processing technologies. The so-called *grid* is a distributed computing infrastructure facilitating the coordinated sharing of computing resources within organizations and across geographically dispersed sites. The main advantages of sharing resources using a grid include (a) pooling of heterogeneous computing resources across administrative domains and dispersed locations, (b) ability to run large-scale applications that outstrip the capacity of local resources, (c) improved utilization of resources and (d) collaborative applications (Kesselman and Foster, 1998).

Essentially, a grid-enabled data mining environment consists of a decentralized high-performance computing platform where data mining tasks and algorithms can be applied on distributed data. Grid-based data mining would allow (a) the distribution of compute-intensive data analysis among a large number of geographically scattered resources, (b) the development of algorithms and new techniques such that the data would be processed where they are stored, thus avoiding transmission and data ownership/privacy issues, and (c) the investigation and potential solution of data mining problems beyond the scope of current techniques (Stankovski *et al.*, 2008).

While grid technology has the potential to address some of the issues of modern data mining applications, the complexity of the grid computing environments themselves gives rise to various issues that need to be tackled. Amongst other things, the heterogeneous and geographically distributed nature of grid resources and the involvement of multiple administrative domains with their local policies make coordinated resource sharing difficult. Ironically, (distributed) data mining technology could offer possible solutions to some of the problems encountered in complex grid computing environments. The basic idea is that the operational data that is generated in grid computing environments (e.g. log files) could be mined to help improve the overall performance and reliability of the grid, e.g. by identifying misconfigured machines.

Hence, there is the potential that both paradigms – grid computing and data mining – could look forward to a future of fruitful, mutually beneficial cooperation.

1.2 Data mining

As already mentioned, data mining refers to the process of extracting useful, non-trivial knowledge from data (Witten and Frank, 2000). The extracted knowledge is typically used in business applications, for example fraud detection in financial businesses or analysis of purchasing behaviour in retail scenarios. In recent years data mining has found its way into many scientific and engineering disciplines (Grossman *et al.*, 2001). As a result the complexity of data mining applications has grown extensively (see Subsection 1.2.1). To address the arising computational requirements distributed and grid computing has been investigated and the notion of a data mining grid has emerged. While the marriage of grid computing and data mining has seen many success stories, many challenges still remain – see Subsection 1.2.2. In the following subsections we briefly hint at some of the current data mining issues and scenarios.

1.2.1 Complex data mining problems

The complexity of modern data mining problems is challenging researchers and developers. The sheer scale of these problems requires new computing architectures, as conventional systems can no longer cope. Typical large-scale data mining applications are found in areas such as molecular biology, molecular design, process optimization, weather forecast, climate change prediction, astronomy, fluid dynamics, physics, earth science and so on. For instance, in high-energy physics, CERN's (European Organization for Nuclear Research) Large Hadron Collider[2] is expected to produce data in the range of 15 petabytes/s generated from the smashing of subatomic particles. These data need to be analysed in four experiments with the aim of discovering new fundamental particles, specifically the Higgs boson or God particle[3].

Another current complex data mining application is found in weather modelling. Here, the task is to discover a model that accurately describes the weather behaviour according to several parameters. The climate*prediction*.net[4] project is the largest experiment to produce forecasts of the climate in the 21st century. The project aims to understand how sensitive weather models are to both small changes and to factors such as carbon dioxide and the sulphur cycle. To discover the relevant information, the model needs to be executed thousands of times.

Sometimes the challenge is not the availability of sheer compute power or massive memory, but the intrinsic geographic distribution of the data. The mining of medical databases is such an application scenario. The challenge in these applications is to mine data located in distributed, heterogeneous databases while adhering to varying security and privacy constraints imposed on the local data sources (Stankovski *et al.*, 2007).

Other examples of complex data mining challenges include large-scale data mining problems in the life sciences, including disease modelling, pathway and gene expression analysis, literature mining, biodiversity analysis and so on (Hirschman *et al.*, 2002; Dubitzky, Granzow and Berrar, 2006; Edwards, Lane and Nielsen, 2000).

1.2.2 Data mining challenges

If data mining tasks, applications and algorithms are to be distributed, some data mining challenges are derived from current distributed processing problems. Nevertheless, data mining has certain special characteristics – such as input data format, pressing steps and tasks – which should be taken into account.

In recent years, two lines of research and development have featured prominently the evolution of data mining in distributed computing environments.

- *Development of parallel or high-performance algorithms, theoretical models and data mining techniques.* Distributed data mining algorithms must support the complete data mining process (pre-processing, data mining and post-processing) in a similar way as their centralized versions do. This means that all data mining tasks, including data cleaning, attribute discretization, concept generalization and so on, should be performed in a parallel way. Several distributed algorithms have been developed according to their centralized versions. For instance, some parallel algorithms have been developed for association rules (Agrawal and Shafer, 1996; Ashrafi, Taniar and Smith, 2004), classification rules (Zaki, Ho and Agrawal, 1999; Cho and Wüthrich, 2002) or clustering algorithms (Kargupta *et al.*, 2001; Rajasekaran, 2005).

[2]http://lhc.web.cern.ch/lhc/

[3]The Higgs boson is the key particle to understanding why matter has mass.

[4]www.climateprediction.net/

- *Design of new data mining systems and architectures to deal with the efficient use of computing resources.* Although some effort has been made towards the development of efficient distributed data mining algorithms, the environmental aspects, such as task scheduling and resource management, are critical aspects to the success of distributed data mining. Therefore, the deployment of data mining applications within high-performance and distributed computing infrastructures becomes a challenge for future developments in the field of data mining. This volume is intended to cover this dimension.

Furthermore, current data mining problems require more development in several areas including data placement, data discovery and storage, resource management and so on, because of the following.

- The high complexity (data size and structure, cooperation) of many data mining applications requires the use of data from multiple databases and may involve multiple organizations and geographically distributed locations. Typically, these data cannot be integrated into a single, centralized database data warehouse due to technical, privacy, legal and other constraints.
- The different institutions have maintained their own (local) data sources using their preferred data models and technical infrastructures.
- The possible geographically dispersed data distribution implies fault tolerance and other issues. Also, data and data model (metadata) updates may introduce replication and data integrity and consistency problems.
- The huge volume of analysed data and the existing difference between computing and I/O access times require new alternatives to avoid the I/O system becoming a bottleneck in the data mining processes.

Common data mining deployment infrastructures, such as clusters, do not normally meet these requirements. Hence, there is a need to develop new infrastructures and architectures that could address these requirements. Such systems should provide (see, for example, Stankovski *et al.*, 2008) the following.

- *Access control, security policies and agreements between institutions to access data.* This ensures seamless data access and sharing among different organizations and thus will support the interoperation needed to solve complex data mining problems effectively and efficiently.
- *Data filtering, data replication and use of local data sets.* These features enhance the efficiency of the deployment of data mining applications. Data distribution and replication need to be handled in a coherent fashion to ensure data consistency and integrity.
- *Data publication, index and update mechanisms.* These characteristics are extremely important to ensure the effective and efficient location of relevant data in large-scale distributed environments required to store the large number of data to be analysed.
- *Data mining planning and scheduling based on the existing storage resources.* This is needed to ensure effective and efficient use of the computing resources within a distributed computing environment.

In addition to the brief outline described above, we highlight some of the key contemporary data mining challenges as identified by (Yang and Wu, 2006). We highlight those that we feel are of particular relevance to ongoing research and development that seeks to combine grid computing and data mining.

(a) Mining complex knowledge from *complex data.*

(b) *Distributed data mining* and mining multi-agent data.

(c) *Scaling up for high-dimensional* and high-speed *data* streams.

(d) *Mining in a network setting.*

(e) *Security, privacy and data integrity.*

(f) Data mining for biological and environmental problems.

(g) Dealing with non-static, unbalanced and cost-sensitive data.

(h) Developing a unifying theory of data mining.

(i) Mining sequence data and time series data.

(j) Problems related to the data mining process.

1.3 Grid computing

Scientific, engineering and other applications and especially grand challenge applications are becoming ever more demanding in terms of their computing requirements. Increasingly, the requirements can no longer be met by single organizations. A cost-effective modern technology that could address the computing bottleneck is grid technology (Kesselman and Foster, 1998).

In the past 25 years the idea of sharing computing resources to obtain the maximum benefit/cost ratio has changed the way we think about computing problems. Expensive and difficult-to-scale supercomputers are being complemented and sometimes replaced by affordable distributed computing solutions.

Cluster computing was the first alternative to multiprocessors, aimed at obtaining a better cost/performance ratio. A cluster can be defined as a set of dedicated and independent machines, connected by means of an internal network, and managed by a system that takes advantage of the existence of several computational elements. A cluster is expected to provide high performance, high availability, load balancing and scalability.

Although cluster computing is an affordable way to solve complex problems, it does not allow us to connect different administration domains. Furthermore, it is not based on open standards, which makes applications less portable. Finally, current grand challenge applications have reached a level of complexity that even cluster environments may not be able to address adequately.

A second alternative to address grand challenge applications is called Internet computing. Its objective is to take advantage of not only internal computational resources (such as the nodes of a cluster) but also those general purpose systems interconnected by a *wide area network* (*WAN*). A WAN is a computer network that covers a broad area, i.e. any network whose communications links cross-metropolitan, regional or national boundaries. The largest and best-known example of a WAN is the Internet. This allows calculations and data analysis to be performed in a highly distributed way by linking geographically widely dispersed resources. In most cases this technology is developed using free computational resources from people or institutions that voluntarily join the system to help scientific research.

A popular example of an Internet-enabled distributed computing solutions is the SETI@home[5] project (University of California, 2007). The goal of the project is the search for extra-terrestrial intelligence through the analysis of radio signals from outer space. It uses Internet-connected computers and a freely available software that that analyses narrow-bandwidth signals from radio telescope data. To participate in this large-scale computing exercise, users download the software and install it on their local systems. Chunks of data are sent to the local computer for processing and the results are sent to a distributor node. The program uses part of the computer's CPU power, disk space and network bandwidth and the user can control how much of the computer resources are used by SETI@Home, and when they can be used.

Similar '@home' projects have been organized in other disciplines under the *Berkeley Open Infrastructure for Network Computing* (*BOINC*)[6] initiative. In biology and medicine, Rosetta@home[7] uses Internet computing to find causes of major human diseases such as the *acquired immunodeficiency syndrome* (*AIDS*), malaria, cancer or Alzheimer's. Malariacontrol.net[8] is another project that adopts an Internet computing approach. Its objective is to develop simulation models of the transmission dynamics (epidemiology) and health effects of malaria.

In spite of its usefulness, Internet computing presents some disadvantages, mainly because most resources are made available by a community of voluntary users. This limits the use of the resources and the reliability of the infrastructure to solve problems in which security is a key factor. Even so, by harnessing immense computing power these projects have made a first step towards distributed computing architectures capable of addressing complex data mining problems in diverse application areas.

One of the most recent incarnations of large-scale distributed computing technologies is *grid computing* (Kesselman and Foster, 1998). The aim of grid computing is to provide an affordable approach to large-scale computing problems. The term grid can be defined as a set of computational resources interconnected through a WAN, aimed at performing highly demanding computational tasks such as grand challenge applications. A grid makes it possible to securely and reliably take advantage of widely dispersed computational resources across several organizations and administrative domains. An administrative domain is a collection of hosts and routers, and the interconnecting network(s), managed by a single administrative authority, i.e. a company, institute or other organization. The geographically dispersed resources that are aggregated within a grid could be viewed as a virtual supercomputer. Therefore, it has no centralized control, as each system still belongs to and is controlled by its original resource provider. The grid automates access to computational resources, assuring security restrictions and reliability.

Ian Foster defines the main characteristics of a grid as follows (Foster, 2002).

- *Decentralized control.* Within a grid, the control of resources is decentralized, enabling different administration policies and local management systems.

[5]http://setiathome.berkeley.edu/

[6]http://boinc.berkeley.edu/

[7]http://boinc.bakerlab.org/rosetta/

[8]http://www.malariacontrol.net/

- *Open technology.* A grid should use of open protocols and standards.
- *High quality of service.* A grid provides high quality of service in terms of performance, availability and security.

Grid solutions are specifically designed to be adaptable and scalable and may involve a large number of machines. Unlike many cluster and Internet computing solutions, a grid should be able to cope with unexpected failures or loss of resources. Commonly used systems (such as clusters) can only grow up to a certain point without significant performance losses. Because of the expandable set of systems that can be attached and adapted, grids can provide theoretical unlimited computational power.

Other advantages of a grid infrastructure can be summarized as follows.

- Overcoming of bottlenecks faced by many large-scale applications.
- Decentralized administration that allows independent administrative domains (such as corporative networks) to join and contribute to the system without losing administrative control.
- Integration of heterogeneous resources and systems. This is achieved through the use of open protocols and standard interconnections and collaboration between diverse computational resources.
- A grid system is able to adapt to unexpected failures or loss of resources.
- A grid environment never becomes obsolete as it may easily assimilate new resources (perhaps as a replacement for older resources) and be adapted to provide new features.
- Provide an attractive cost/performance ratio making high-performance computing affordable.

Current grids are designed so as to serve a certain purpose or community. Typical grid configurations (or types of grid) include the following.

- *Computing (or computational) grid.* This type of grid is designed to provide as much computing power as possible. This kind of environment usually provides services for submitting, monitoring and managing jobs and related tools. Typically, in a computational grid most machines are high-performance servers. Sometimes two types of computational grid are distinguished: distributed computing grids and high-throughput grids (Krauter, Buyya and Maheswaran, 2002).
- *Data grid.* A data grid stores and provides reliable access to data across multiple organizations. It manages the physical data storage, data access policies and security issues of the stored data. The physical location of the data is normally transparent to the user.
- *Service grid.* A service grid (Krauter, Buyya and Maheswaran, 2002) provides services that are not covered by a single machine. It connects users and applications into collaborative workgroups and enables real-time interaction between users and applications via a virtual workspace. Service grids include on-demand, collaborative and multimedia grid systems.

While grid technology has been used in productive setting for a while, current research still needs to address various issues. Some of these are discussed below.

1.3.1 Grid computing challenges

Although grid computing allows the creation of comprehensive computing environments capable of addressing the requirements of grand challenge applications, such environments can often be very complex. Complexity arises from the heterogeneity of the underlying software and hardware resources, decentralized control, mechanisms to deal with faults and resource losses, grid middleware such as resource broker, security and privacy mechanisms, local policies and usage patterns of the resources and so on. These complexities need to be addressed in order to fully exploit the grid's features for large-scale (data mining) applications.

One way of supporting the management of a grid is to monitor and analyse all information of an operating grid. This may involve information on system performance and operation metrics such as throughput, network bandwidth and response times, but also other aspects such as service availability or the quality of job–resource assignment[9]. Because of their complexity, distributed creation and real-time aspects, analyzing and interpreting the 'signals' generated within an operating grid environment can become a very complex data analytical task. Data mining technology is turning out to be the methodology of choice to address this task, i.e. to *mine grid data*.

1.4 Data mining grid – mining grid data

From the overview on data mining and grid technology, we see two interesting developments, the concept of a *data mining grid* and *mining grid data*. A data mining grid could be viewed as a grid that is specifically designed to facilitate demanding data mining applications. In addition, grid computing environments may motivate a new form of data mining, *mining grid data*, which is geared towards supporting the efficient operation of a grid by facilitating the analysis of data generated as a by-product of running a grid. These two aspects are now briefly discussed.

1.4.1 Data mining grid: a grid facilitating large-scale data mining

A *data mining application* is defined as the use of data mining technology to perform data analysis tasks within a particular application domain. Basic elements of a data mining application are the *data* to be mined, the data mining *algorithm*(s) and methods used to mine the data, and a *user* who specifies and controls the data mining process. A data mining process may consist of several data mining algorithms, each addressing a particular data mining task, such as feature selection, clustering or visualization. A given data mining algorithm may have different software implementations. Likewise, the data to be mined may be available in different implementations, for instance as a database in a database management system, a file in a particular file format or a data stream.

A data mining grid is a system whose main function is to facilitate the sharing and use of *data*, *data mining programs* (implemented algorithms), *processing units* and *storage devices*

[9] A grid job could be anything that needs a grid resource, e.g. a request for bandwidth or disk space, an application or a set of application programs.

in order to improve existing, and enable novel, data mining applications (see Subsection 1.2.2). Such a system should take into account the unique constraints and requirements of data mining applications with respect to the data management and data mining software tools, and the users of these tools (Stankovski *et al.*, 2008). These high-level goals lead to a natural breakdown of some basic requirements for a data mining grid. We distinguish *user*, *application* and *system requirements*. The user requirements are dictated by the need of end users to define and execute data mining tasks, and by developers and administrators who need to evolve and maintain the system. Application program and system requirements are driven by technical factors such as resource type and location, software and hardware architectures, system interfaces, standards and so on. Below we briefly summarize what these requirements may be.

Ultimately, a data mining grid system facilitating advanced data mining applications is operated by a human user – an end user wanting to solve a particular data mining task or a system developer or administrator tasked with maintaining or further developing the data mining grid. Some of the main requirements such users may have include the following.

- *Effectiveness and efficiency.* A data mining grid should facilitate more effective (solution quality) and/or more efficient (higher throughput, which relates to speed-up) solutions than conventional environments.
- *Novel use/application.* A data mining grid should facilitate novel data mining applications currently not possible with conventional environments.
- *Scalability.* A data mining grid should facilitate the seamless adding of grid resources to accommodate increasing numbers of users and growing application demands without performance loss.
- *Scope.* A data mining grid should support data mining applications from different application domains and should allow the execution of all kinds of data mining task (pre-processing, analysis, post-processing, visualization etc.)
- *Ease of use.* A data mining grid should hide grid details from users who do not want to concern themselves with such details, but be flexible enough to facilitate deep, grid-level control to those users wish to operate on this level. Furthermore, mechanisms should be provided by a data mining grid that allow users to search for grid-wide located data mining applications and data sources. Finally, a data mining grid should provide tools that help users to define complex data mining processes.
- *Monitoring and steering.* A data mining grid should provide tools that allow users to monitor and steer (e.g. abort, provide new input, change parameters) data mining applications running on the grid.
- *Extensibility, maintenance and integration.* Developers should be able to port existing data mining applications to the data mining with little or no modification to the original data mining application program. System developers should be able to extend the features of the core data mining grid system without major modifications to the main system components. It should be easy to integrate new data mining applications and core system components with other technology (networks, Web services, grid components, user interfaces etc).

To meet the user requirements presented above, a data mining grid should meet additional technical requirements relating to *data mining application* software (data, programs) and the underlying data mining *grid system* components. Some basic requirements of this kind are as follows.

- *Resource sharing and interoperation.* A data mining grid should facilitate the seamless interoperation and sharing of important data mining resources and components, in particular, data mining application programs (implemented algorithms), data (different standard data file formats, database managements systems, other data-centric systems and tools), storage devices and processing units.
- *Data mining applications.* A data mining grid should accommodate a wide range of data mining application programs (algorithms) and should provide mechanisms that take into account the requirements, constraints and user-defined settings associated with these applications.
- *Resource management.* A data mining grid system should facilitate resource management to match available grid resources to job requests (resource broker), schedule the execution of the jobs on matched resources (scheduler) and manage and monitor the execution of jobs (job execution and monitoring). In particular, a data mining grid resource manager should facilitate data-oriented scheduling and parameter sweep applications, and take into account the type of data mining task, technique and method or algorithm (implementation) in its management policies

1.4.2 Mining grid data: analysing grid systems with data mining techniques

Grid technology provides high availability of resources and services, making it possible to deal with new and more complex problems. But it is also known that a grid is a very heterogeneous and decentralized environment. It presents different kinds of security policy, data and computing characteristic, system administration procedure and so on. Given these complexities, the management of a grid, any grid not just a data mining grid, becomes a very important aspect in running and maintaining grid systems. Grid management is the key to providing high reliability and quality of service. The complexities of grid computing environments make it almost impossible to have a complete understanding of the entire grid. Therefore, a new approach is needed. Such an approach should pool, analyse and interpret all relevant information that could be obtained from a grid. The insights provided should then be used to support resource management and system administration. Data mining has proved to be a remarkably powerful tool, facilitating the analysis and interpretation of large volumes of complex data. Hence, given the complexities involved in operating and maintaining grid environments efficiently and the ability of data mining to analyse and interpret large volumes of data, it is evident that 'mining grid data' could be a solution to improving the performance, operation and maintenance of grid computing environments.

Nowadays, most management techniques consider the grid as a set of independent, complex systems, building together a huge pool of computational resources. Therefore, the administration procedures are subjected to a specific analysis of each computer system, organizational units, etc. Finally, the decision making is based on a detailed knowledge of each of the elements that make up a grid. However, if we consider how more commonly used systems (such as regular desktop computers or small clusters) are managed, it is easy to realize that resource administration is very often based on more general parameters such as CPU or memory usage, not directly related to the specific architectural characteristics, although it is affected by them. This can be considered as an abstraction method that allows administrators to generalize and apply their knowledge to different systems. This abstraction is possible thanks to a set of underlying procedures, present in almost every modern computer.

Nevertheless, in complex systems such as a grid, this level of abstraction is not enough. The heterogeneous and distributed nature of grids implies a new kind of architectural complexity. Data mining techniques can contribute to observe and analyse the environment as a single system, offering a new abstraction layer that reduces grid observation to a set of representative generic parameters. This approach represents a new perspective for management, allowing consideration of aspects regarding the whole system activity, instead of each subsystem's behaviour.

The complexity of this formulation makes it hard to face grid understanding directly as a single problem. It is desirable to focus on a limited set of aspects, trying to analyse and improve them first. This can provide insight on how to deal with the abstraction of grid complexity, which can be extended to more complete scenarios. The great variety of elements that can be found in the grid offers a wide range of information to process. Data from multiple sources can be gathered and analysed using data mining techniques to learn new useful information about different grid features. The nature of the information obtained determines what kind of knowledge is going to be obtained.

Standard monitoring parameters such as CPU or memory usage of the different grid resources can provide insight on a grid's computational behaviour. A better knowledge of the grid variability makes it possible to improve the environment performance and reliability. A deep internal analysis of the grid can reveal weak points and other architectural issues.

From a different point of view, user behaviour can be analysed, focusing on access patterns, service request, the nature of these requests etc. This would make it possible to refine the environment features and capabilities, trying to effectively fit user needs and requirements.

The grid's dynamic evolution can also be analysed. Understanding the grid's present and past behaviour allows us to establish procedures to predict its evolution. This would help the grid management system to anticipate future situations and optimize its operation.

1.5 Conclusions

With the advance of computer and information technology, increasingly complex and resource-demanding applications have become possible. As a result, even larger-scale problems are envisaged and in many areas so-called *grand challenge problems* (Wah, 1993) are being tackled. These problems put an even greater demand on the underlying computing resources. A growing class of applications that need large-scale resources is modern data mining applications in science, engineering and other areas (Grossman *et al.*, 2001). Grid technology (Kesselman and Foster, 1998) is an answer to the increasing demand for affordable large-scale computing resources.

The emergence of grid technology and the increasingly complex nature of data mining applications have led to a new synergy of *data mining* and *grid*. On one hand, the concept of a *data mining grid* is in the process of becoming a reality. A data mining grid facilitates novel data mining applications and provides a comprehensive solution for affordable high-performance resources satisfying the needs of large-scale data mining problems. On the other hand, *mining grid data* is emerging as a new class of data mining application. Mining grid data could be understood as a methodology that could help to address the complex issues involved in running and maintaining large grid computing environments. The dichotomy of these

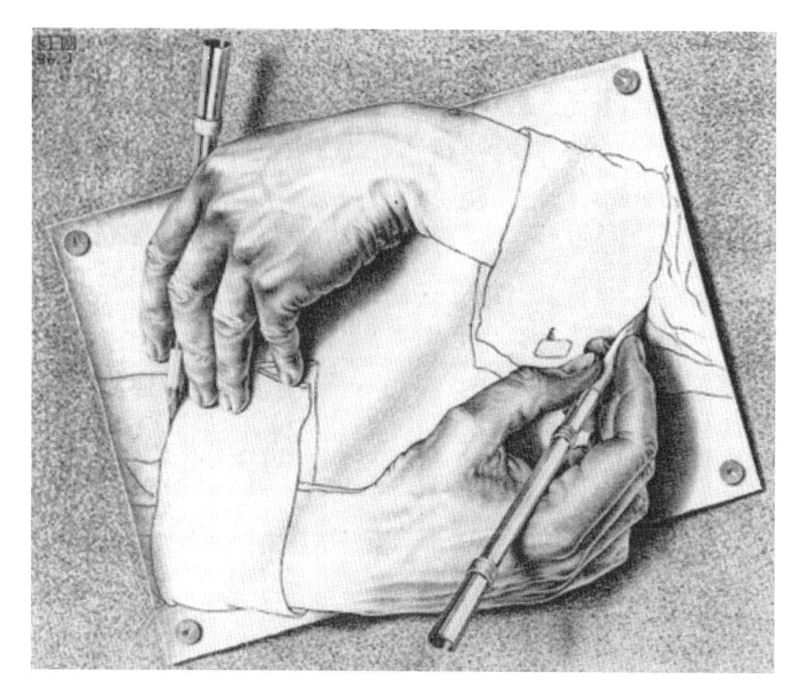

Figure 1.1 Analogy symbolizing the new synergy between a data mining grid and the mining of grid data. *'M. C. Escher : The Graphic Work'* (with permission from Benedikt-Taschen Publishers)

concepts – a data mining grid and mining grid data – is the subject of this volume and is beautifully illustrated in Figure 1.1. The two paradigms should go hand in hand and benefit from each other – a data mining grid can efficiently deploy large-scale data mining applications and data mining techniques can be used to understand and reduce the complexity of grid computing environments.

However, both areas are relatively new and demand further research and development. This volume is intended to be a contribution to this quest. What seems clear, though, is that the two areas are looking forward to a great future. The time has come to face the music and dance!

1.6 Summary of chapters in this volume

Chapter 1 is entitled 'Data mining meets grid computing: time to dance?'. The title indicates that there is a great synergy afoot, a synergy between data mining and grid technology. The chapter describes how the two paradigms – data mining and grid computing – can benefit from each other: data mining techniques can be efficiently deployed in a grid environment and operational grids can be mined for patterns that may help to optimize the effectiveness and efficiency of the grid computing infrastructure.

Chapter 2 is entitled 'Data analysis services in the knowledge grid'. It describes a grid-based architecture supporting distributed knowledge discovery called Knowledge Grid. It discusses how the Knowledge Grid framework has been developed as a collection of grid services and how it can be used to develop distributed data analysis tasks and knowledge discovery processes exploiting the service-oriented architecture model.

Chapter 3 is entitled 'GridMiner: an advanced support for e-science analytics'. It describes the architecture of the GridMiner system, which is based on the Cross Industry Standard Process for Data Mining. GridMiner provides a robust and reliable high-performance data mining and OLAP environment, and the system highlights the importance of grid-enabled applications in terms of e-science and detailed analysis of very large scientific data sets.

Chapter 4 is entitled 'ADaM services: scientific data mining in the service-oriented architecture paradigm'. The ADaM system was originally developed in the early 1990s with the goal of mining large scientific data sets for geophysical phenomena detection and feature extraction. The chapter describes the ADaM system and illustrates its features and functions on the basis of two applications of ADaM services within a SOA context.

Chapter 5 is entitled 'Mining for misconfigured machines in grid systems'. This chapter describes the Grid Monitoring System (GMS) – a system that adopts a distributed data mining approach to detection of misconfigured grid machines.

Chapter 6 is entitled 'FAEHIM: federated analysis environment for heterogeneous intelligent mining'. It describes the FAEHIM toolkit, which makes use of Web services composition, with the widely deployed Triana workflow environment. Most of the Web services are derived from the Weka data mining library of algorithms.

Chapter 7 is entitled 'Scalable and privacy-preserving distributed data analysis over a service-oriented platform'. It reviews a recently proposed scalable and privacy-preserving distributed data analysis approach. The approach computes abstractions of distributed data, which are then used for mining global data patterns. The chapter also describes a service-oriented realization of the approach for data clustering and explains in detail how the analysis process is deployed in a BPEL platform for execution.

Chapter 8 is entitled 'Building and using analytical workflows in Discovery Net'. It describes the experience of the authors in designing the Discovery Net platform and maps out the evolution paths for a workflow language, and its architecture, that address the requirements of different scientific domains.

Chapter 9 is entitled 'Building workflows that traverse the bioinformatics data landscape'. It describes how the myGrid supports the management of the scientific process in terms of *in silico* experimentation in bioinformatics. The approach is illustrated through an example from the study of trypanosomiasis resistance in the mouse model. Novel biological results obtained from traversing the 'bioinformatics landscape' are presented.

Chapter 10 is entitled 'Specification of Distributed data mining workflows with DataMiningGrid'. This chapter gives an evaluation of the benefits of grid-based technology from a data miner's perspective. It is focused on the DataMiningGrid, a standard-based and extensible environment for grid-enabling data mining applications.

Chapter 11 is entitled 'Anteater: service-oriented data mining'. It describes SOA-based data mining platform Anteater, which relies on Anthill, a runtime system for irregular, data intensive, iterative distributed applications, to achieve high performance. Anteater is operational and being used by the Brazilian Government to analyse government expenditure, public health and public safety policies.

Chapter 12 is entitled 'DMGA: a generic brokering-based data mining grid architecture'. It describes DMGA (Data Mining Grid Architecture), a generic brokering-based architecture for deploying data mining services in a grid. This approach presents two different composition models: horizontal composition (offering workflow capabilities) and vertical composition (increasing performance of inherently parallel data mining services). This scheme is especially significant to those services accessing a large volume of data, which can be distributed through diverse locations.

Chapter 13 is entitled 'Grid-based data mining with the environmental scenario search engine (ESSE)'. The natural environment includes elements from multiple domains such as space, terrestrial weather, oceans and terrain. The environmental modelling community has begun to develop several archives of continuous environmental representations. These archives contain a complete view of the Earth system parameters on a regular grid for a considerable period of time. This chapter describes the ESSE for data grids, which provides uniform access to heterogeneous distributed environmental data archives and allows the use of human linguistic terms while querying the data. A set of related software tools leverages the ESSE capabilities to integrate and explore environmental data in a new and seamless way.

Chapter 14 is entitled 'Data pre-processing using OGSA-DAI'. It explores the Open Grid Services Architecture – Data Access and Integration (OGSADAI) software, which is a uniform framework for providing data services to support the data mining process. It is shown how the OGSA-DAI activity framework already provides powerful functionality to support data mining, and that this can be readily extended to provide new operations for specific data mining applications. This functionality is demonstrated by two application scenarios and compares OGSA-DAI with other available data handling solutions.

References

Agrawal, R. and Shafer, J. C. (1996), 'Parallel mining of association rules', *IEEE Transactions on Knowledge and Data Engineering* **8** (6), 962–969.

Ashrafi, M. Z., Taniar, D. and Smith, K. (2004), 'ODAM: An optimized distributed association rule mining algorithm', *IEEE Distributed Systems Online* **5** (3), 2–18.

Butler, D. (1999), 'Computing 2010: from black holes to biology', *Nature* C67–C70.

Cho, V. and Wüthrich, B. (2002), 'Distributed mining of classification rules', *Knowledge and Information Systems* **4** (1), 1–30.

Clifton, C., Kantarcioglu, M., Vaidya, J., Lin, X. and Zhu, M. Y. (2002), 'Tools for privacy preserving distributed data mining', *SIGKDD Explorer Newsletter* **4** (2), 28–34.

Dubitzky, W., Granzow, M. and Berrar, D. P. (2006), *Fundamentals of Data Mining in Genomics and Proteomics*, Springer, Secaucus, NJ.

Edwards, J., Lane, M. and Nielsen, E. (2000), 'Interoperability of biodiversity databases: biodiversity information on every desktop', *Science* **289** (5488), 2312–2314.

Fayyad, U., Piatetsky-Shapiro, G. and Smyth, P. (1996), From data mining to knowledge discovery, *in* U. Fayyaad et al., ed., 'Advances in Knowledge Discovery and Data Mining', AAAI Press, pp. 1–34.

Foster, I. (2002), 'What is the Grid? A three point checklist', *Grid Today*.

Frawley, W., Piatetsky-Shapiro, G. and Matheus, C. (1992), 'Knowledge discovery in databases: An overview', *AI Magazine* 213–228.

Gomes, C. P. and Selman, B. (2005), 'Computational science: Can get satisfaction', *Nature* **435**, 751–752.

Grossman, R. L., Kamath, C., Kumar, V. and Namburu, R. R., eds (2001), *Data Mining for Scientific and Engineering Applications*, Kluwer.

Hirschman, L., Park, J. C., Tsujii, J., Wong, L. and Wu, C. H. (2002), 'Accomplishments and challenges in literature data mining for biology', *Bioinformatics* **18** (12), 1553–1561.

Kargupta, H., Huang, W., Sivakumar, K. and Johnson, E. (2001), 'Distributed clustering using collective principal component analysis', *Knowledge and Information Systems Journal* **3**, 422–448.

Kargupta, H., Kamath, C. and Chan, P. (2000), Distributed and parallel data mining: Emergence, growth, and future directions, *in* 'Advances in Distributed and Parallel Knowledge Discovery', AAAI/MIT Press, pp. 409–416.

Kesselman, C. and Foster, I. (1998), *The Grid: Blueprint for a New Computing Infrastructure*, Kaufmann.

Krauter, K., Buyya, R. and Maheswaran, M. (2002), 'A taxonomy and survey of grid resource management systems for distributed computing', *Software – Practice and Experience* **32**, 135–164.

Rajasekaran, S. (2005), 'Efficient parallel hierarchical clustering algorithms', *IEEE Transactions on Parallel and Distributed Systems* **16** (6), 497–502.

Sánchez, A., Peña, J. M., Pérez, M. S., Robles, V. and Herrero, P. (2004), Improving distributed data mining techniques by means of a grid infrastructure, *in* R. Meersman, Z. Tari and A. Corsaro, eds, 'OTM Workshops', Vol. 3292 of *Lecture Notes in Computer Science*, Springer, pp. 111–122.

Stankovski, V., May, M., Franke, J., Schuster, A., McCourt, D. and Dubitzky, W. (2004), A service-centric perspective for data mining in complex problem solving environments, *in* H. R. Arabnia and J. Ni, eds, 'Proceedings of International Conference on Parallel and Distributed Processing Techniques and Applications', Vol. 2, pp. 780–787.

Stankovski, V., Swain, M., Kravtsov, V., Niessen, T., Wegener, D., Kindermann, J. and Dubitzky, W. (2008), 'Grid-enabling data mining applications with DataMiningGrid: An architectural perspective', *Future Generation Computer Systems* **24**, 259–279.

Stankovski, V., Swain, M., Stimec, M. and Mis, N. F. (2007), Analyzing distributed medical databases on DataMiningGrid, *in* T. Jarm, P. Kramar and A. Zupanic, eds, '11th Mediterranean Conference on Medical and Biomedical Engineering and Computing', Springer, Berlin, pp. 166–169.

Talia, D. (2006), Grid-based distributed data mining systems, algorithms and services, *in* 'HPDM 2006: 9th International Workshop on High Performance and Distributed Mining', Bethesda, MD.

University of California (2007), 'SETI@Home. The Search for ExtraTerrestrial Inteligence (SETI)', http://setiathome.ssl.berkeley.edu

Wah, B. (1993), 'Report on workshop on high performance computing and communications for grand challenge applications: computer vision, speech and natural language processing, and artificial intelligence', *IEEE Transactions on Knowledge and Data Engineering* **5** (1), 138–154.

Witten, I. and Frank, E. (2000), *Data Mining: Practical Machine Learning Tools and Techniques with Java Implementations*, Kaufmann.

Wright, A. (2007), *Glut: Mastering Information Through the Ages*, Henry, Washington, D.C.

Yang, Q. and Wu, X. (2006), '10 challenging problems in data mining research', *International Journal of Information Technology and Decision Making* **5**, 597–604.

Zaki, M. J., Ho, C. T. and Agrawal, R. (1999), Parallel classification for data mining on shared-memory multiprocessors, *in* 'Proceedings International Conference on Data Engineering'.

2

Data analysis services in the Knowledge Grid

Eugenio Cesario, Antonio Congiusta, Domenico Talia and **Paolo Trunfio**

ABSTRACT

Grid environments were originally designed for dealing with problems involving compute-intensive applications. Today, however, grids have enlarged their horizon as they are going to manage large amounts of data and run business applications supporting consumers and end users. To face these new challenges, grids must support adaptive data management and data analysis applications by offering resources, services and decentralized data access mechanisms. In particular, according to the service-oriented architecture model, data mining tasks and knowledge discovery processes can be delivered as services in grid-based infrastructures. By means of a service-based approach it is possible to define integrated services supporting distributed business intelligence tasks in grids. These services can address all the aspects involved in data mining and knowledge discovery processes: from data selection and transport to data analysis, knowledge model representation and visualization. We worked along this direction for providing a grid-based architecture supporting distributed knowledge discovery named Knowledge Grid. This chapter discusses how the Knowledge Grid framework has been developed as a collection of grid services and how it can be used to develop distributed data analysis tasks and knowledge discovery processes exploiting the service-oriented architecture model.

2.1 Introduction

Computer science applications are becoming more and more network centric, ubiquitous, knowledge intensive and computing demanding. This trend will result soon in an ecosystem of pervasive applications and services that professionals and end users can exploit everywhere. A long-term perspective can be envisioned where a collection of services and applications will be accessed and used as public utilities, as water, gas and electricity are used today.

Key technologies for implementing this perspective vision are SOA and Web services, semantic Web and ontologies, pervasive computing, P2P systems, grid computing, ambient intelligence architectures, data mining and knowledge discovery tools, Web 2.0 facilities,

Data Mining Techniques in Grid Computing Environments Edited by Werner Dubitzky

mashup tools and decentralized programming models. In fact, it is mandatory to develop solutions that integrate some or many of these technologies to provide future knowledge-intensive software utilities.

In the area of grid computing a proposed approach in accordance with the trend outlined above is the *service-oriented knowledge utilities (SOKU)* model (NGG, 2006), which envisions the integrated use of a set of technologies that are considered as a solution to the information, knowledge and communication needs of many knowledge-based industrial and business applications. The SOKU approach stems from the necessity of providing knowledge and processing capabilities to everybody, thus promoting the advent of a competitive knowledge-based economy. The SOKU model captures three key notions.

- *Service oriented.* Services that must be instantiated and assembled dynamically, so that the structure, behaviour and location of software is changing at run-time.
- *Knowledge.* SOKU services are knowledge assisted to facilitate automation and advanced functionality.
- *Utility.* Directly and immediately useable services with established functionality, performance and dependability; the emphasis is on user needs and issues such as trust, ease of use and standard interface. Although the SOKU model is not yet implemented, grids are increasingly equipped with data management tools, semantic technologies, complex workflows, data mining features and other Web intelligence approaches. These technologies can facilitate the process of having grids as a strategic element for pervasive knowledge intensive applications and utilities.

Grids were originally designed for dealing with problems involving large numbers of data or compute-intensive applications. Today, however, grids have enlarged their horizon as they are going to run business applications supporting consumers and end users (Cannataro and Talia, 2004). To face these new challenges, grid environments must support adaptive data management and data analysis applications by offering resources, services and decentralized data access mechanisms. In particular, according to the service-oriented architecture model, data mining tasks and knowledge discovery processes can be delivered as services in grid-based infrastructures.

Through a service-based approach it is possible to define integrated services for supporting distributed business intelligence tasks in grids. These services can address all the aspects involved in data mining and in knowledge discovery processes: from data selection and transport to data analysis, knowledge model representation and visualization. Along this direction, a grid-based architecture supporting distributed knowledge discovery is the *Knowledge Grid* (Cannataro and Talia, 2003; Congiusta, Talia and Trunfio, 2007). This chapter discusses how the Knowledge Grid framework has been developed as a collection of grid services and how it can be used to develop distributed data analysis tasks and knowledge discovery processes using the SOA model.

2.2 Approach

The Knowledge Grid framework makes use of basic grid services to build more specific services supporting distributed *knowledge discovery in databases* (*KDD*) on the grid (Cannataro and

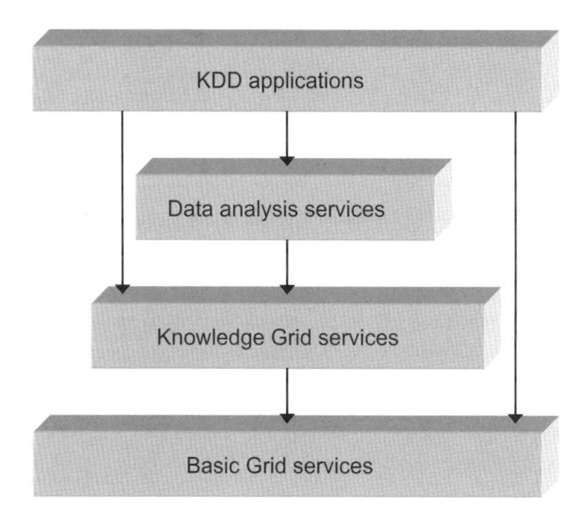

Figure 2.1 The Knowledge Grid layered approach

Talia, 2003). Such services allow users to implement knowledge discovery applications that involve data, software and computational resources available from distributed grid sites. To this end, the Knowledge Grid defines mechanisms and higher-level services for publishing and searching information about resources, representing and managing KDD applications, as well as managing their output results.

Previous work on the Knowledge Grid has been focused on the development of the system by using early grid middleware (Cannataro *et al.*, 2004*b*), as well as the design and evaluation of distributed KDD applications (Cannataro *et al.*, 2004*b*; Bueti, Congiusta and Talia, 2004; Cannataro *et al.*, 2004*a*). The *Web Services Resource Framework* (*WSRF*) has been adopted for re-implementing the Knowledge Grid services following the SOA approach (Congiusta, Talia and Trunfio, 2007). Such services will be used either to execute distributed data mining applications or to build more specific data analysis services, which can be in turn exploited to compose and execute higher-level KDD applications.

Such an approach can be described through the layered architecture shown in Figure 2.1, which represents the relationship among basic grid services, Knowledge Grid services, data analysis services and KDD applications.

- *Basic grid services.* Core functionality provided by standard grid environments such as Globus Toolkit (Foster, 2005), UNICORE (Erwin and Snelling, 2001) and gLite (Erwin and Snelling, 2004). Such functionalities include *security services* (mechanisms for authentication, authorization, cryptography etc.), *data management services* (data access, file transfer, replica management etc.), *execution management services* (resource allocation, process creation etc.) and *information services* (resource representation, discovery and monitoring).

- *Knowledge Grid services.* Services specifically designed to support the implementation of data mining services and applications. They include *resource management services*, which provide mechanisms to describe, publish and retrieve information about data sources, data mining algorithms and computing resources, and *execution management services*, which allow users to design and execute distributed KDD applications. Both resource and execution management services include core-level components and higher-level services, as described in Section 2.3.

- *Data analysis services.* Ad hoc services that exploit the Knowledge Grid services to provide high-level data analysis functionalities. A data analysis service can expose either a single data pre-processing or data mining task (e.g. classification, clustering, etc.), or a more complex knowledge discovery task (e.g. parallel classification, meta-learning etc.). Data analysis services are discussed in Section 2.4.
- *KDD applications.* Knowledge discovery applications built upon the functionalities provided by the underlying grid environments, the Knowledge Grid framework or higher-level data analysis services. Different models, languages and tools can be used to design distributed KDD applications in this framework, as discussed in Section 2.5.

The discussed architectural approach is flexible in allowing KDD applications to be built upon data analysis services and upon Knowledge Grid services. Moreover, these services can be composed together with basic services provided by a generic grid toolkit to develop KDD applications.

2.3 Knowledge Grid services

The Knowledge Grid is a software framework for implementing knowledge discovery tasks in a wide range of high-performance distributed applications. It offers to users high-level abstractions and a set of services by which they can integrate grid resources to support all the phases of the knowledge discovery process.

The Knowledge Grid architecture is designed according to the SOA, which is a model for building flexible, modular and interoperable software applications. The key aspect of SOA is the concept of a *service*, a software block capable of performing a given task or business function. Each service operates by adhering to a well defined interface, defining required parameters and the nature of the result. Once defined and deployed, services are like: 'black boxes' that is, they work independently of the state of any other service defined within the system, often cooperating with other services to achieve a common goal. The most important implementation of SOA is represented by Web services, whose popularity is mainly due to the adoption of universally accepted technologies such as XML, SOAP and HTTP. Also the grid provides a framework whereby a great number of services can be dynamically located, balanced and managed, so that applications are always guaranteed to be securely executed, according to the principles of on-demand computing.

The grid community has adopted the *Open Grid Services Architecture (OGSA)* as an implementation of the SOA model within the grid context. In OGSA every resource is represented as a Web service that conforms to a set of conventions and supports standard interfaces. OGSA provides a well defined set of *Web service* (*WS*) interfaces for the development of interoperable grid systems and applications (Foster *et al.*, 2003). Recently the WSRF has been adopted as an evolution of early OGSA implementations (Czajkowski *et al.*, 2004). WSRF defines a family of technical specifications for accessing and managing stateful resources using Web services. The composition of a Web service and a stateful resource is termed as Web service resource. The possibility to define a state associated with a service is the most important difference between WSRF-compliant Web services and pre-WSRF services. This is a key feature in designing grid applications, since Web service resources provide a way to represent, advertize and access properties related to both computational resources and applications.

2.3.1 The Knowledge Grid architecture

Basically, the Knowledge Grid architecture is composed of two groups of services, classified on the basis of their roles and functionalities. Indeed, two main aspects characterize a knowledge discovery process performed in accordance with the Knowledge Grid philosophy. The first is the management of data sources, data sets and tools to be employed in the whole process. The second is concerned with the design and management of a knowledge flow, that is the sequence of steps to be executed in order to perform a complete knowledge discovery process by exploiting the advantages coming from a grid environment. Notice that such a division has also a functional meaning, because a Knowledge Grid user first looks for data sets and tools to be used in the knowledge discovery process, then defines the knowledge flow to be executed (by using the resources found in the first step). On the basis of such a rationale, there are two distinct groups of services composing the Knowledge Grid architecture: the *Resource Management Services* and the *Execution Management Services*.

The *Resource Management Service (RMS)* group is composed of services that are devoted to the management of resources, such as data sets, tools and algorithms involved in a KDD process performed by the Knowledge Grid, as well as models inferred by previous analysis. Due to the heterogeneity of such resources, metadata represent a suitable descriptive model for them, in order to provide information about their features and their effective use. In this way, all the relevant objects for knowledge discovery applications, such as data sources, data mining software, tools for data pre-processing and results of computation, are described by their metadata. RMSs comprise services that enable a user to store resource metadata and perform different kinds of resource discovery task. Such tasks may typically include queries on resource availability resource location, as well as their access properties.

The *Execution Management Service (EMS)* group is concerned with the design and management of the knowledge discovery flow. A knowledge flow in the Knowledge Grid is modelled by a so-called *execution plan* that can be represented by a direct graph describing interactions and data flows between data sources, extraction tools, data mining tools and visualization tools. The execution graph describes the whole flow of the knowledge discovery process, the resources involved, the data mining steps etc. The EMS module contains services that are devoted to create an execution plan and to elaborate and to execute it, on the basis of the available resources and the basic grid services running on the host. It is worth noticing that when the user submits a knowledge discovery application to the Knowledge Grid, she has no knowledge about all the low-level details needed by the execution plan. More precisely, the client submits to the Knowledge Grid a high-level description of the KDD application, named a *conceptual model*, more targeted to distributed knowledge discovery aspects than to grid-related issues. A specific function of the EMS is to create an execution plan on the basis of the conceptual model received from the user, and execute it by using the resources effectively available. To realize this logic, the EMS follows a two-step approach: it initially models an *abstract execution plan* that in a second step is resolved into a *concrete execution plan*.

Figure 2.2 shows the architecture of the Knowledge Grid. A client application that wants to submit a knowledge discovery computation to the Knowledge Grid does not need to interact with all of these services, but just with a few of them. Inside each group, in fact, there are two layers of services: *high-level services* and *core-level services*. The design idea is that user-level applications directly interact with high-level services, which, in order to perform client requests, invoke suitable operations exported by the core-level services. In turn,

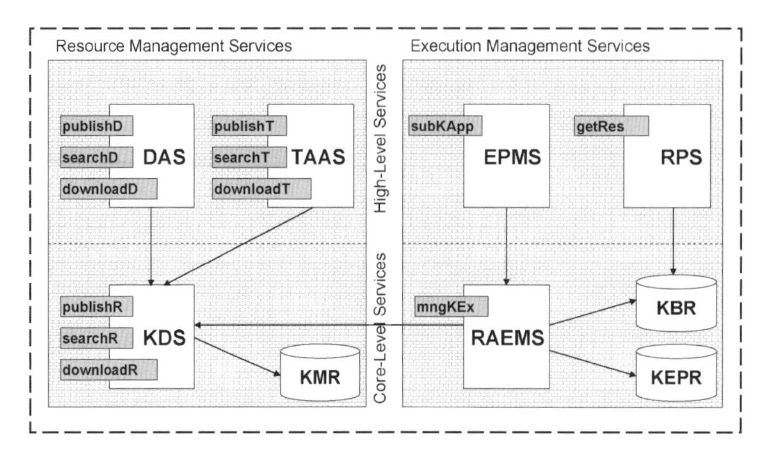

Figure 2.2 The Knowledge Grid architecture

core-level services perform their operations by invoking basic services provided by available grid environments running on the specific host, as well as by interacting with other core-level services. In other words, operations exported by high-level services are designed to be invoked by user-level applications, whereas operations provided by core-level services are thought to be invoked both by high-level and core-level services.

Client interface

Even though the client component is not a module integrated in the Knowledge Grid, it is useful to introduce its role and interaction with the architecture in order to better describe functionalities of the whole framework. A user of the Knowledge Grid can request the execution of basic tasks (e.g. searching of data and software, data transfers, simple job executions), as well as distributed data mining applications described by arbitrarily complex conceptual models or execution plans. As previously mentioned, a user-level application performs its tasks by invoking the appropriate operations provided by the different high-level services. Generally, a client running on a grid node may invoke services that are running on a different grid node. Therefore, the interactions between the client interface and high-level services are possibly remote.

Resource Management Services

This group of services includes basic and high-level functionalities for managing different kinds of Knowledge Grid resource. Among such resources, data sources and algorithms are of main importance; for this reason the Knowledge Grid architecture provides ad hoc components, namely *DAS* and *TAAS*, for dealing with data, tools and algorithms.

DAS The *Data Access Service* (*DAS*) is concerned with the publishing, searching and transferring of data sets to be used in a KDD application, as well as the search for inferred models (resulting from the mining process). The DAS exports the `publishData`, `searchData` and `downloadData` operations. The `publishData` is invoked by a user-level application to publish metadata about a data set; as soon as a publish operation is requested, it passes the corresponding metadata to the local *KDS* by invoking the `publishResource` operation. The `searchData` operation is invoked by a client interface that needs to locate a data set on the basis of a specified set of criteria. The DAS submits its request to the local KDS, by invoking the corresponding `searchResource` and, as soon as the search is completed, it receives the result from the KDS. Such a result consists in a set of references to the data sets matching the specified search criteria. It is worth noticing that the search operation is not just handled on the local host, but it is also remotely forwarded to other hosts. Finally, the `downloadData` operation works similarly to the previous ones: it receives the reference of the data set to be download and forwards the request to the `downloadResource` of the local KDS component. The DAS is, together with the TAAS, a high-level service of the RMS group.

TAAS The *Tools and Algorithms Access Service* (*TAAS*) is concerned with the publishing, searching and transferring of tools to be used in a knowledge discovery application. Such tools can be extraction tools, data mining tools and visualization tools. It has the same basic structure as the DAS and performs the main tasks by interacting with a local instance of the KDS, which in turn may invoke one or more other remote KDS instances. The operations that are exported by the TAAS are `publishTool`, `searchTool` and `downloadTool`. They have similar functionalities and similar behaviour to DAS operations, with the difference that TAAS operations are concerned with tools and not with data.

KDS The *Knowledge Directory Service* (*KDS*) is the only core-level service of the RMS group. It manages metadata describing Knowledge Grid resources. Such resources comprise hosts, repositories of data to be mined, tools and algorithms used to extract, analyse and manipulate data, knowledge models obtained as results of mining processes. Such information is stored in a local repository, the *Knowledge Metadata Repository* (*KMR*). The KDS exports the `publishResource`, `searchResource` and `downloadResource` operations. As previously noticed, they are invoked by the high-level services DAS or TAAS (to perform operations on data sets or tools respectively). The `publishResource` operation is invoked to publish information (metadata) about a resource; the publish operation is made effective by storing their metadata in the local KMR. The `searchResource` operation is invoked to retrieve a resource on the basis of a given set of criteria represented by a query. An important aspect to be pointed out is that the KDS performs such a searching task both locally, by accessing the local KMR, and remotely, by querying other remote KDSs (that in turn will access their local KMR). The `downloadResource` operation asks that a given resource is downloaded to the local site; also in this case, if the requested resource is not stored on the local site, the KDS will forward such a request to other remote KDS instances.

Execution Management Services

Services belonging to this group have the functionality of allowing the user to design and execute KDD applications, as well as delivering and visualising the computation result.

EPMS The *Execution Plan Management Service* (*EPMS*) allows for defining the structure of an application by building the corresponding execution graph and adding a set of constraints about resources. This service, on the basis of a *conceptual model* received from the client, generates its corresponding *abstract execution plan*, which is a more formal representation of the structure of the application. Generally, it does not contain information on the physical grid resources to be used, but rather constraints and other criteria about them. Nevertheless, it can include both well identified resources and abstract resources, i.e. resources that are defined through logical names and quality-of-service requirements. The EPMS exports the `submitKApplication` operation, through which it receives a conceptual model of the application to be executed and transforms it into an abstract execution plan for subsequent processing by the RAEMS. The mapping between the abstract execution plan and a more concrete representation of it, as well as its execution, is delegated to the RAEMS.

RPS The *Results Presentation Service* (*RPS*) offers facilities for presenting and visualizing the extracted knowledge models (e.g. association rules, clustering models and decision trees), as well as to store them in a suitable format for future reference. It is pointed out that the result stage-out, often needed for delivering results to the client, is an important task of this service. Indeed, a user can remotely interacts with the Knowledge Grid and, as soon as the final result is computed, it is delivered to the client host and made available for visualization.

RAEMS The *Resource Allocation and Execution Management Service* (*RAEMS*) is used to find a suitable mapping between an abstract execution plan (received from the EPMS) and available resources, with the goal of satisfying the constraints (CPU, storage, memory, database, network bandwidth requirements) imposed by the execution plan. The output of this process is a concrete execution plan, which explicitly defines the resource requests for each data mining process. In particular, it matches requirements specified in the abstract execution plan with real names, location of services, data files etc. From the interface viewpoint, the RAEMS exports the `manageKExecution` operation, which is invoked by the EPMS and receives the abstract execution plan. Starting from this, the RAEMS queries the local KDS (through the `searchResource` operation) to obtain information about the resources needed to create a concrete execution plan from the abstract execution plan. As soon as the concrete execution plan is obtained, the RAEMS coordinates the actual execution of the overall computation. For this purpose, the RAEMS invokes the appropriate services (data mining services and basic grid services) as specified by the concrete execution plan. As soon as the computation terminates, the RAEMS stores its results in the *Knowledge Base Repository (KBR)*, while the execution plans are stored in the *Knowledge Execution Plan Repository (KEPR)*. As a final operation, in order to publish results obtained by the computation, the RAEMS publishes their metadata (result metadata) into the KMR (by invoking the `publishResource` operation of the local KDS).

2.3.2 Implementation

This section describes the WSRF implementation of the Knowledge Grid, pointing out the adopted technologies and the service structure, along with the main implemented mechanisms.

The Knowledge Grid architecture uses grid functionalities to build specific knowledge discovery services. Such services can be implemented by using different available grid environments, such as Globus Toolkit[1], UNICORE[2] and gLite[3]. The Knowledge Grid framework

[1] http://www.globus.org
[2] http://www.unicore.org
[3] http://www.cern.ch/glite

and its services are being developed by using the Globus Toolkit (Foster, 2005). The Globus Toolkit, currently at Version 4, is an open source framework provided by the Globus Alliance for building computational grids based on WSRF. Currently, a growing number of projects and companies are using it to develop grid services and to exploit grid potentials. Such a phenomenon is contributing to the diffusion of the framework within the international community.

As a preliminary consideration, it has to be noticed that the implementation of each Knowledge Grid service follows the *Web service resource factory pattern*, a well known design pattern in software design and especially in object-oriented programming. Such a pattern, recommended by the WSRF specifications, requires having one service, the *factory service*, in charge of creating the resources and another one, the *instance service*, to actually operate on them. A resource maintains the state (e.g. results) of the computation performed by the service specific operations. The factory service provides a `createResource` operation that is used (by a client) to create a stateful resource. As soon as a new resource has been created, the factory service returns an *endpoint reference* that identifies univocally the resource for successive accesses to it. The instance service exports all the other methods useful for the functionalities of the service, by accessing the resources through their endpoint reference. Both services interact with a third external entity (not a service), the *Resource Home*, in charge of managing the resources (creates them and returns their reference). For completeness, it has to be specified that each service of the Knowledge Grid, in addition to the service specific operations and the aforementioned `createResource`, exports two other operations: `destroy` and `subscribe`. Even though they are not reported in Figure 2.2 and were not mentioned before, such operations are part of the service interface as well as other methods explicitly reported. The `destroy` operation is used to destroy a resource (previously created by `createResource`). The `subscribe` operation subscribes a client for notifications about a specific published topic.

In order to implement services in a highly decentralized way, a common design pattern is adopted. It allows a client to be notified when interesting events occur on the service side. The *WS-Notification* specification defines a *publish–subscribe* notification model for Web services, which is exploited to notify interested clients and services about changes that occur to the status of a WS resource. In other words, a service can *publish* a set of topics that a client can subscribe to; as soon as the topic changes, the client receives a notification of the new status.

Resource Management Services

The Resource Management Services are services devoted to the publishing, searching and downloading of resources (data sets, tools etc.) involved in Knowledge Grid applications. In particular, the high-level DAS and TAAS services interact, through the core-level KDS service, with the KMR repository to manage metadata about data sets and algorithms respectively.

KDS, DAS, TAAS The KDS, DAS and TAAS have been implemented according to the factory-instance pattern. Therefore, the development of each Knowledge Grid service is split into two basic services: factory service and instance service. All the services are implemented in Java. The service interfaces are described in Section 2.3.1, as well as their qualitative functionalities.

The interactions between a client, the DAS (or TAAS), the KDS and the KMR to manage a publish, search or download operation are depicted in Figure 2.3. When a client needs to publish a data set it invokes the `publishData` operation of the DAS, passing the URL of the metadata describing the data set. In turn, the DAS forwards such a request to the KDS by invoking its

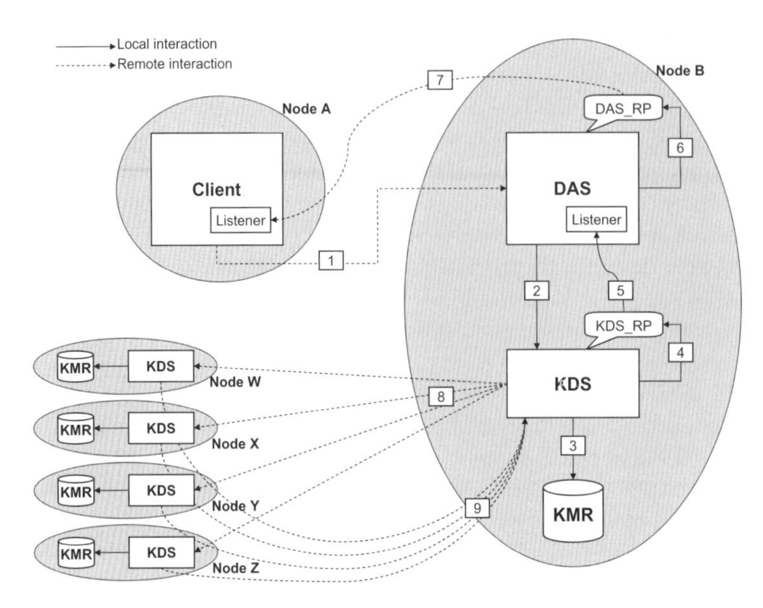

Figure 2.3 Services interactions for resource management

`publishResource` operation. The KDS stores the metadata info in the KMR, by adding an entry corresponding to the newly available resource. There is one resource property associated with each service: the *DAS_RP* is the resource property of the DAS, whereas the *KDS_RP* is the resource property of the KDS. Both such resource properties are string variables. Additionally, they are also published in the *topic list*, in order to be available as 'items of interest for subscription' for notification purposes. The sample scenario reproduces a general case where the client is running on Node A, while a Knowledge Grid instance (the client directly interacts with) is running on Node B. Moreover, there are some Knowledge Grid instances running on different nodes (W, X, Y and Z).

Whenever a client submit the request, it invokes the `createResource` operation of the DAS to create a resource property DAS_RP, subscribes a listener (a Java class inside the client) as consumer of notifications on the topic DAS_RP (through the DAS `subscribe` method) and invokes the `publishData` operation exported by the DAS (Step 1). In turn, the DAS dispatches this operation in a similar way: it subscribes a listener as consumer of notifications on the resource property KDS_RP (after having created it), and invokes the local KDS `publishResource` operation (Step 2). As a consequence of this invocation, the metadata file referred to by the URL (and describing the data set to be published) is inserted into the KMR (Step 3). As soon as such operation is terminated, the value of the KDS_RP is set by the *kdsURL* of the published resource (Step 4), or it assumes an error message, if an error occurs (e.g. the metadata file is corrupted, the resource has been already published etc.). The kdsURL is a reference identifying a published resource in the KMR. The change in the KDS_RP value is automatically notified to the *DAS Listener* (Step 5), which in turn stores such a value on the DAS_RP (Step 6) and asks for the destroying of KDS_RP. The value change of DAS_RP throws a notification for the *Client Listener* (Step 7), and finally to the client.

The operation of publishing a tool is similar to that described above, with the difference that the TAAS is involved, instead of the DAS. Analogously, also the search and download

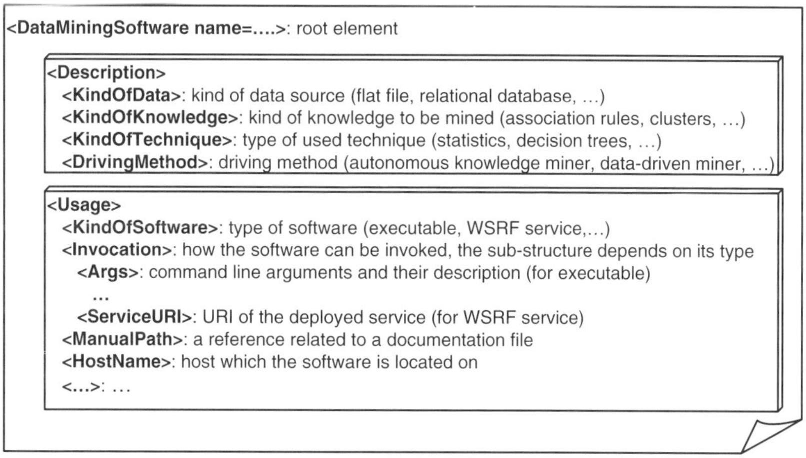

<DataMiningSoftware name=....>: root element

<Description>
<KindOfData>: kind of data source (flat file, relational database, ...)
<KindOfKnowledge>: kind of knowledge to be mined (association rules, clusters, ...)
<KindOfTechnique>: type of used technique (statistics, decision trees, ...)
<DrivingMethod>: driving method (autonomous knowledge miner, data-driven miner, ...)

<Usage>
<KindOfSoftware>: type of software (executable, WSRF service,...)
<Invocation>: how the software can be invoked, the sub-structure depends on its type
<Args>: command line arguments and their description (for executable)
...
<ServiceURI>: URI of the deployed service (for WSRF service)
<ManualPath>: a reference related to a documentation file
<HostName>: host which the software is located on
<...>: ...

Figure 2.4 Software metadata schema

operations are executed in a similar way, with the addition that the KDS forwards (Step 8) the requests also to other remote KDSs and waits for their response (Step 9). For completeness, it is worth noticing that in the case of a download request the data transferring is handled by a request to the *RFT service*, provided as a basic service by the Globus Toolkit.

KMR The KMR stores metadata needed to identify and classify all the heterogeneous resources used in data mining applications on grids. They include data mining software and data sources.

Data mining software There are two main aspects in the definition of metadata describing a data mining algorithm: its *description* (what the algorithm does) and its *usage* (how it can be invoked). Data mining software can be categorized on the basis of the following parameters: the kind of input data source, the kind of knowledge to be discovered, the types of technique used and the driving method of the mining process (Mastroianni, Talia and Trunfio, 2004). As concerns its usage, metadata should specify whether it is an executable file or an available WSRF service, and the parameters to be used for its invocation. Such information can be mapped on an XML schema that defines the format and the syntax of the XML file that is used to describe the features of a generic data mining software. In Figure 2.4 is reported a sketch of the XML file containing such metadata information. It is composed of two parts. The first part is the software description, containing some tags specifying the features of the data mining algorithm as mentioned above. The second part is the software usage, describing how to invoke or execute the algorithm. In particular, the `<invocation>` subsection specifies the way the software can be invoked. This feature depends on the kind of software: if it is in the form of an executable, this section specifies details of the software invocation (executable name and its arguments), whereas in the case of a software available through a service, within the tag is specified the service URI (from which it is possible to retrieve the WSDL (Web Services Description Language) file and information about the service interface).

Data source Data sources are analysed by data mining algorithms to extract knowledge from them. They can originate from text files, relational databases, webpages and other semi-structured documents. Metadata information describing a data source should specify its *structure* (what attributes describe it, their types etc.) and its *access* (how it is retrieved). This allows the Knowledge Grid to deal with different kinds of data source as long as they are properly characterized by appropriate metadata descriptions. For instance, data sources represented by flat files (*attribute-relation file format* (*ARFF*) and *comma-separated values* (*CSV*) formats) are described by metadata information as detailed by Mastroianni, Talia and Trunfio (2004).

After having briefly described the structure of the metadata modeling algorithms and data sources, let us briefly sketch the implementation of the KMR. Currently it is implemented as a directory of the file system in which metadata about resources are stored as XML files. In such a directory an XML file for each resource published on the local host is stored. In addition to all these files, there is also a *registry* file, containing an entry <*LogicalName,PhysicalName*> for each published resource. The *LogicalName* (the so called kdsURL) is in the form *kds://<hostname>/<relative path of the document in the KMR of hostname>*, while the *PhysicalName* is in the form *<absolute path of the document in the KMR of hostname>*. As soon as a new resource has to be published on a host, its metadata file is stored in the KMR and an entry is inserted in the registry which makes the publishing operation effectively concluded). As an evolution of such a current solution based on a file system, there is an ongoing implementation of a storage back-end based on a database. Both relational (PostgreSQL, MySQL) and XML (XIndice, Exist) databases have been chosen, in order to evaluate and compare their suitability and efficiency.

Execution Management Services

Interfaces and communication patterns involving such services have been defined and the implementation of the software is in a preliminary status. As previously mentioned, applications are described at high level using a *conceptual model* generated by the user (e.g. through design facilities provided by the user-level module). It must specify, at some level of abstraction, the tasks that compose the process, as well as the logic ruling their integration. In the current approach *UML activity diagrams* are used to represent the conceptual model of an application. Such an activity diagram represents the high-level flow of service invocations that constitute the application logic. The *conceptual model* is then passed to the EPMS, which is in charge of transforming it into an *abstract execution plan* for subsequent processing by the RAEMS. The execution plan is represented by *Business Process Execution Language (BPEL)* (Andrews *et al.*, 2003), that is an XML-based language to express the business logic of the application being modelled. The abstract execution plan may not refer to specific implementations of services, application components and input data. The RAEMS receives the abstract execution plan and creates a *concrete execution plan*. In order to carry out such a task, the RAEMS needs to evaluate and resolve a set of grid resources, by contacting the KDS and choosing the most appropriate ones. Another important difference between the abstract and concrete execution plans is that in the first the WSDL of services is used without specifying the service location, while in the second the location on which to execute the service is concretely decided (w.r.t. locations currently available, processor speed, quality of service etc.). In other words, the RAEMS attempts to locate suitable service instances that match the service requirements specified as WSDL definitions. As soon as the RAEMS has built the concrete execution plan, it is in charge of submitting it to the workflow engine that coordinates services execution. The RAEMS and the workflow engine cooperate for managing the execution of the overall application, taking into account relevant scheduling decisions. Such decisions are generated by specific sub-components of the RAEMS that monitor the status of the involved grid resources and jobs, and enforce optimization policies also based on appropriate heuristics (more details are described by Pugliese and Talia (2004)).

Whenever a legacy software is to be integrated within the execution plan, this is obtained by including the submission of a grid job to the GRAM service of the Globus Toolkit. In order to do this, the RAEMS takes care of generating the needed *resource specification language* document to be referred by the concrete execution plan as a GRAM service parameter. Finally, the notification mechanism will be used to send back information to the client about the execution status of the running jobs.

2.4 Data analysis services

Distributed data mining includes a variety of application scenarios relying on many different patterns, techniques and approaches. Their implementation may involve multiple independent or correlated tasks, as well as sharing of distributed data and knowledge models (Kargupta and Chan, 2000).

An example of a distributed data mining technique is *meta-learning*, which aims to build a global classifier from a set of inherently distributed data sources (Prodromidis and Chan, 2000). Meta-learning is basically a two-step process: first, a number of independent classifiers are generated by applying learning programs to a collection of distributed and homogeneous data sets in parallel. Then, the classifiers computed by learning programs are collected in a centralized site and combined to obtain the global classifier.

Another example of distributed data mining is *collective data mining* (Kargupta *et al.*, 2000). Instead of combining incomplete local models, collective data mining builds the global model through the identification of significant sets of local information. In other terms, the local blocks directly compose the global model. This result is based on the fact that any function can be expressed in a distributed fashion using a set of appropriate basis functions. If the basis functions are orthogonal the local analysis generates results that can be correctly used as components of the global model.

When such kinds of application are to be deployed on a grid, in order to fully exploit the advantages of the environment, many issues must be faced, including heterogeneity, resource allocation and execution coordination. The Knowledge Grid provides services and functionalities that permit us to cope with most of the above-mentioned issues and to design complex applications using high-level abstractions such as service composition facilities (i.e. workflows).

In some cases, however, end users are not familiar with distributed data mining patterns and complex workflow composition. Therefore, higher-level data analysis services, exposing common distributed data mining patterns as single services, might be useful to broaden the accessibility of the Knowledge Grid features.

In general, data analysis services may implement either

- single steps of the overall KDD process or
- complex KDD patterns involving multiple independent or correlated data mining tasks.

The former include tasks such as data filtering, classification, clustering and association rule discovery. The latter refer to patterns such as parallel classification, meta-learning, collective data mining and so on. Both types of data analysis service can be in turn used as basis for other data analysis services, or directly employed in the implementation of distributed KDD applications over the Knowledge Grid.

As an example scenario in which data analysis services may be profitably employed, let us consider the case of a user willing to perform a meta-learning analysis over distributed data. In order to execute this application on the Knowledge Grid, the user must design the overall application by defining a workflow that coordinates the execution of different classifiers on multiple nodes, followed by a final combining of the models to obtain a global classifier.

While this task can be performed using the Knowledge Grid facilities (see Section 2.5), it requires some expertize that in some cases is not owned by the final user. Thus, a specific data analysis service, here called a *meta-learning service (MLS)*, could be provided to support seamless execution of meta-learning applications over the Knowledge Grid.

Like the underlying Knowledge Grid services, the MLS data analysis service should be implemented as a WSRF-compliant Web service. Among the other WSRF-specific operations, the MLS interface should provide a `submitMLApplication` operation for submitting the meta-learning task, with parameters specifying the URLs of the data sources to be analysed, the names of the classifiers algorithms to be used and the name of the combining algorithm.

Note that the MLS `submitMLApplication` is different from the EPMS `submitKApplication`, because the latter receives a conceptual model of generic KDD application, while the former receives parameters exclusively targeted to the execution of a meta-learning task. The role of the MLS is thus to translate the request submitted to its `submitMLApplication` into a request for the `submitKApplication` of the EPMS, releasing the user from the task of designing the conceptual model of the application.

By exploiting the MLS, the meta-learning application could be thus executed through the following steps.

(a) The user invokes `submitMLApplication` of the MLS passing the URLs of the data sources to be analysed and the names of the algorithms to be used.

(b) The MLS defines the appropriate conceptual model for the application, and passes it to EPMS of a Knowledge Grid node.

(c) The Knowledge Grid services then proceed by: (i) generating the abstract execution plan, (ii) creating the concrete execution plan by finding the needed algorithms and resources, (iii) coordinating the overall execution, (iv) passing the results to the MLS.

(d) The MLS then returns the obtained results (i.e. the global model inferred by the meta-learning process) to the submitting user.

While in the example above the MLS acts as a broker between end users and the Knowledge Grid services, in other cases data analysis services may in turn invoke lower-level data analysis services (e.g. simple classification services). In general, the use of data analysis services can foster the reuse of KDD applications and patterns over the Knowledge Grid, by fully exploiting the service composition paradigm.

2.5 Design of Knowledge Grid applications

The knowledge discovery design phase is the process of selecting data sets, algorithms and related configurations for performing the needed steps leading to the extraction of knowledge models from data. To such a purpose, a task-flow involving the selected data sets and algorithms can be used for specifying how the several activities of the knowledge discovery process are organized. Such a task-flow typically involves several atomic tasks (such as pre-processing, classification etc.), to be performed sequentially and/or concurrently (and in general on different grid nodes) with specified relationships among them, as well as data constraints.

The application design is also a fundamental process through which it is possible to exploit the facilities offered by the underlying grid environment. The application design process in the Knowledge Grid can be described as follows.

- Starting points are the resource descriptions provided by the Knowledge Grid information system (namely DAS and TAAS services).
- It is necessary to integrate resources in a task-flow (or workflow) representing the structure of the application. This step includes the possible addition of abstract tasks. The task-flow is expressed in an abstract and generic way, to assure flexibility, easiness of design and comprehension, and portability over time and space of the application model. Such a model is referred to as a conceptual model.

As already discussed, the design phase is followed by a phase of application management, in which the application model is processed by the EPMS and the RAEMS for refining the specification, generating execution plans and submitting the application for execution.

Designing and executing a distributed KDD application within the Knowledge Grid is a multi-step process that involves interactions and information flows between services at the different levels of the architecture. A key aspect in the Knowledge Grid is how applications are modelled, and how the application models are represented and processed through the different services.

As already mentioned, applications are described at a high level using a conceptual model, generated, for instance, by the user through the design facilities provided by the client interface. The conceptual model (or application model) must specify, at some level of abstraction, the tasks that compose the application, as well as the logic ruling their integration.

The Knowledge Grid embraces user-friendliness as a basic design principle. For this reason a significant effort has been devoted to the provision of design tools and abstractions able to fill the gap between the user perspective and the grid infrastructure. The application design support offered by the Knowledge Grid has evolved over time from a basic visual language to a UML-based grid-workflow formalism.

2.5.1 The VEGA visual language

VEGA was the first programming interface of the Knowledge Grid (Cannataro *et al.*, 2004*b*). The graphical interface provided by VEGA proposes to users a set of graphical objects representing grid resources. These objects can be manipulated using visual facilities allowing for inserting links among them, thus composing a graphical representation of the set of jobs included in the application. A complex application is generally composed by several jobs, some of which must be executed concurrently and others sequentially. To reflect this, VEGA

Table 2.1 Composition rules in VEGA

Resource A	Resource B	Link type	Meaning
Data	Software	Input	Computation input
Data	Software	Output	Computation output
Data	Host	File transfer	Data transfer to the host
Software	Host	File transfer	Software staging to the host
Software	Host	Execute	Computation execution on the host

provides the *workspace* abstraction. An application is designed as a sequence of workspaces that are executed sequentially, while each workspace contains a set of jobs to be executed concurrently.

The insertion of links between resources must follow precise rules, in order to produce logically correct application models. Table 2.1 summarizes such basic composition rules.

Before the generation of abstract and instantiated execution plans the application model is validated by VEGA in order to assert both basic composition rules and other consistency properties.

2.5.2 UML application modelling

In the latest implementation of the Knowledge Grid, the conceptual model of an application is expressed under the form of a grid workflow, since it represents a more general way for expressing grid computations able to overcome some limitations of the VEGA model.

Many formalisms have been traditionally used for modeling workflows, such as directed acyclic graphs, Petri nets and UML activity diagrams. Many grid workflow systems adopt standard coordination languages such as BPEL and WSCI (Arkin *et al.*, 2002), or XML-based ad hoc solutions.

Within the Knowledge Grid, the UML-activity-diagram formalism is used to represent the conceptual model of the application, while BPEL is used for representing execution plans. The activity diagram represents the high-level flow of service invocations that constitute the application logic, whereas the BPEL expresses how the various services are actually invoked and coordinated.

The UML activity diagrams are built using typical task-graph operators such as *start*, *fork*, *join*, which rule the execution flow. As mentioned earlier, abstract and concrete execution plans are distinguished. At the abstract level, the execution plan may not refer to specific implementations of services, application components, and input data. All these entities are identified through logical names and, in some cases, by means of a set of constraints about some of their properties, possibly expressing quality of service requirements. For example, requirements on processor speed, amount of main memory or disk space can be used to single out grid nodes, while requirements on grid software may concern input data or target platforms.

Prior the application execution, all of the resource constraints need to be evaluated and resolved on a set of available grid resources, to choose the more appropriate ones with respect to the current status of the grid environment. Of course, due to the dynamic nature of the grid, an abstract execution plan can be instantiated into different execution plans at different times. Concrete execution plans include real names and locations of services, data files, etc. According to this approach the workflow definition is decoupled from the underlying grid configuration.

The translation of the conceptual model (represented by the UML formalism) into the abstract execution plan (represented by a BPEL document) is performed by the EPMS. To this end, the EPMS incorporates an algorithm for mapping the UML operators into corresponding BPEL notations. The BPEL notation is explicitly targeted to service invocations and thus richer with respect to the UML one. It includes several constructs for interacting with services according to different patterns, as well as means for manipulating and accessing service input-messages and responses (both of them through explicit variable-manipulation operators and XPath expressions).

2.5.3 Applications and experiments

The Knowledge Grid framework has been successfully used to implement several applications in different application areas. The applications developed in such contexts have been useful for evaluating the overall system under different aspects, including performance.

The main issues and problems that have been addressed include the use of massive data sets and the distribution and coordination of the computation load among different nodes. In the remaining part of this section two applications are summarized: the clustering of human protein sequences, and the integration of a query-based data mining system within the Knowledge Grid. More detailed descriptions, along with performance evaluations, are presented in previous work (Bueti, Congiusta and Talia, 2004; Cannataro *et al.*, 2004*a*).

Protein folding

The implemented application carries out the clustering of human protein sequences using the TribeMCL method (Enright, Dongen and Ouzounis, 2002). TribeMCL is a clustering method through which it is possible to cluster correlated proteins into groups termed protein families. The clustering is achieved by analysing similarity patterns between proteins in a given data set, and using these patterns to assign proteins to related groups. TribeMCL uses the Markov Clustering algorithm.

The application comprises four phases: (i) data selection, (ii) data pre-processing, (iii) clustering, and (iv) results visualization. The measurement of application execution times has been performed in two different cases: (a) using only 30 human proteins, and (b) using all the human proteins classified in the Swiss-Prot database [4] (a well known and widely used repository containing extensively annotated protein sequence information).

Comparing the execution times, it is possible to note that the execution of the clustering phase is a computationally intensive operation; consequently, it takes much more time when all the proteins have to be analysed. The grid execution of these phases using three nodes has been performed in a time reduced by a factor of 2.5 with respect to the sequential execution.

KDD Markup Language and the Knowledge Grid

KDDML-MQL is an environment for the execution of complex KDD processes expressed as high-level queries (Alcamo, Domenichini and Turini, 2000). The environment is composed of two layers, the bottom layer called KDDML and the top layer is referred to as MQL. KDDML (KDD Markup Language) is an XML-based query language for performing knowledge extrac-

[4] http://www.ebi.ac.uk/swissprot/

tion from databases. It combines, through a set of operators, data manipulation tasks such as data acquisition, pre-processing, mining and post-processing, employing specific algorithms from several suites of tools. The composition of the algorithms is obtained by appropriate wrappers between algorithm internal representations and the KDDML representation of data and models.

Objectives of this effort have been

- combining the possibility to express a data mining application as a query with the advantages offered by the Knowledge Grid environment for executing the data mining query on a grid and
- extending the Knowledge Grid environment with a new class of knowledge discovery applications.

KDDML operators can be applied both to a data set, a leaf in the tree representing the overall query, and to the results produced by another operator. KDDML can support the parallel execution of the KDD application thanks to the possibility to split the query representing the overall KDD process into a set of sub-queries to be assigned to different processing nodes.

The distributed execution of KDDML has been modelled according to the master worker paradigm. In particular, the query entering component acts as the master, while each instance of the query executor is a worker. The main objectives of the execution of such queries on the grid are

- distributing them to a number of grid nodes to exploit parallel execution of sub-queries and
- avoiding an excessive fragmentation of the overall task, i.e. allocating a number of nodes so that the exchanges of intermediate (partial) results are minimized.

The grid execution of a KDDML application on a four-node grid has shown that the benefits from the Knowledge Grid environment become profitable as soon as the data set size increases. For instance, the extraction of a classification model from a data set of 320 000 tuples, according to a two-step process (that is, first applying a clustering algorithm and next a classification one), resulted in an average speed-up of about 3.70, reducing execution time from about 3 hours to about 45 minutes.

2.6 Conclusions

To support complex data-intensive applications, grid environments must provide adaptive data management and data analysis tools and techniques through the offer of resources, services and decentralized data access mechanisms. In this chapter, we have argued that, according to the service-oriented architecture model, data mining tasks and knowledge discovery processes can be delivered as services in grid-based infrastructures. Through a service-based approach it is possible to design basic and complex services for supporting distributed business intelligence tasks in grids. These services can address all the aspects that must be considered in data mining tasks and in knowledge discovery processes from data selection and transport to data analysis, knowledge model representation and visualization. We have discussed this approach and illustrated how it is implemented in the Knowledge Grid framework starting from basic services to data analysis services and KDD applications expressed as composition of lower level services.

References

Alcamo, P., Domenichini, F. and Turini, F. (2000), An XML-based environment in support of the overall kdd process, *in* 'International Conference on Flexible Query Answering Systems (FQAS'00)', Physica, Warsaw, pp. 413–424.

Andrews, T., Curbera, F., Dholakia, H., Goland, Y., Klein, J., Leymann, F., Liu, K., Roller, D., Smith, D., Thatte, S., Trickovic, I. and Weerawarana, S. (2003), 'BPEL for Web services version 1.1'. http://www-128.ibm.com/developerworks/library/specification/ws-bpel

Arkin, A., Askary, S., Fordin, S., Jekeli, W., Kawaguchi, K., Orchard, D., Pogliani, S., Riemer, K., Struble, S., Takacsi-Nagy, P., Trickovic, I. and Zimek, S. (2002), 'Web Service Choreography Interface (WSCI) 1.0'. http://www.w3.org/TR/wsci

Bueti, G., Congiusta, A. and Talia, D. (2004), Developing distributed data mining applications in the Knowledge Grid framework, *in* 'International Conference on High-Performance Computing for Computational Science (VECPAR'04)', Vol. 3402 of *LNCS*, Springer, Valencia, pp. 156–169.

Cannataro, M., Comito, C., Congiusta, A. and Veltri, P. (2004*a*), 'PROTEUS: a bioinformatics problem solving environment on grids', *Parallel Processing Letters* **14** (2), 217–237.

Cannataro, M., Congiusta, A., Pugliese, A., Talia, D. and Trunfio, P. (2004*b*), 'Distributed data mining on grids: services, tools, and applications', *IEEE Transactions on Systems, Man, and Cybernetics: Part B* **34** (6), 2451–2465.

Cannataro, M. and Talia, D. (2003), 'The Knowledge Grid', *Communications of the ACM* **46** (1), 89–93.

Cannataro, M. and Talia, D. (2004), 'Semantics and knowledge grids: building the next-generation grid', *IEEE Intelligent Systems and Their Applications* **19** (1), 56–63.

Congiusta, A., Talia, D. and Trunfio, P. (2007), 'Distributed data mining services leveraging WSRF', *Future Generation Computer Systems* **23** (1), 34–41.

Czajkowski, K., Ferguson, D. F., Foster, I., Frey, J., Graham, S., Sedukhin, I., Snelling, D., Tuecke, S. and Vambenepe, W. (2004), 'The WS-Resource Framework Version 1.0'. http://www.ibm.com/developerworks/library/ws-resource/ws-wsrf.pdf

Enright, A. J., Dongen, S. V. and Ouzounis, C. A. (2002), 'An efficient algorithm for large-scale detection of protein families', *Nucleic Acids Research* **30** (7), 1575–1584.

Erwin, D. W. and Snelling, D. F. (2001), UNICORE: a grid computing environment, *in* 'International Conference on Parallel and Distributed Computing (Euro-Par'01)', Vol. 2150 of *LNCS*, Springer, Manchester, UK, pp. 825–834.

Erwin, D. W. and Snelling, D. F. (2004), Middleware for the next generation grid infrastructure, *in* 'International Conference on Computing in High Energy and Nuclear Physics (CHEP'04)', Interlaken.

Foster, I. (2005), Globus toolkit version 4: software for service-oriented systems, *in* 'International Conference on Network and Parallel Computing (NPC'05)', Vol. 3779 of *LNCS*, Springer, Beijing, pp. 2–13.

Foster, I., Kesselman, C., Nick, J. M. and Tuecke, S. (2003), The physiology of the grid, *in* F. Berman, G. Fox and A. Hey, eds, 'Grid Computing: Making the Global Infrastructure a Reality', Wiley.

Kargupta, H. and Chan, P., eds (2000), *Advances in distributed and parallel knowledge discovery*, AAAI Press.

Kargupta, H., Park, B. H., Hershbereger, D. and Johnson, E. (2000), Collective data mining: a new perspective toward distributed data mining, *in* H. Kargupta and P. Chan, eds, 'Advances in Distributed and Parallel Knowledge Discovery', AAAI Press.

Mastroianni, C., Talia, D. and Trunfio, P. (2004), 'Metadata for managing grid resources in data mining applications', *Journal of Grid Computing* **2** (1), 85–102.

NGG (2006), 'Future for European Grids: Grids and Service-Oriented Knowledge Utilities'. Next Generation Grids (NGG) Expert Group Report 3.

Prodromidis, A. and Chan, P. (2000), Meta-learning in distributed data mining systems: issues and approaches, *in* H. Kargupta and P. Chan, eds, 'Advances in Distributed and Parallel Knowledge Discovery', AAAI press.

Pugliese, A. and Talia, D. (2004), Application-oriented scheduling in the knowledge grid: a model and architecture, *in* 'International Conference on Computational Science and its Applications (ICCSA'04)', Vol. 3044 of *LNCS*, Springer, Assisi, pp. 55–65.

3

GridMiner: An advanced support for e-science analytics

Peter Brezany, Ivan Janciak and **A. Min Tjoa**

ABSTRACT

Knowledge discovery in data sources available on computational grids is a challenging research and development issue. Several grid research activities addressing some facets of this process have already been reported. This chapter introduces the GridMiner framework, developed within a research project at the University of Vienna. The project's goal is to deal with all tasks of the knowledge discovery process on the grid and integrate them in an advanced service-oriented grid application. The GridMiner framework consists of two main components: technologies and tools, and use cases that show how the technologies and tools work together and how they can be used in realistic situations. The innovative architecture of the GridMiner system is based on the Cross-Industry Standard Process for Data Mining. GridMiner provides a robust and reliable high-performance data mining and OLAP environment, and the system highlights the importance of grid-enabled applications in terms of e-science and detailed analysis of very large scientific data sets. The interactive cooperation of different services – data integration, data selection, data transformation, data mining, pattern evaluation and knowledge presentation– within the GridMiner architecture is the key to productive e-science analytics.

3.1 Introduction

The term e-science refers to the future large-scale science that will increasingly be carried out through distributed global collaborations enabled by the Internet. This phenomenon is a major force in the current e-science research and development programmes (as pioneered in the UK) and NSF cyberstructure initiatives in the US. Typically, the individual user scientists require their collaborative scientific enterprises to have features such as access to very large data collections and very large scale computing resources. A key component of this development is *e-science analytics*, which is a dynamic research field that includes rigorous and sophisticated scientific methods of data pre-processing, integration, analysis, data mining and visualization

Data Mining Techniques in Grid Computing Environments Edited by Werner Dubitzky

associated with information extraction and knowledge discovery from scientific data sets. Unlike traditional business analytics, e-science analytics has to deal with huge, complex, heterogeneous, and very often geographically distributed data sets which contain volumes measured in terabytes and will soon total petabytes. Because of the huge volume and high dimensionality of data, the associated analytic tasks are often both input/output and compute intensive and consequently dependent on the availability of high-performance storage, computing hardware, software resources and software solutions. This is one of the main reasons why high-performance computing analytics has become a regular topic of the programmes at the Supercomputing Conference Series since 2005. Further, the user community is often at different geographically distributed locations, and finally, a high level of security has to be guaranteed for many analytical applications, e.g. in finance and medical sectors. The grid is an infrastructure proposed to bring all these issues together and make a reality of the vision for e-science analytics outlined above. It enables flexible, secure, coordinated resource (computers, storage systems, equipments, database, and so forth) sharing among dynamic collections of individuals and institutions. Over the past four years, several pioneering grid-based analytic system prototypes have been developed. In most of these systems, analytic tasks are implemented as grid services, which are combined into interactive workflows orchestrated and executed by special services called workflow engines. In the following, we characterize some relevant developments.

The myGrid project is developing high-level middleware for data and legacy application resource integration to support *in silico* experiments in biology. Their developed workflow system Taverna (Wolstencroft, *et al.*, 2005) provides semantic support for the composition and enactment of bioinformatics workflows. Cannataro and Talia (2003) proposed a design of the Knowledge Grid architecture based on the Globus Toolkit (Foster and Kesselman, 1998). Discovery Net (Curcin, *et al.*, 2002) provides a service-oriented computing model for knowledge discovery allowing users to connect to and use data analysis software and data sources that are available on-line. The *Science Environment for Ecological Knowledge (SEEK)* project (Jones, *et al.*, 2006) aims to create a distributed data integration and analysis network for environmental, ecological and taxonomy data. The SEEK project uses the Kepler (Berkley, *et al.*, 2005) workflow system for service orchestration. The recently finished EU-IST project DataMiningGrid (Stankovski, *et al.*, 2008) developed advanced tools and services for data mining applications on a grid. It uses Web Services Resource Framework-compliant technology with Triana (Churches, *et al.*, 2006), an open source problem solving environment developed at Cardiff University, that combines an intuitive visual interface with powerful data analysis tools in its user interface.

In all the above-mentioned projects, the main focus has been put on the functionality of the developed prototypes and not on the optimization of their runtime performance. Moreover, the end user productivity aspects associated with their use have not been sufficiently investigated. Their improvement can have significant impact on many real-life spheres, e.g. it can be a crucial factor in achievement of scientific discoveries, optimal treatment of patients, productive decision making, cutting costs and so forth. These issues, which pose serious research challenges for continued advances in e-science analytics, motivated our work on a grid-technology-based analytics framework called GridMiner[1], which is introduced in this chapter. Our aim was to develop a core infrastructure supporting all the facets of e-science analytics for a wide spectrum of applications. The framework consists of two main components.

[1] http://www.gridminer.org

- *Technologies and tools.* They are intended to effectively assist application developers to develop grid-enabled high-performance analytics applications. GridMiner includes services for sequential, parallel and distributed data, text mining, on-line analytical processing, data integration, data quality monitoring and improvement based on data statistics and visualization of results. These services are integrated into interactive workflows, which can be steered from desk-top or mobile devices. A tool called workflow composition assistant allows semi-automatic workflow construction based on the SemanticWeb technology to increase the productivity of analytic tasks.
- *Use cases.* They show how the above technologies and tools work together, and how they can be used in realistic situations.

The presented research and development was conducted in cooperation with some of the worldwide leading grid research and application groups. The set of pilot applications directly profiting from the project results includes the medical domain (cancer research, traumatic brain injuries, neurological diseases and brain informatics) and the ecological domain (environment monitoring and event prediction).

3.2 Rationale behind the design and development of GridMiner

Since the 1990s, grid technology has traversed different phases or generations. In 2002 when the research on GridMiner started, all major grid research projects were built on the Globus Toolkit (Foster and Kesselman, 1998) and UNICORE (Romberg, 2002). Therefore, our initial design of the GridMiner architecture was based on the state-of-the-art research results achieved by Globus and its cooperating research projects, which provided grid software development kits exposing 'library-style APIs'. Very soon after the start of the project, the grid research programme and industry activities focused on architecture and middleware development aligned with the Web services standards. Further, we realized that our effort had to consider the achievements in the Semantic Web, workflow management and grid database technologies.

As a first step, we designed and prototyped a runtime environment and framework called GridMiner-Core based on the *Open Grid Service Architecture* (*OGSA*) concepts built on top of the Globus Toolkit Version 3 (Foster, *et al.*, 2002) and grid database access services provided by the *OGSA Data Access and Integration* (*OGSA-DAI*) (Antonioletti, *et al.*, 2005) middleware. This solution allowed the integration and execution of data pre-processing, data mining tools, applications and algorithms in a grid-transparent manner, i.e. the algorithm contributors could focus on knowledge discovery problems without the necessity to handle grid specifics. Thus, the framework abstracted from grid details such as platform, security, failure, messaging and program execution. The design decisions, scalability behaviour with respect to the data set size and number of users working concurrently with the infrastructure and performance of the prototype were evaluated. Based on the results and experience from the GridMiner-Core research tasks, a full service-oriented GridMiner architecture has been investigated (Hofer and Brezany, 2004).

Since the birth of the OGSA technology, based on the Web service concepts, the grid community has been focusing on the service-oriented architectures. This trend was confirmed by the latest OASIS specification, named *Web Service Resource Framework* (*WSRF*) (Czajkowski, *et al.*, 2004). Therefore, our goal was also to develop an infrastructure supporting all the phases of the knowledge discovery process in the service-oriented grid environments with respect to the scientific workflows and the latest available technologies. Furthermore, we have devoted significant effort to the investigation of appropriate modelling mechanisms. We adopted and extended the existing *CRoss-Industry Standard Process for Data Mining* (*CRISP-DM*) Reference Model (Chapman, *et al.*, 2000) and its phases as essential steps for the service-oriented scientific workflows. The model is discussed in Section 3.4.

3.3 Use case

Before describing the GridMiner framework, we present a practical use case taken from a medical application addressing management of patients with traumatic brain injuries (Brezany, *et al.*, 2003a). A traumatic brain injury typically results from an accident in which the head strikes an object. Among all traumatic causes of death, it is the most important one, besides injuries of the heart and great vessels. Moreover, survivors of a traumatic brain injury may be significantly affected by a loss of cognitive, psychological or physical function. Below, we discuss how data mining and its implementation in GridMiner can support treatment of traumatic brain injury patients.

At the first clinical examination of a traumatic brain injury patient, it is very common to assign the patient into a category, which allows one to plan the treatment of the patient and also helps to predict the final outcome of the treatment. There are five categories of the final outcome defined by the *Glasgow Outcome Scale* (*GOS*): dead, vegetative, severely disabled, moderately disabled and good recovery. It is obvious that the outcome is influenced by several factors, which are usually known and are often monitored and stored in a hospital data warehouse; frequently used factors include Injury Severity Score, Abbreviated Injury Scale and Glasgow Coma Score. It is evident that if we want to categorize a patient then there must be a prior knowledge based on cases and outcomes of other patients with the same type of injury. This knowledge can be mined from the historical data and represented as a classification model. The model can be then used to assign the patient to one of the outcome categories. In particular, using the model, one of the values from the GOS can be assigned to a concrete patient.

One of the basic assumptions in classification is that by considering a larger number of cases, the accuracy of the final model can be improved. Therefore, access to the data for similar traumatic brain injury cases stored in other hospitals would help to create a more accurate classification model. In the grid environment, a group of hospitals can share their data resources such as anonymized patients records or some statistical data related to the management of the hospitals. The group of hospitals can be then seen as a *virtual organization* (*VO*) (Foster, *et al.*, 2002). There can be also other partners in the VO offering their shareable resources, for example, analytical services or high-performance computing resources, as illustrated in Figure 3.1. In such a scenario, it is necessary to deal with other challenges such as secure access to the distributed data, cleaning and integration of the data and its transformation into a format suitable for data mining. There are also many other tasks related to this problem such as legal aspects of data privacy that have to be solved, especially for such sensitive data patients' records.

There exist several possibilities for how to deal with the above-mentioned challenges. Our approach is based on the principle that the movement of the data should be reduced as much

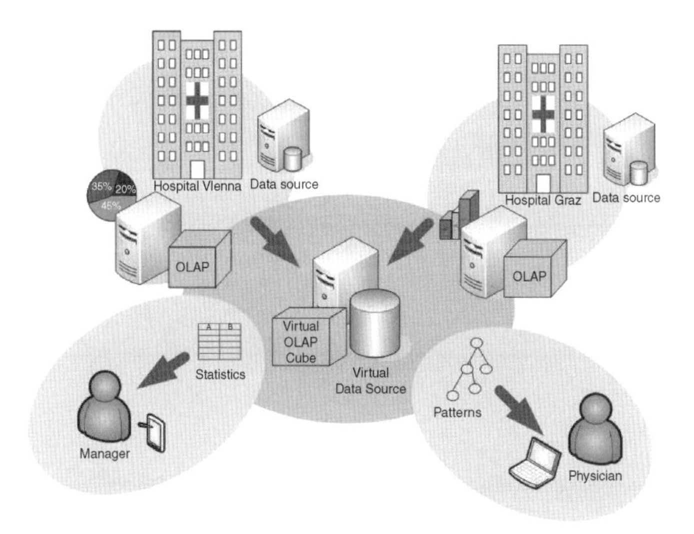

Figure 3.1 Use case scenario

as possible to avoid the leak of valuable information. This means that some tasks, such as data pre-processing, should be performed as close as possible to the data source. If the data movement is unavoidable then only patterns or aggregated data should be moved out of the hospital's data centre.

Another situation involves data integration prior to mining that requires all data records. To address this requirement problem we adopted an approach based on data virtualization and data mediation. This means that in order to reach a particular record a distributed query mechanism supported by the mediator service (Brezany, *et al.*, 2003b) is applied. The data integration problem has also several other aspects, which should not be ignored. The first one is semantic inconsistency of the integrating data sources, and the second one is the data format inconsistency.

Before a data mining technique can be applied, the data need to be preprocessed to correct or eliminate erroneous records. Problematic records may contain missing values or noise such as typos or outlier values that can negatively impact the quality of the final model.

From a health care control and governance point of view, hospital management data about the health care services and their quality are more interesting. The data of interest in this case include statistics, which can be represented as aggregated data. This can be supported by *online analytical processing* (*OLAP*) techniques.

3.4 Knowledge discovery process and its support by the GridMiner

Knowledge discovery in databases (*KDD*) can be defined as a non-trivial process of identifying valid, novel, potentially useful and ultimately understandable patterns in data (Fayyad, Piatetsky-Shapiro and Smyth, 1996). KDD is a highly interactive process and to achieve useful results the user must permanently have the possibility to influence this process by applying different algorithms or adjusting their parameters. Initiatives such as CRISP-DM try to define the KDD project, its corresponding phases, their respective tasks and relationships between these

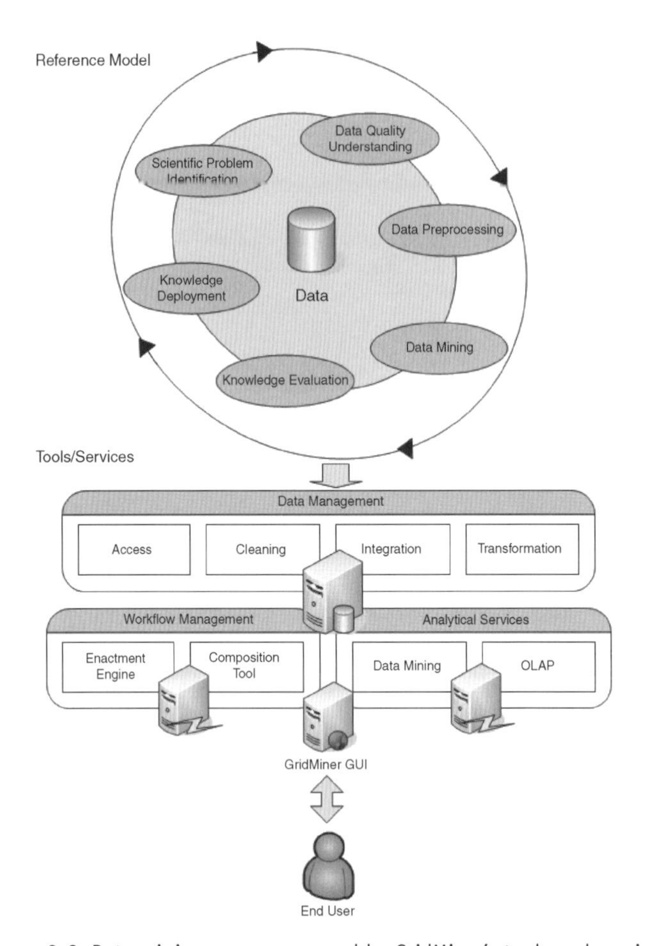

Figure 3.2 Data mining process covered by GridMiner's tools and services

tasks. Within the GridMiner project, we have adopted and extended the CRISP-DM model and its phases as essential steps to the service-oriented scientific workflows. The whole concept is depicted in Figure 3.2. The top part corresponds to our new reference model, and the bottom part shows an illustrative GridMiner framework and its supporting tools and services.

3.4.1 Phases of knowledge discovery

In the following paragraphs, we discuss the phases at the highest level of our new reference model for data mining projects as implemented in the GridMiner framework.

Scientific problem identification This initial phase focuses on clearly establishing goals and requirements of the scientific problem that is going to be solved, the role of data mining in the solution approach and selecting suitable data sets in the given data space and mining techniques. Within this phase, the main objectives of the data mining process are identified and their basic properties are specified in a preliminary workplan. The ways of solving the problem can include different methods and approaches, for example a scientific experiment, statistical confirmation of a hypothesis and so forth.

Data quality understanding The data and its quality understanding phase starts with an initial data collection and proceeds with activities allowing the user to become familiar with the data, to identify data quality problems, to obtain first insights into the data or to detect interesting subsets to form hypotheses.

Data pre-processing According to Pyle (1999), data pre-processing is usually the most challenging and time consuming step in the whole knowledge discovery process. The aim of this phase is to improve data quality, which has a significant impact on the quality of the final model and, therefore, on the success of the whole process. Data in operational databases are typically not clean. This means that they can contain erroneous records due to wrong inputs from users or application failures. Besides, these data may be incomplete, and essential information may not be available for some attributes. Hence, for data mining and other analysis tasks, it is necessary to clean data and to integrate the data from multiple sources such as databases, XML files and flat files into a coherent single data set. Because of the large size of typical grid data sources, building traditional data warehouses is impractical. Therefore, data integration is performed dynamically at the point of data access or processing requests.

Moreover, the size of the input data for modelling can enhance the accuracy of the predictive model as well as having significant impact on the time of model building. The data selection step allows the choice of an appropriate subset of the whole data set, which can be used as a training set to build a model as well as a test set for the model evaluation.

Data mining This phase deals with selecting, applying and tuning a modelling technique on the preprocessed data set. It involves the application and parameterization of a concrete data mining algorithm to search for structures and patterns within the data set. Typically, data mining has the two high-level goals of *prediction* and *description*, which can be achieved using a variety of data mining methods such as association rules, sequential patterns, classification, regression, clustering, change and deviation detection and so forth. Besides these most fundamental methods, the data mining process could, for example in bioinformatics (Wang, *et al.*, 2005), refer to finding motifs in sequences to predict folding patterns, to discover genetic mechanisms underlying a disease, to summarize clustering rules for multiple DNA or protein sequences and so on. An overview of other modern and future data mining application areas is given by Kargupta, *et al.* (2003).

The large size of the available data sets and their high dimensionality in many emerging applications can make knowledge discovery computationally very demanding to an extent that parallel computing can become an essential component of the solution. Therefore, we investigated new approaches enabling development of highly optimized services for management of data quality, data integration, data mining text mining and OLAP. These services are able to run on parallel and distributed computing resources and process large, voluminous data within a short period of time, which is crucial for decision making. There were several concurrency levels investigated as depicted in Figure 3.3.

(a) *Workflows.* Speculative parallelism can be used for parameter study of analytical tasks (for example, concurrent comparison of accuracy of neural network, decision tree and regression classification methods).

(b) *Inter-service parallelism.* Services can perform coordinated distributed processing.

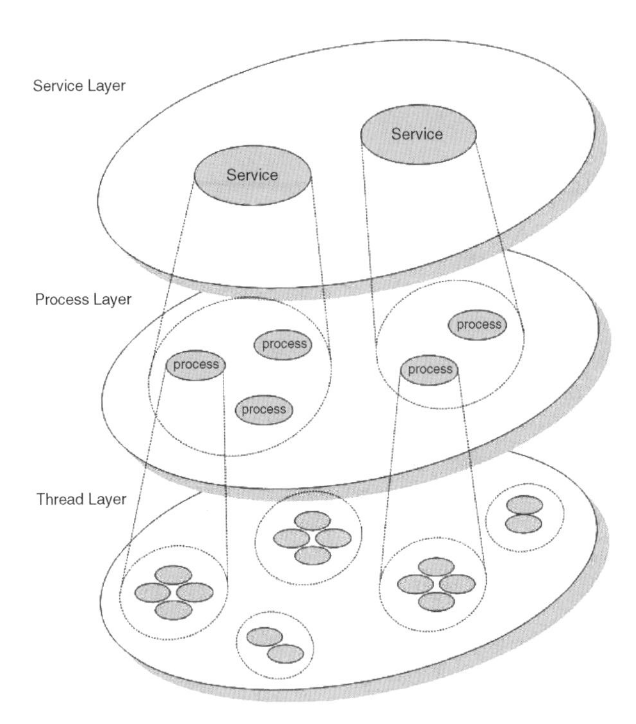

Figure 3.3 Inter and intra-service parallelism

(c) *Intra-service parallelism.* Service implementation is based on process and thread mechanisms, which are the concepts of parallel programming technology.

Within the data exploration processes we also investigated the impact of the data size on the models' accuracy to answer the question 'When is it necessary to mine complete data sets, as opposed to mining a sample of the data?' This opens a new dimension for data exploration methodologies. We also included OLAP in our research focus because we consider OLAP and data mining as two complementary technologies, which, if applied in conjunction, can provide efficient and powerful data analysis solutions on the grid.

Evaluation Data mining algorithms discover patterns in data, not all of which may be new or be of interest to the user. The evaluation phase is designed to address this issue and discriminate relevant from irrelevant patterns. The main goal of the presentation task is to give a user all discovered information in an appropriate form. In GridMiner, we addressed different visualization techniques for representing data mining models in, for example, tables, graphs and charts. These models are represented by the PMML so they can be imported to the other applications and further evaluated.

Knowledge deployment It is obvious that the purpose of a data mining model, as a final product of the data mining process, is to increase knowledge about the analysed data. A model itself, without an appropriate description, can be of limited utility for the end user who is carrying out the deployment phase. Therefore, a kind of report describing used data and all the steps leading to the model is required. The *Predictive Markup Model Language* (*PMML*)

(Data Mining Group, 2004) facilitates not only the characterization of data mining models but also the description of data, metadata, statistics and data transformations. Hence, together with visualization, the PMML is a suitable format for knowledge representation. However, the representation of the models is usually not enough. Its interpretation by a domain expert is extremely helpful and can point to the weakness of the model to eliminate its misuse.

In the following sections we will introduce the services and tools supporting the phases of the knowledge discovery process. We will not go too deep into implementation details, which can be found in the cited papers, but rather introduce the design concepts we have used to develop the presented tools.

3.4.2 Workflow management

With respect to the discussion about the phases of the KDD, it is obvious that the GridMiner architecture highly depends on appropriate workflow management mechanisms. We came to the conclusion that GridMiner needs a dynamic workflow concept where a user can compose the workflow according to his individual needs. As a first step, an XML-based Dynamic Service Composition Language (Kickinger, *et al.*, 2004) was proposed; it serves as input to the *Dynamic Service Control Engine (DSCE)* (Kickinger, *et al.*, 2003), which was also developed within the project. The DSCE was implemented as an application-independent, stateful and transient OGSA (Foster, *et al.*, 2002) service consisting of a set of other OGSA services. Additionally, the engine allows the starting, stopping, resuming, modifying and cancelling of workflows and notifies a client about the status of the execution.

Workflow Enactment Engine The new scientific workflow concepts (Taylor, *et al.*, 2007), the acceptance of the WS-BPEL language (Sarang, Mathew and Juric, 2006) in the scientific community and standardization of WSRF have become the leading motivation for the development of a new workflow engine. A newly established subproject of the GridMiner project called *Workflow Enactment Engine Project (WEEP)* (Janciak, Kloner and Brezany, 2007) is aiming to implement a workflow engine for WSRF services using WS-BPEL 2.0 as a formalism to specify, process and execute workflows. The engine follows up ideas of its ancestor DSCE and is being developed as a central component for data mining workflow orchestration in GridMiner. The core of the engine provides a run-time environment in which process instantiation and activation occurs, utilizing a set of internal workflow management components responsible for interpreting and activating the process definition and interacting with the external resources necessary to process the defined activities. The main functionality of the engine can be summarized as follows.

- *Validation and interpretation.* The validation and interpretation of a BPEL process definition and its representation as a new resource with a unique identifier. The resource can be seen as a workflow template representing a stateless process.
- *Instantiation.* The instantiation of the workflow template to the stateful process represented by a service.
- *Execution.* The execution of the process instance together with controlling and monitoring of the whole workflow and its particular activities.

With the above-mentioned functionality, the engine is able to reuse deployed workflows, create new instances with different input parameters and represent the process as a single service.

GridMiner Assistant To support a semi-automatic composition of the data mining workflows we have developed a specialized tool called GridMiner Assistant (Brezany, Janciak and Tjoa, 2007), which assists the user in the workflow composition process. The GridMiner Assistant is implemented as a Web application able to navigate a user in the phases of the knowledge discovery process and construct a workflow consisting of a set of cooperating services aiming to realize concrete data mining objectives. The GridMiner Assistant provides support in choosing particular objectives of the knowledge discovery process and manages the entire process by which properties of data mining tasks are specified and results are presented. It can accurately select appropriate tasks and provide a detailed combination of services that can work together to create a complex workflow based on the selected outcome and its preferences. The GridMiner Assistant dynamically modifies the task composition depending on the entered values, defined process preconditions and effects, and existing description of services.

3.4.3 Data management

GridMiner needs to interoperate with data access and management technologies that are being developed by other communities. On top of these technologies, we developed services for data access, data integration, data preprocessing, computation of statistics and a novel concept of a grid-based data space. All these services and concepts are described in the following paragraphs.

Data access In order to avoid building a new proprietary solution and reimplementing solved aspects of *Grid Data Services* (*GDS*) (Antonioletti, *et al.*, 2003), we have decided to integrate our developed concepts into OGSA-DAI (Antonioletti, *et al.*, 2005). OGSA-DAI (see also Chapter 14 in this volume) is a middleware implementation of GDS for supporting access and integration of data from separate data sources within a grid. Moreover, the OGSA-DAI allows us to define extensions (called activities), which can internally transform data into the required form and deliver them as a file or as a data stream. The middleware uses XML as its native format for data representation. Therefore, XML was also adopted by data mining services implemented in GridMiner as standard input data format so no further conversion of the data is necessary. The latest release of the OGSA-DAI is implemented as a WSRF service and is also available as a core component for data access and integration in the Globus Toolkit. For all these reasons the OGSA-DAI was included in the GridMiner architecture as an essential service supporting data mining services with the access to the data sources on the grid.

Data integration Modern science, business, industry and society increasingly rely on global collaborations, which are very often based on large-scale linking of databases that were not expected to be used together when they were originally developed. We can make decisions and discovery with data collected within our business or research. But we improve decisions or increase the chance and scope of discoveries when we combine information from multiple sources. Then correlation and patterns in the combined data can support new hypotheses, which can be tested and turned into useful knowledge. Our approach to data integration is based on the data mediation concept, which allows for the simplification of work with multiple, federated and usually geographically distributed data sources. The realization of the 'wrapper/mediator' approach is supported by the *Grid Data Mediation Service* (*GDMS*) (Woehrer, Brezany and Tjoa, 2005), which creates a single, virtual data source. The GDMS allows integration of heterogeneous relational databases, XML databases and comma-separated value files into one logically homogeneous data source.

Data statistics, data understanding, data pre-processing The D^{3}G framework (Wöhrer, *et al.*, 2006), developed within the GridMiner project, describes an architecture for gathering data statistics on the fly and uses them in remote data pre-processing methods on query results. Additionally, it provides access to incrementally maintained data statistic of pre-defined database tables via the GDS. The implementation of the framework is based on an extension of OGSA-DAI with a set of new activities allowing for the computation of different kinds of data statistic. The framework provides a set of OGSA-DAI activities for extracting descriptive and advanced statistics, together with a specialized monitoring tool for continuously updating statistics.

- *Descriptive statistics.* Could be viewed as a metadata extraction activity providing basic information about the selected tables for nominal and numerical attributes.
- *Advanced statistics.* Allow for the enhancement of basic statistics into advanced data statistics based on additional input information about a data set, for example statistics for intervals.
- *Data source monitoring.* Provides an interface to the incrementally maintained statistics about whole tables returning more recent statistics.

All data statistics are presented in the PMML format to support service interoperability. The gathered statistical data are useful in deciding what data pre-processing technique to use in the next phase or whether it is necessary at all, which is especially interesting for very expensive pre-processing methods.

DataEx The aim of this research task is to develop a new model for scientific data management on the grid, which extends our earlier work on structural and semantic data integration (Woehrer, Brezany and Tjoa, 2005) to provide a new class of data management architecture that responds to the rapidly expanding demands of large-scale data exploration. DataEx follows and leverages the visionary ideas of *dataspaces*, a concept introduced by Franklin, Halevy and Maier (2005). A dataspace consists of a set of participants and a set of relationships. Participants are single data sources (elements that store or deliver data). Relationships between them should be able to model traditional correlations, for example one is a replica of another one, as well as novel future connections, such as two independently created participants contain information about the same physical object. Dataspaces are not a data integration approach – rather more a data co-existence approach.

3.4.4 Data mining services and OLAP

Sequential, parallel and distributed data mining algorithms for building different data mining and text mining models were proposed and implemented as grid services using the multi-level concurrency approach. Each service can be used either autonomously or as a building block to construct distributed and scalable services that operate on data repositories integrated into the grid and can be controlled by the workflow engine. Within the project, the following services providing data and text mining tasks were developed.

- *Decision trees.* Distributed version of grid service able to perform high-performance classification based on the SPRINT algorithm (Shafer, Agrawal and Mehta, 1996). The service is implemented as a stateful OGSA service able to process and distribute data provided by data sources represented by the OGSA-DAI interface. The algorithm and the service implementation details are described by Brezany, Kloner and Tjoa (2005).
- *Sequence patterns.* Implementation of the out-of-core sequence mining algorithm SPADE producing a sequence model and its evaluation. The work on the service is in progress.
- *Text classification.* Parallel implementation of the tree classification algorithm operating over text collections represented in the XML format. Implementation details and performance results can be found elsewhere (Janciak, *et al.*, 2006).
- *Clustering.* Service enabling discovery of a clustering model using the k-means algorithm and evaluation of the final model. The work on the service is in progress.
- *Neural networks.* Parallel and distributed version of the back-propagation algorithm implemented as a high-performance application in the Titanium language (a parallel Java dialect) and fully tested on HPC clusters. The used algorithm and implementation details together with performance results can be found elsewhere (Brezany, Janciak and Han, 2006).
- *Association rules.* A specialized service operating on top of OLAP using its data cube structure as a fundamental data source for association rule discovery. A detailed description of the design and implementation of the service can be found elsewhere (Elsayed and Brezany, 2005).

All the services were fully tested in our local test bed, and the performance results were compared with other implementations. Moreover, the visualization application was integrated into the graphical user interface and can visualize the models of the developed services, which are represented in the PMML format.

OLAP So far, in the grid community, no significant effort has been devoted to data warehousing and associated OLAP, which are kernel parts of modern decision support systems. Our research effort on this task resulted in several original solutions for scalable OLAP on the grid.

Because there was no appropriate open source OLAP system for proof-of-concept prototyping of developed concepts available, we decided to develop appropriate design and implementation patterns from scratch. First, we designed the architecture of a sequential grid OLAP Engine (Fiser, *et al.*, 2004) as a basic building block for distributed OLAP on the grid. This architecture includes the data cube structure, an index database and function blocks for cube construction, querying and connection handling. During investigation of existing cube storage and indexing schemes, we discovered that there was no suitable method available for management of sparse OLAP cubes on the grid. The existing methods, for example *Bit-Encoded Binary Structure* (*BESS*) (Goil and Choudhary, 1997), require that the number of positions within each cube dimension is known before the records are imported from data repositories into the cube. However, such exact information is often not available in grid applications. Therefore, we proposed a method called *Dynamic Bit Encoding* (*DBE*) as an extension of BESS. DBE is based on basic bit logic operations; coding and indexing are fully extendable during the cube construction.

With this concept it was possible to focus our research on a parallel and distributed OLAP solution. In our approach, the OLAP cube is assumed to be built on top of distributed computational and data storage resources – the goal is to merge independent systems into one large virtual federation to gain high computational power and storage capacity. However, the OLAP Management System has to virtualize all these resources. Therefore, for the end user and other applications, we consider this data cube as one virtual cube, which is managed by resources provided by cooperating partners. Cube segments (sets of chunks) are assigned to grid nodes, which are responsible for their management.

A most helpful programming model for the development of scalable scientific applications for distributed-memory architectures is the SPMD model (SPMD: single program–multiple data). This model can be applied to data parallel problems in which one and the same basic code executes against partitioned data. We took an analogous approach to the development of scalable OLAP applications on the grid. We proposed the *SSMC model (SSMC: single service–multiple cube data)*. In this model, there is only one OLAP service code, which is used by the grid OLAP Service Factory for generation of a set service instances. The instances are then initiated on appropriate grid nodes due to the configuration specification, which can be smartly designed in the administration domain of the GridMiner graphical user interface and can be arbitrarily extended. For each service instance, it is only necessary to get information about the locations of its immediate children's services due to the hierarchical system architecture, data distribution and indexing strategies. The cube data aggregation operations and query processing are performed in each node by a pool of threads, which can really run in parallel if the node includes multiple processors.

The sequential and parallel OLAP engines were implemented in Java and integrated into GridMiner as a grid service. A special data mining service for discovery of association rules from OLAP cubes created and accessed by the OLAP service was proposed and implemented. The results of data mining and OLAP are represented by the *OLAP Model Markup Language* (*OMML*), which was proposed by our project.

3.4.5 Security

Security of the data and service access is extremely important in GridMiner, which is designed for distributed data access and manipulation. In cooperation with the *Cancer Bioinformatics Grid* (*caBIG*) project, we integrated *Dorian* (Langella, *et al.*, 2006) as a federated identity management service into the GridMiner framework. Dorian provides a complete grid-enabled solution, based on public key certificates and the *Security Assertion Markup Language* (*SAML*), for managing and federating user identities in a grid environment. SAML has been developed as a standard for exchanging authentication and authorization statements between security domains.

To obtain grid credentials or a proxy certificate, a digitally signed SAML assertion, vouching that the user has been authenticated by his/her institution, has to be sent to Dorian in exchange for grid credentials. The user authorization is accomplished through the user's identity. Dorian can only issue grid credentials to users that supply an SAML assertion from a Trusted Identity Provider. Dorian also provides a complete graphical user interface.

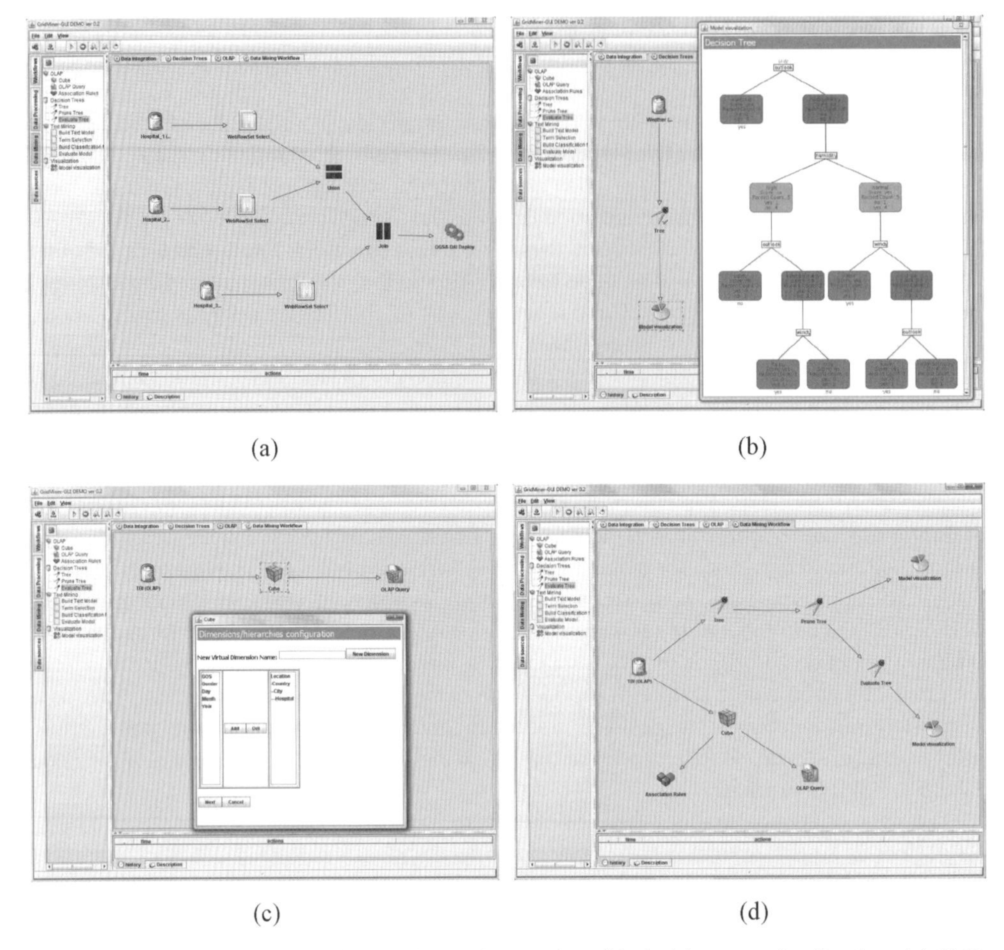

Figure 3.4 Graphical user interface. (a) Data integration, (b) decision tree visualization, (c) OLAP cube configuration, (d) workflow composition

Secure data mining services have their own credentials consisting of a private key and certificate assigned by a Grid Certificate Authority. The services are trusted based on the service identity; thus, services have full access authorization to other secure services.

Data access security The software architecture of the grid database access service OGSA-DAI provides the basic security mechanisms, which can be specified in a special security configuration file (RoleMap file). The application developers can extend the functionality of this approach. This idea is also followed in the GridMiner application of OGSA-DAI.

3.5 Graphical user interface

The aim of the *graphical user interface* (*GUI*) is to hide the complexity of the system architecture and offer an easy-to-use front end for the data mining experts as well as for the system administrators.

The GUI was designed as a stand alone Java application able to be remotely started by 'Java Web Start' practically in any operating system supporting Java. It provides a browser window allowing a user to interact with several Web applications used to configure and parameterize data mining tasks and interact with a specialized Web application for the workflow execution control. Another important contribution of the GUI is the possibility to interactively construct data mining workflows in a graphical representation on abstract and concrete levels (see Figure 3.4(d)). The workflow is created using drag-and-drop functionality and can be presented as a graph as well as being specified in the workflow language. Additionally, for the data integration tasks it is possible to construct mediation schemas used by GDMS (see Figure 3.4(a)). In summary, the main functionality of the GUI includes the following.

- Training and test data set selection based on query preparation and execution.
- Parameterization of tasks and configuration of the grid services.
- Data mining workflow composition and its execution.
- Controlling and monitoring of workflow execution.
- Visualizing data mining models, statistics and results of OLAP queries.

The layout of the GridMiner GUI is divided into three main panels as follows.

- *Resource panel.* The panel on the left hand side is used to manage grid services. They are clustered due to their relationship with the phases of the KDD process and are therefore categorized into data processing and data mining groups. The panel allows for the addition and selection of new data sources and management of created workflow instances.
- *Workflow composition panel.* This central panel of the GUI allows for the creation and modification of data mining workflow instances consisting of the resources selected from the resource panel using drag-and-drop mechanisms.
- *Log panel.* The bottom panel displays log and debug messages as well as notification messages during the workflow execution.

All the services of the workflow instance can be configured and parameterized by the user. Double-clicking on the resource icon in the *Workflow composition panel* opens a window displaying the Web application responsible for configuration of the particular service (see Figure 3.4(c)). Double-clicking on the visualization icon corresponding to the mined model opens a visualization application (see Figure 3.4(b)).

Visualization Several visualization methods for different data mining models have been applied in the GridMiner project, namely visualization for decision trees, association rules, neural networks, sequences and clustering. The visualization is supported by a Web application converting data mining models or statistics stored in PMML documents to their graphical or textual representations. The graphical representation is made by converting the models into the widely supported *Scalable Vector Graphics* (*SVG*) format, which can be displayed by standard Web browsers. A specialized interface for OLAP query visualization is also supported by a dynamic webpage displaying aggregated values and dimensions of the OLAP cube.

3.6 Future developments

From the previous paragraphs, we can see that we are already witnessing various grid-based data mining research and development activities. There are many potential extensions of this work towards comprehensive, productive and high-performance analysis of scientific data sets. Below we outline three promising future research avenues.

3.6.1 High-level data mining model

The most popular database technologies, like the relational one, have precise and clear foundations in the form of uniform models and algebraic and logical frameworks on which significant theoretical and system research has been conducted. The emergence of the relational query language SQL and its query processors is an example of this development. A similar state has to be achieved in data mining in order for it to be leveraged as a successful technology and science. The research in this field will include three main steps: (i) identifying the properties required of such a model in the context of large-scale data mining, (ii) examining the currently available models that might contribute (e.g. CRISP-DM) and (iii) developing a formal notation that exposes the model features.

3.6.2 Data mining query language

From a user's point of view, the execution of a data mining process and the discovery of a set of patterns can be considered as either the result of an execution of a data mining workflow or an answer to a sophisticated database query. The first is called the procedural approach, while the second is the descriptive approach. GridMiner and other grid-based data mining developments are based on this model. To support the descriptive approach several languages have been developed. The *Data Mining Query Language* (*DMQL*) (Han, 2005) and OLE DB for Data Mining (Netz, *et al.*, 2001) are good examples. A limitation of these languages is their poor support for data preparation, transformation and post-processing. We believe that the design of such a language has to be based on an appropriate data mining model; its features will be reflected by the syntactic and semantic structure of the language. Moreover, implementation of queries expressed in the language on the grid, which represents a highly volatile distributed environment, will require investigation of dynamic approaches able to sense the query execution status and grid status in specified time intervals and appropriately adapt the query execution plan.

3.6.3 Distributed mining of data streams

In the past five years with advances in data collection and generation technologies, a new class of application has emerged that requires managing data streams, i.e. data composed of continuous, real time sequence of items. A significant research effort has already been devoted to stream data management (Chaudhry, Shaw and Abdelguerfi, 2005) and data stream mining (Aggarwal, 2007). However, in advanced applications there is a need to mine multiple data streams. Many systems use a centralized model (Babcock, *et al.*, 2002). Here, the distributed data streams are directed to one central location before they are mined. Such a model is limited in many aspects. Recently, several researchers have proposed a distributed model considering distributed data sources and computational resources–an excellent survey was provided by

Parthasarathy, Ghoting and Otey (2007). We believe that investigation of grid-based distributed mining of data streams is also an important future research direction.

3.7 Conclusions

The characteristics of data exploration in modern scientific applications impose unique requirements for analytical tools and services as well as their organization into scientific workflows. In this chapter, we have introduced our approach to the e-science analytics that has been incorporated into the GridMiner framework. The framework aims to exploit modern software engineering and data engineering methods based on service-oriented Web and grid technologies. An additional goal of the framework is to give the data mining expert powerful support to achieve appealing results. There are two main issues driving the development of such an infrastructure. The first is the increasing volumes and complexity of involved data sets, and the second is the heterogeneity and geographic distribution of these data sets and the scientists who want to analyse them. The GridMiner framework attempts to address both challenges, and we believe successfully. Several data mining services have been already deployed and are ready to perform the knowledge discovery tasks and OLAP.

References

Aggarwal, C. (2007), *Data Streams: Models and Algorithms*, Advances in Database Systems, Springer.

Antonioletti, M., Atkinson, M., Baxter, R., Borley, A., Hong, N. P. C., Collins, B., Hardman, N., Hume, A. C., Knox, A., Jackson, M., Krause, A., Laws, S., Magowan, J., Paton, N. W., Pearson, D., Sugden, T., Watson, P. and Westhead, M. (2005), 'The design and implementation of grid database services in OGSA-DAI: Research articles', *Concurrency and Computation: Practice and Experience* **17** (2–4), 357–376.

Antonioletti, M., Hong, N. C., Atkinson, M., Krause, A., Malaika, S., McCance, G., Laws, S., Magowan, J., Paton, N. and Riccardi, G. (2003), 'Grid data service specification', http://www.gridpp.ac.uk/papers/DAISStatementSpec.pdf

Babcock, B., Babu, S., Datar, M., Motwani, R. and Widom, J. (2002), Models and issues in data stream systems, *in* 'Proceedings of the 21st ACM SIGMOD–SIGACT–SIGART Symposium on Principles of Database Systems (PODS'02)', ACM, New York, NY, pp. 1–16.

Berkley, C., Bowers, S., Jones, M., Ludaescher, B., Schildhauer, M. and Tao, J. (2005), Incorporating semantics in scientific workflow authoring, *in* 'Proceedings of the 17th International Conference on Scientific and Statistical Database Management (SSDBM'05)', Lawrence Berkeley Laboratory, Berkeley, CA, pp. 75–78.

Brezany, P., Janciak, I. and Han, Y. (2006), Parallel and distributed grid services for building classification models based on neural networks, *in* 'Second Austrian Grid Symposium', Innsbruck.

Brezany, P., Janciak, I. and Tjoa, A. M. (2007), Ontology-based construction of grid data mining workflows, *in* H. O. Nigro, S. E. G. Císaro and D. H. Xodo, eds, 'Data Mining with Ontologies: Implementations, Findings, and Frameworks', Hershey, New York, pp. 182–210.

Brezany, P., Kloner, C. and Tjoa, A. M. (2005), Development of a grid service for scalable decision tree construction from grid databases, *in* 'Sixth International Conference on Parallel Processing and Applied Mathematics', Poznan.

Brezany, P., Tjoa, A. M., Rusnak, M. and Janciak, I. (2003a), Knowledge grid support for treatment of traumatic brain injury victims, *in* '2003 International Conference on Computational Science and its Applications', Vol. 1, Montreal, pp. 446–455.

Brezany, P., Tjoa, A. M., Wanek, H. and Woehrer, A. (2003b), Mediators in the architecture of grid information systems, *in* 'Fifth International Conference on Parallel Processing and Applied Mathematics, PPAM 2003', Springer, Czestochowa, Poland, pp. 788–795.

Cannataro, M. and Talia, D. (2003), 'The knowledge grid', *Communications of the ACM* **46** (1), 89–93.

Chapman, P., Clinton, J., Kerber, R., Khabaza, T., Reinartz, T., Shearer, C. and Wirth, R. (2000), *CRISP-DM 1.0: Step-by-Step Data Mining Guide*, CRISP-DM Consortium: NCR Systems Engineering Copenhagen (USA and Denmark) DaimlerChrysler AG (Germany), SPSS Inc. (USA) and OHRA Verzekeringen en Bank Groep BV (The Netherlands).

Chaudhry, N., Shaw, K. and Abdelguerfi, M. (2005), *Stream Data Management (Advances in Database Systems)*, Springer.

Churches, D., Gombas, G., Harrison, A., Maassen, J., Robinson, C., Shields, M., Taylor, I. and Wang, I. (2006), 'Programming scientific and distributed workflow with Triana services: Research articles', *Concurrency and Computation: Practice and Experience* **18** (10), 1021–1037.

Ćurčin, V., Ghanem, M., Guo, Y., Köhler, M., Rowe, A., Syed, J. and Wendel, P. (2002), Discovery Net: towards a grid of knowledge discovery, *in* 'Proceedings of the 8th ACM SIGKDD International Conference on Knowledge Discovery and Data Mining (KDD'02)', ACM Press, New York, NY, pp. 658–663.

Czajkowski, K., Ferguson, D., Foster, I., Frey, J., Graham, S., Maguire, T., Snelling, D. and Tuecke, S. (2004), 'From Open Grid Services Infrastructure to WS-Resource Framework: Refactoring and Evolution, version 1.1', Global Grid Forum.

Data Mining Group(2004), 'Predictive Model Markup Language', http://www.dmg.org/

Elsayed, I. and Brezany, P. (2005), Online analytical mining of association rules on the grid, Technical Report Deliverable of the TU/Uni-Vienna GridMiner Project, Institute for Software Science, University of Vienna.

Fayyad, U., Piatetsky-Shapiro, G. and Smyth, P. (1996), 'From data mining to knowledge discovery in databases', *Ai Magazine* **17**, 37–54.

Fiser, B., Onan, U., Elsayed, I., Brezany, P. and Tjoa, A. M. (2004), Online analytical processing on large databases managed by computational grids, *in* 'DEXA 2004'.

Foster, I. and Kesselman, C. (1998), The Globus Project: a status report, *in* 'Proceedings of the 7th Heterogeneous Computing Workshop (HCW'98)', IEEE Computer Society, Washington, DC, p. 4.

Foster, I., Kesselman, C., Nick, J. M. and Tuecke, S. (2002), 'The physiology of the grid: an open grid services architecture for distributed systems integration', http://www.globus.org/research/papers/ogsa.pdf

Franklin, M., Halevy, A. and Maier, D. (2005), 'From databases to dataspaces: a new abstraction for information management', *SIGMOD Record* **34** (4), 27–33.

Goil, S. and Choudhary, A. (1997), 'High performance OLAP and data mining on parallel computers', *Journal of Data Mining and Knowledge Discovery* **1** (4), 391–417.

Han, J. (2005), *Data Mining: Concepts and Techniques*, Morgan Kaufmann, San Francisco, CA.

Hofer, J. and Brezany, P. (2004), DIGIDT: distributed classifier construction in the grid data mining framework GridMiner-Core, *in* 'Workshop on Data Mining and the Grid (DM-Grid 2004), held in conjunction with the 4th IEEE International Conference on Data Mining (ICDM'04)', Brighton, UK.

Janciak, I., Kloner, C. and Brezany, P. (2007), 'Workflow Enactment Engine Project', http://weep.gridminer.org

Janciak, I., Sarnovsky, M., Tjoa, A. M. and Brezany, P. (2006), Distributed classification of textual documents on the grid, *in* 'The 2006 International Conference on High Performance Computing and Communications, LNCS 4208', Munich, pp. 710–718.

Jones, M. B., Ludaescher, B., Pennington, D., Pereira, R., Rajasekar, A., Michener, W., Beach, J. H. and Schildhauer, M. (2006), 'A knowledge environment for the biodiversity and ecological sciences', *Journal of Intelligent Information Systems* **29** (1), 111–126.

Kargupta, H., Joshi, A., Sivakumar, K. and Yesha, Y., eds(2003), *Data Mining: Next Generation Challenges and Future Directions*, AAAI/MIT Press, Menlo Park, CA.

Kickinger, G., Brezany, P., Tjoa, A. M. and Hofer, J. (2004), Grid knowledge discovery processes and an architecture for their composition, *in* 'IASTED Conference', Innsbruck, pp. 76–81.

Kickinger, G., Hofer, J., Tjoa, A. M. and Brezany, P. (2003), Workflow management in GridMiner, *in* '3rd Cracow Grid Workshop', Cracow.

Langella, S., Oster, S., Hastings, S., Siebenlist, F., Kurc, T. and Saltz, J. (2006), Dorian: grid service infrastructure for identity management and federation, *in* 'Proceedings of the 19th IEEE Symposium on Computer-Based Medical Systems (CBMS'06)', IEEE Computer Society, Washington, DC, pp. 756–761.

Netz, A., Chaudhuri, S., Fayyad, U. M. and Bernhardt, J. (2001), Integrating data mining with SQL databases: OLE DB for data mining, *in* 'Proceedings of the 17th International Conference on Data Engineering', pp. 379–387.

Parthasarathy, S., Ghoting, A. and Otey, M. E. (2007), A survey of distributed mining of data streams, *in* C. C. Aggarwal, ed., 'Data Streams: Models and Algorithms', Springer, pp. 289–307.

Pyle, D. (1999), *Data Preparation for Data Mining*, Morgan Kaufmann San Francisco, CA.

Romberg, M. (2002), 'The UNICORE grid infrastructure', *Scientific Programming* **10** (2), 149–157.

Sarang, P., Mathew, B. and Juric, M. B. (2006), *Business Process Execution Language for Web Services, 2nd Edition. An Architects and Developers Guide to BPEL and BPEL4WS*, Packt.

Shafer, J. C., Agrawal, R. and Mehta, M. (1996), SPRINT: A scalable parallel classifier for data mining, *in* T. M. Vijayaraman, A. P. Buchmann, C. Mohan and N. L. Sarda, eds, 'Proceedings of the 22nd International Conference on Very Large Databases (VLDB'96)', Morgan Kaufmann, pp. 544–555.

Stankovski, V., Swain, M., Kravtsov, V., Niessen, T., Wegener, D., Kindermann, J. and Dubitzky, W. (2008), 'Grid-enabling data mining applications with DataMiningGrid: an architectural perspective', *Future Generation Computer Systems* **24**, 259–279.

Taylor, I. J., Deelman, E., Gannon, D. B. and Shields, M. (2007), *Workflows for e-Science: Scientific Workflows for Grids*, Springer, Secaucus, NJ.

Wang, J. T.-L., Zaki, M. J., Toivonen, H. and Shasha, D., eds(2005), *Data Mining in Bioinformatics*, Springer.

Woehrer, A., Brezany, P. and Tjoa, A. M. (2005), 'Novel mediator architectures for grid information systems', *Future Generation Computer Systems* **21** (1), 107–114.

Wöhrer, A., Novakova, L., Brezany, P. and Tjoa, A. M. (2006), D^3G: novel approaches to data statistics, understanding and pre-processing on the grid, *in* 'IEEE 20th International Conference on Advanced Information Networking and Applications', Vol. 1, Vienna, pp. 313–320.

Wolstencroft, K., Oinn, T., Goble, C., Ferris, J., Wroe, C., Lord, P., Glover, K. and Stevens, R. (2005), Panoply of utilities in Taverna, *in* 'Proceedings of the First International Conference on e-Science and Grid Computing (E-SCIENCE'05)', IEEE Computer Society, Washington, DC, pp. 156–162.

4

ADaM services: Scientific data mining in the service-oriented architecture paradigm

Rahul Ramachandran, Sara Graves, John Rushing, Ken Keizer, Manil Maskey, Hong Lin and **Helen Conover**

ABSTRACT

The *Algorithm Development and Mining System* (*ADaM*) was originally developed in the early 1990s with the goal of mining large scientific data sets for geophysical phenomena detection and feature extraction. The original design was a comprehensive system that comprised key software components for distributed computing including a mining daemon to handle mining requests, a mining database to fetch and stage appropriate data, a mining scheduler to schedule different mining jobs, a set of mining operations and a mining engine to parse mining plans or workflows and to execute the right mining operations in the right sequence. ADaM has over 100 algorithms and has been used extensively in different applications. With the advent of the *service-oriented architecture* (*SOA*) paradigm, ADaM was redesigned and refactored as a toolkit which provides both image processing and pattern recognition algorithms for science data sets as standalone components. These components can be easily packaged as Web or grid services using standard-based protocols. The services can be scripted together to form mining workflows using different composers. The use of standard protocols allows a mining workflow to be deployed as a solutions package at different sites to be reused by others or to be integrated into other applications using service composition. This chapter will describe the transition of ADaM from a system using tightly coupled client/server technology and non-standard protocols into a toolkit that can be moulded to fit both Web and grid service computing paradigms. The chapter will describe two different applications of ADaM services within the SOA context. In the first application, the ADaM Toolkit has been used as the basis for Web services to provide mining capability to scientists at a NASA data archive. The second application will employ grid mining services created using grid middleware. Mining examples are also presented to illustrate the use of ADaM Web and grid services to mine science data in the SOA paradigm.

Data Mining Techniques in Grid Computing Environments Edited by Werner Dubitzky

4.1 Introduction

There are two main problem areas that drive the need for data mining functionality at organizations such as NASA and the Earth Science community at large. First, since the Earth functions as a complex system, typical information system infrastructure and science practices are inadequate to address the many interrelated and interdisciplinary science problems. Solving these complex problems requires more advanced approaches, such as data mining, for analysing science data sets. Second, the volume of data being collected and stored in different science archives today defies even partial manual examination or analysis. Data mining addresses both of these problem areas by integrating techniques and algorithms from a multitude of disciplines to analyse and explore these massive data sets by providing automated methods. These techniques are borrowed from the diverse fields of machine learning, statistics, databases, expert systems, pattern recognition and data visualization. Data mining is an iterative, multistep process which includes data preparation, cleaning, pre-processing, feature extraction, selection and application of the mining algorithm(s), and, finally, analysis and interpretation of the result or the discovered patterns (Dunham, 2003).

Service-oriented architecture is a new paradigm for building client/server systems based on loose coupling between interacting software components. A service is a base unit of work made available by a provider to produce the desired end results for a consumer. SOA emphasizes the notion of 'separation of concern' – that most software systems are complex and it is more effective to consume a service rather than build one. SOA components do have to adhere to a set of constraints. They must have simple and ubiquitous interfaces and must communicate to each other via descriptive messages constrained by an extensible schema.

Data mining systems such as the ADaM system (Hinke *et al.*, 2000) were originally designed and developed as a tightly coupled application for specific requirements. With the new distributed computing landscape consisting of Web and grid services, it became clear that ADaM users required functional flexibility. For example, some users required the complete end-to-end mining capability at an archive whereas others only wanted to leverage a specific mining algorithm within their data analysis workflow. Based upon these user requirements, the ADaM system was redesigned as a comprehensive toolkit that can be used to build customized solutions and general applications and provide mining services. This chapter provides an overview of the ADaM system and its transition into a simple and effective toolkit being used to provide mining services for both Web and grid computing environments.

4.2 ADaM system overview

The *Information Technology and Systems Center* (*ITSC*) at the University of Alabama in Huntsville originally developed the ADaM system under a research grant from NASA to investigate new methods of processing large volumes of Earth Observing System remote sensing data sets. In order to allow for the distributed use of the data mining functionality, the ADaM system was designed as a client/server architecture, which supported remote client applications communicating with the data mining server. This allowed the server system to be co-located with archived data stores while being driven by either remote or local clients. Figure 4.1 depicts the connections between the components of the ADaM client/server architecture. Each component performed a specific well defined task.

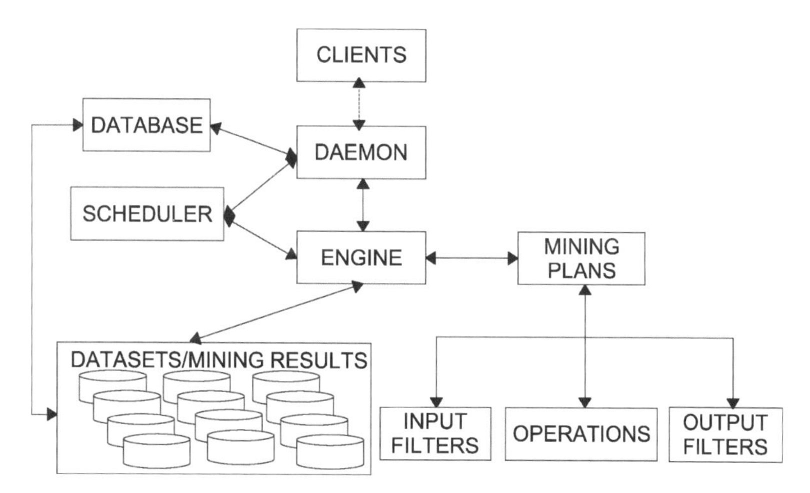

Figure 4.1 ADaM system architecture

The *mining engine* was the software component that managed the processing of data through a series of specified operations. The input, processing and output modules were dynamically loaded as needed at execution time, allowing the addition of newly developed modules without the need to rebuild the engine. The mining engine interpreted a mining plan script that contained the details about each specified operation and the order that they should be executed. Other communication with the mining engine was managed through the mining daemon process. The *mining daemon* was the gateway to the mining engine. All network communications with the mining system were handled by the daemon through a message handling protocol. Upon installation, the daemon was configured to listen on a specific port for any socket communications. The daemon was capable of handling a fairly rich set of messages that allowed it to perform file management duties, command the mining engine and provide user security screening. The daemon could also determine at run time which processing modules were available on the server. The database component was used to store information required for the smooth operation of the system and the interaction of its components. This information included the names, locations and related metadata for input data sets available on the server. The database also stored information about users, jobs, mining results and other related information. The *scheduler* component examined the list of jobs to be executed on the server and determined which job or jobs to execute at any given time. The scheduler invoked the mining engine for each job and monitored its progress, updating the job status in the database whenever it changed. Processing components included operations or mining algorithms and input/output filters (data translation modules). These operations and data set filters were implemented as a shared library. The libraries were loaded dynamically by the mining engine, allowing new modules to be added to the system without recompiling or relinking. Each of the operations and filters were completely independent of all the others and produced or operated on an internal data model representing scientific data. The mining plan script conveyed the processing instructions to the mining engine. The plan contained the number and sequence of processing steps as well as the detailed parameters (tokens) describing how to perform each step, such as where to find the input data and where to store the output and configuration parameters for all the various operations.

The ADaM data mining system is being used to extract content-based metadata from large Earth science data archives. It detects phenomena or events of interest to scientists (such as hurricanes or tropical cyclones[1]) and then stores this information to facilitate the data search and order process (Ramachandran *et al.*, 2000). Some mining results are also stored in a spatial database to find coincidences between mining-generated phenomena, climatological events and static information such as country and river basin boundaries (Hinke *et al.*, 1997). ADaM has also been used in other mining applications (Hinke *et al.*, 2000; Rushing *et al.*, 2001).

In 2000, ADaM was one of the first applications to run on NASA's Information Power Grid. For this early grid implementation, the ADaM Engine was modified to deploy as mining agents with required data to different grid processing nodes (Hinke and Novotny, 2000). More recently, ADaM has been redesigned as a toolkit to take advantage of the evolution of grid computing toward service-oriented architecture in recent years.

4.3 ADaM toolkit overview

The key issues for ADaM's redesign involved the packaging of the mining capabilities, the processing model for the toolkit, the data models used by the toolkit and the strategy for importing and exporting data. The ADaM Toolkit (Rushing *et al.*, 2005) is structured so that its components (individual algorithm implementations) are completely independent of one another. This avoids coupling and interdependence and allows easier integration with external tools. The ADaM Toolkit layers are depicted in Figure 4.2. The ADaM Toolkit itself consists of a layer of data models and a layer of tools constructed on top of these models.

The capabilities provided by ADaM can be categorized into two basic types: image processing capabilities and pattern analysis capabilities. The two types of capability are distinct in the types of data on which they operate. To address both types of capability, two different data models were developed: one for images and another for pattern data. Components share data models only when it is practical and efficient. The toolkit includes utilities to translate back and forth between the two models. The image data model represents an image as a three-dimensional array of pixel values, which are referenced by x-, y- and z-coordinates.

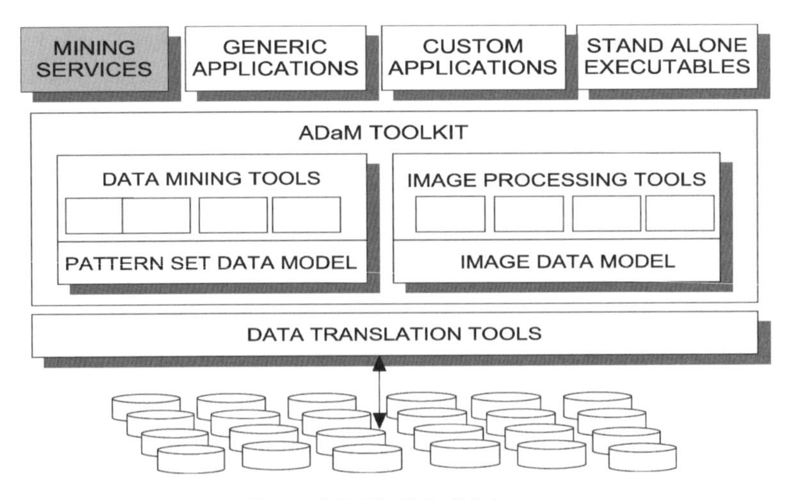

Figure 4.2 ADaM Toolkit layers

[1] http://pm-esip.msfc.nasa.gov/cyclone/

Two-dimensional images have z- size of one. Multispectral image data can be represented using arrays of 2D images along the z-coordinate. Methods to obtain and set pixel values, find the size of the image, and read and write binary image files are made available in the data model. The pattern set data model is implemented as the ADaM pattern set class. The pattern set is represented as an array of pattern vectors, with associated descriptors for each attribute. The attribute descriptors have the names of the attributes, their types and their range of legal values. One attribute may be designated as a class attribute. A pattern set may have any number of attributes associated with it, and may also contain an arbitrarily large number of patterns. Both numeric and categorical attributes are allowed in the pattern set and it may consist of mixed types of attribute. Methods to add or remove attributes, add or remove pattern vectors, obtain and set vector values, find attribute names and properties and read and write pattern data files are made available in the data model.

The data mining algorithms in the ADaM Toolkit are packaged as independent executables, allowing for maximum flexibility and efficiency. The executables are small (typically less than 100 KB each) and have no external dependences, making it possible to deliver arbitrary subsets of the toolkit. The actual functionality for a component is implemented in a single, statically linked C++ callable library, which can be called by the executable driver or a user application. This toolkit architecture provides users with flexibility of use. Individual modules can be used as single standalone executables, components may be mixed and matched to build customized mining solutions for specific problems and workflows can be constructed from the modules using a scripting or service orchestration language. The entire toolkit can be used to construct a general mining application such as ADaM-IVICS (Berendes *et al.*, 2007), which integrates data mining capabilities into a data visualization and analysis package.

4.4 Mining in a service-oriented architecture

The motivation to implement the data mining and image processing components of the ADaM Toolkit as services is that these services can be deployed with ease by various distributed online data providers and community projects, and be used immediately in data processing and analysis activities. These services form the building blocks for distributed, flexible processing in an SOA built around Web or grid services. This architecture allows different groups that focus on science data processing, basic research or the application arena to chain the mining services together with other services and community-based toolkits in many different ways. The underlying principle of an SOA is to minimize the need for writing new code each time a new processing capability is required, and to provide optimal means to reuse existing computing assets. In the context of science data processing and analysis, this architecture can provide a suite of data processing and analysis functions to scientists and end-users in a flexible processing system. In addition, different user algorithms packaged as services can easily be integrated with other processing and analysis services for quick testing and comparison. An SOA promotes the notion of composite applications that combine different existing services from various systems within the SOA. This means that not only can data processing and analysis functions interoperate horizontally across platforms, but lower level services can also be combined to create multi-layered, vertically abstracted services. Thus, a composite application is a new set of functionality based on well defined and established functions already present in the SOA. Building composite applications leverages quick reuse of existing services in the SOA, creates multiple layers of abstraction for the services, requires minimal coding to build new functionality and leverages existing computing assets to provide quick delivery of new capabilities to the users. The composite applications are not just leveraging existing services, but are also exploiting implicit domain expertise. For example, by utilizing ADaM data mining services, users will be able to build specialized composite applications using these services without having to be experts in the area of data mining. Likewise, these mining services can be combined with services from other domains to create powerful data processing and analysis solutions.

4.5 Mining web services

Because a sequence of operations is often necessary to provide an appropriate data mining solution to a complex problem, the SOA approach allows users to develop a workflow that employs various Web services for pre-processing to prepare the data for mining, one or more mining steps and post-processing to present the results. For use within an SOA, ADaM Toolkit components support Web service interfaces based on *World Wide Web Consortium* (*W3C*) specifications including SOAP interfaces and WSDL service descriptions. The primary customers for these services are data centres that provide online data to their community such as NASA Data Pools and would like to provide additional processing features to their end users. Because of the size and volume of data at different online science data providers, pushing data to distributed mining service locations often incurs a large network load. This burden can be overcome if the services are available at these online data provider sites along with the data. Therefore, the ADaM mining Web services have to be deployable and able to reside at many different distributed online data repositories. The need to be deployable places implementation constraints on these Web services, in that they must support different Web servers and computing platforms, thus allowing distribution to different groups with minimal site-specific support. The general architecture for mining Web services is presented in Figure 4.3. The functionality of the various components is the same as the original ADaM architecture (Figure 4.1) but implementation of the components has changed. The mining daemon is now a Web service. In the implementation described in this chapter, the *Business Process Execution Language* (*BPEL*) standard-based engine and associated service workflow have replaced the earlier mining engine and plan. BPEL provides a standard XML schema for workflow composition

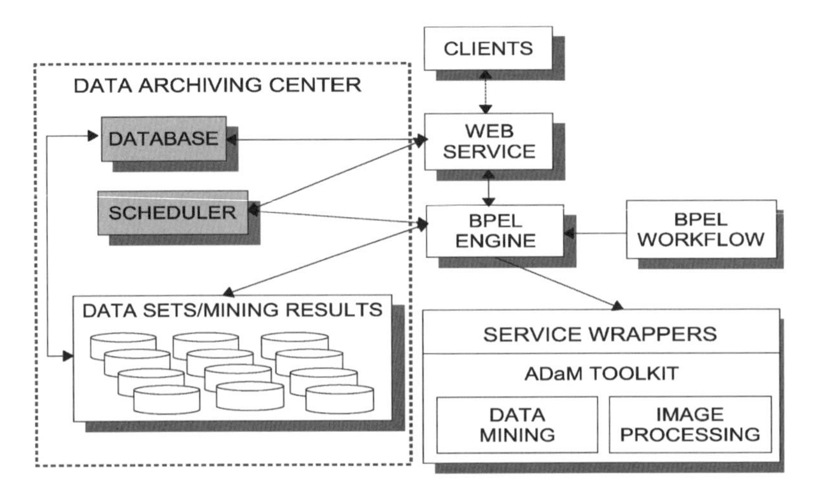

Figure 4.3 Generic Web service architecture to provide ADaM mining services at data archiving centres

of Web services that are based on SOAP. The workflow describes how tasks are orchestrated, including what components perform them, what their relative order is, how they are synchronized and how information flows to support the tasks. This standardized composition description is deployable on BPEL-compliant engines. Deploying a workflow translates into taking a BPEL description and asking a BPEL engine to expose the workflow as a Web service. Being able to expose a composition of Web services makes the workflow itself a Web service and thus reusable. A BPEL engine is also responsible for invoking the individual services.

The advantage of a standard-based approach such as BPEL is interoperability: a common execution language allows for the portability of workflow definitions across multiple systems. Any BPEL engine will now be able to take an ADaM workflow described in BPEL and invoke the individual services. The database component and the scheduler implementations in the Web service architecture are left to the individual data archives, as they might have different requirements and restrictions.

4.5.1 Implementation architecture

For one implementation of a data mining SOA, shown in Figure 4.4, ITSC has teamed with *NASA Goddard Earth Sciences Data and Information Services Center* (*GES DISC*). The GES DISC provides a large online repository of many NASA scientific data sets, together with a processing environment called the *Simple, Scalable Script-Based Science Processor for Measurements* (*S4PM*) (Lynnes, 2006), for applying user-supplied processing components to the data. ITSC provides an external 'sandbox' that allows the user to experiment with a small subset of the data and a variety of ADaM Toolkit components and parameter settings. A mining-specific workflow composition application hides the complexities of the BPEL language from users. Note that the user client, toolkit of mining services and BPEL service orchestration engine may be co -located at the data center or distributed across the Web. Once the user is satisfied with a workflow, it is deployed to a BPEL engine, which generates a WSDL description for the composite service (workflow) and returns its URL. This URL is then transmitted to the GES DISC via a Web service request along with a specification of the data to be mined, such as the data set identifier and temporal or spatial constraints. This request is provided to the GES DISC's processing engine S4PM for mining data at the data center. The S4PM engine is

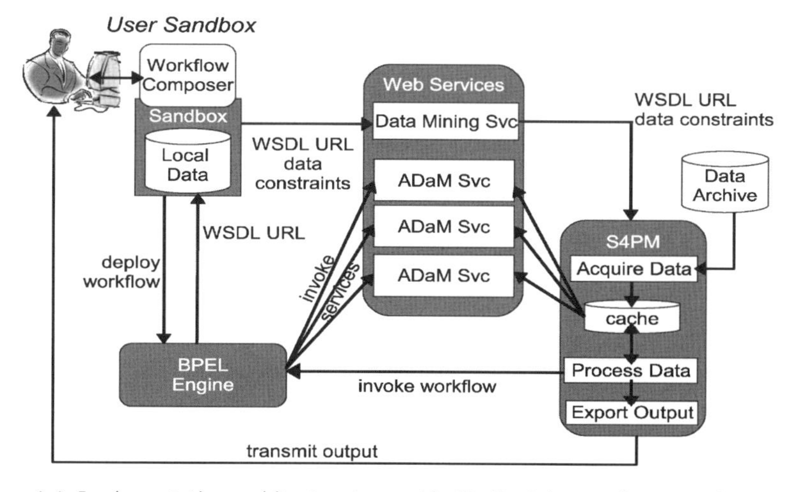

Figure 4.4 Implementation architecture to provide ADaM mining services at NASA GES DISC

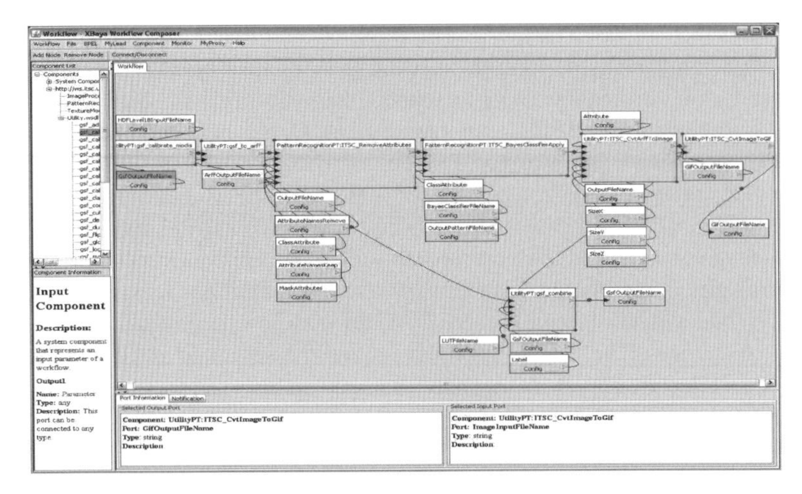

Figure 4.5 Example mining workflow to detect airborne dust in MODIS imagery

responsible for acquiring the right data from the archive, staging the data and then executing the requested workflow on each input file. However, note that the workflow itself is not provided to S4PM. Instead, S4PM uses the supplied WSDL URL to fetch the WSDL document and then invoke the corresponding composite Web service in the BPEL engine, supplying the full path of the input file. The BPEL engine invokes the atomic Web services in the proper order at the GES DISC. The result is then staged to an FTP directory, from which it can be retrieved and verified.

4.5.2 Workflow example

An example mining workflow to detect and classify air-borne dust in NASA's *Moderate Resolution Imaging Spectrometer* (*MODIS*) imagery is presented in Figure 4.5. The workflow consists of pre-processing steps to calibrate the MODIS data, perform the data model translation and subset the spectral bands to select the spectral bands most sensitive to dust classification. A naïve Bayes classifier is used for the detection and the result is converted into an image for visualization. The original data (Figure 4.6(a)) show dust blowing off the northwest coast of Africa. The classification result is presented as a thematic map in Figure 4.6(b). The pixels coloured black represent dust over water whereas the pixels coloured grey represent dust over land. Details of the dust detection study are discussed by Ramachandran *et al.* (2007). This example dust detection using classifier and data processing components demonstrates the different kinds of mining application that scientists will be able to perform with the ADaM services at data centres such as GES DISC.

4.5.3 Implementation issues

Several issues were encountered while implementing ADaM Web services and are listed below.

SOAP wrappers The Web service implementation of existing ADaM modules included placing a SOAP wrapper around each ADaM operation. Many options of the SOAP implementation were evaluated before deciding on SOAP::Lite[2], the Perl implementation of SOAP. Although

[2] http://www.soaplite.com/

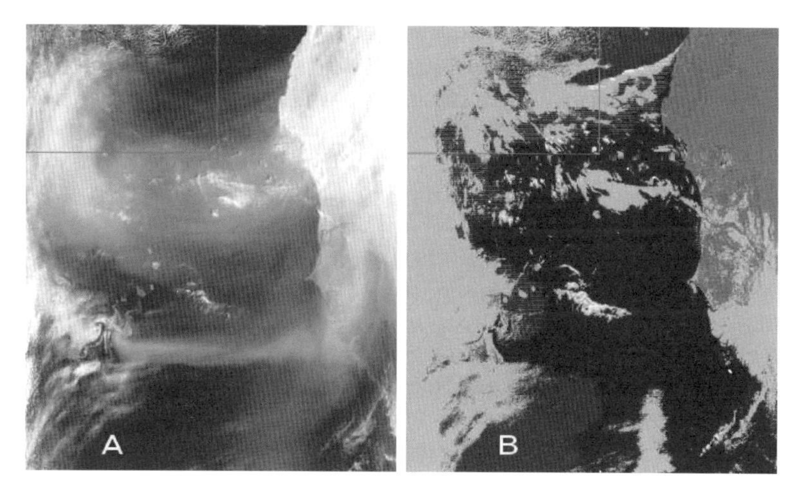

Figure 4.6 (*A*) Example MODIS true colour imagery depicting airborne dust off the coast of Africa. (*B*) Classification mask generated by the workflow (black, dust over water; grey, dust over land)

Java's implementation of SOAP, Axis[3], provides a wide variety of tools that assist in implementing services and clients, Java was not appropriate in this case because the ADaM Toolkit was implemented in C++, and interfacing *Java Native Interface* (*JNI*) to C++ can be problematic. The Axis C++ offers a C++ implementation of SOAP; however, it does not appear to be well accepted by the software industry so far. The Python implementation of SOAP, pySOAP, has not been updated or maintained in recent years. The *Web Service Definition Language* (*WSDL*) for each of the ADaM services has been published and each of the Web services was tested for interoperability using Perl and Java clients.

Security issues Even though this initial implementation did not focus on security issues associated with Web services, some basic rules were enforced. To allow the data centre full control over enforcing local policies for file creation, the SOAP wrappers specify the full path of the input file only, plus additional parameters as necessary. The data centre constructs the output directory and filename and returns it the calling program in the SOAP response. The output of one service may then be used as the input file to the next step in the workflow. This interface mechanism allows the data centre to control where files are created on its system and to implement safeguards against inappropriate usage.

Data staging There are two types of architecture available for service orchestration: data-centric and Web service-centric. In a Web service-centric architecture, the services are deployed at one centralized location. Data to be processed are made available using some sort of file handling protocol such as *File Transfer Protocol* (*FTP*), *UDP-Based Data Transfer Protocol* (*UDT*) (Gu and Grossman, 2007), or GridFTP[4]. These services transfer the data to the host of the Web services. Additional Web services are created to wrap these file handlers, so that these services can be incorporated into a workflow. In any workflow, these file transfer services will

[3]Apache Axis SOAP implementation: http://ws.apache.org/axis/

[4]Globus Toolkit 4.0, GridFTP: http://www.globus.org/ toolkit/docs/4.0/data/gridftp/.

always precede the actual processing components. Use of this architecture also makes maintaining the services easier, as changes to the services can be handled rather quickly. However, implementations using this architecture are limited by the rate at which data transfers occur.

In a data-centric architecture, the Web services are deployed at the data centers, so that all the data to be processed are available locally. Thus, the overhead of data transfer is avoided. The BPEL engine that hosts the workflow services can be hosted separately. A user can deploy a workflow to any BPEL engine using the service WSDLs that are exposed by the Web service hosts (in this case the data centres). The BPEL engine exposes a new WSDL to the composite service workflow such that data centre users can invoke the workflow to process specified data by calling the composite service. The processed data are not transferred to and from the BPEL engine every time a service in the workflow is invoked; instead a file handle is being passed in/out as parameters. Advantages of such third party transfers are highlighted by Ludäscher *et al.* (2005). The implementation architecture described in this chapter is data-centric as the primary target customers are large online data repositories interested in providing data processing and analysis services to their user community.

Deployment descriptors for BPEL engine A BPEL engine is required to process the instructions described in BPEL. There are many BPEL engines and visual workflow composers that are available as open source software. The problem with having different BPEL engines is that, even though BPEL is a standard language, these engines require an additional deployment descriptor that describes engine-specific additional dependences. The deployment descriptor varies from engine to engine. The ActiveBPEL[5] engine was selected because of its popularity.

4.6 Mining grid services

The *Linked Environments for Atmospheric Discovery* (*LEAD*) project[6] is building a comprehensive cyber infrastructure in mesoscale meteorology using grid computing technologies. This multidisciplinary effort involving nine US institutions and some 80 scientists and students is addressing the fundamental IT and meteorology research challenges needed to create an integrated, scalable framework for identifying, accessing, decoding, assimilating, predicting, managing, analysing, mining and visualizing a broad array of meteorological data and model output, independent of format and physical location (Droegemeier *et al.*, 2005). One of the major goals of LEAD is to develop and deploy technologies that will allow people (students, faculty, research scientists, operational practitioners) and atmospheric tools (radars, numerical models, data assimilation systems, data mining engines, hazardous weather decision support systems) to interact with weather. Even though mesoscale meteorology is the driver for LEAD, the functionalities and infrastructure being developed are extensible to other domains such as biology, oceanography and geology (Droegemeier *et al.*, 2005). An overview of the LEAD architecture and middleware components can be found in the work of (Gannon *et al.*, 2007).

The grid mining service approach for data mining, shown in Figure 4.7, is similar to the Web service implementation (Figure 4.3), with modifications as necessitated by the complexities and capabilities of the grid middleware. Both grid and Web mining services are built on ADaM executable modules. A data mining workflow can be created by using a workflow composition

[5]The ActiveBPEL Open Source Engine: http://www. activebpel.org/
[6]https://portal.leadproject.org/gridsphere/gridsphere

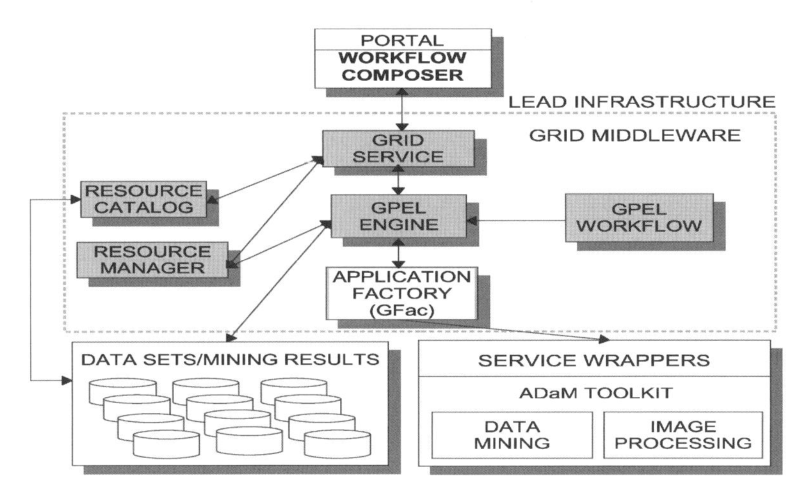

Figure 4.7 Grid Mining Service architecture

tool, which is part of the LEAD portal. A variant of BPEL, called GPEL (Kandaswamy *et al.*, 2006), provides additional grid-specific features used to describe workflows in LEAD. A GPEL engine is used to parse the workflow. One of differences in the Web and grid service implementations is that LEAD has only a small set of primary persistent services. 'Application services' including the mining modules are virtual and are created on demand by an application factory. The GPEL engine queries the application factory service to determine the best instance of the mining service and uses the query result to invoke the individual service at different host computers.

4.6.1 Architecture components

The *Generic Application Factory* (*GFac*) service (Slominski, 2007) provides a mechanism to convert any application executable such as an ADaM Toolkit component into a service. The GFac requires a wrapper around the application executable with the shell script meeting specific GFac Service Toolkit specifications. Use of GFac to convert ADaM modules to LEAD grid services entails wrapping an application executable, registering the service, creating an instance of the service from the registry and finally invoking the service. The ADaM Toolkit components can reside on separate hosts (computing resources) from the Factory service within the grid. Detailed descriptions for registration, creation and invoking of ADaM grid services are as follows.

Service registration Shell wrapped ADaM modules are registered as services in a GFac registry. Registering a service is a process of capturing metadata about the application server's host and the ADaM module's input and output parameters. A GFac specific XML descriptor is created for the registration of each service. GFac also provides a Web interface portal for registration of these services. During workflow creation the registry is automatically loaded for the discovery of these ADaM services.

Service creation An instance of any ADaM registered service can be created using the information in the GFac registry. After the mining service instance is created, a URL to the corresponding WSDL is returned. Many instances of the same service can be created. Each of these services will have a unique WSDL. Usually, this action is performed using a grid portal. The GFac portlet contacts the GFac registry and creates a service.

Service execution An instance of an ADaM service can be invoked as a standalone service from GFac or as a component of a workflow from a workflow composer via a portal. The input and output data for the service are staged using GridFTP for data transfer. The access to the GFac and the workflow portlets is via the portal and the grid credential used for the portal is also used to instantiate mining workflows at the application server.

Workflow composer Mining workflows are composed using XBaya, a graphical user interface for composing services. The composed workflows are deployed to a *Grid Process Execution Language* (*GPEL*) engine. During service registration, the user can select the input and output data to be staged, and any data to be published to the data catalogue.

GPEL GPEL is an XML based workflow process description language. A set of XML schemas is defined that closely follows the industry standard process execution language, BPEL. To utilize the Grid resources, GPEL has added additional attributes to split the services into multiple tasks and eventually combine the results. In addition, GPEL allows parallel execution of workflow components which is essential for mining large volumes of data.

4.6.2 Workflow example

A mining workflow to generate a cloud mask by applying a unsupervised classifier (k-means clustering algorithm) on images from the *Geostationary Operational Environmental Satellite* (*GOES*) can be created using the XBaya composer in the LEAD portal. The workflow contains additional processing services required to convert the input data to the required ADaM data model and to convert the result to an image file for visualization. The original data and the workflow results are presented in Figure 4.8.

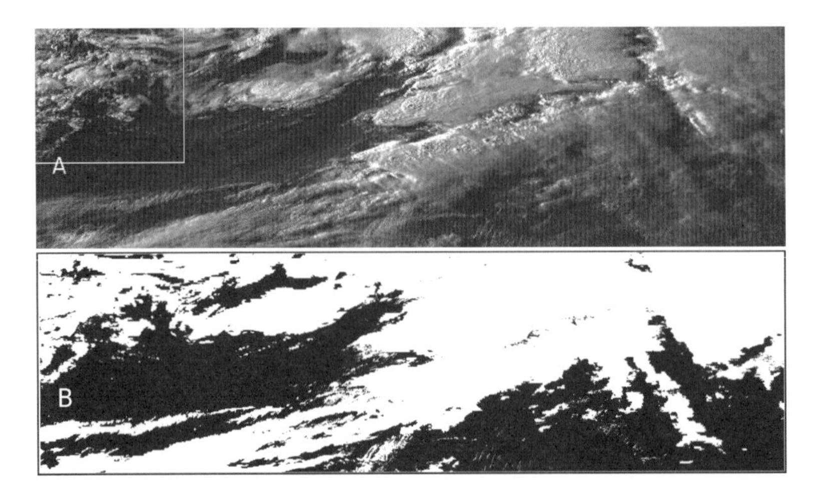

Figure 4.8 Clustering results from the sample workflow. (*A*) Original GOES image. (*B*) Two class cluster mask generated by the k-means algorithm

4.7 Summary

The redesigned ADaM Toolkit has been adapted to provide a suite of Web and grid services. Even though the basic architecture design and component functionality has not changed from the original ADaM system design, standards-based specifications and protocols have replaced internal specifications and protocols. The use of these standards now maximizes the interoperability of these services with other applications and tools. ADaM services allow scientists to utilize the components in the mining and image processing toolkit in a distributed mode. Using different workflow composers, scientists can mix and match various ADaM services with other available services to create complex mining and analysis workflows to solve their specific problem. Even though applications of these services presented in this chapter have focused on Earth Science problems, these mining services can and are being effectively applied to solve problems in various other domains.

Acknowledgements

The authors would like to acknowledge the contributions of Chris Lynnes and Long Pham at the NASA Goddard Earth Sciences Data and Information Services Center for the ADaM Web services effort. Dennis Gannon at Indiana University and his students have been instrumental in incorporating the ADaM Toolkit as grid services within the LEAD project. This research work was supported by NASA grant NNG06GG18A and NSF grant ATM-0331579.

References

Berendes, T., Ramachandran, R., Graves, S. and Rushing, J. (2007), ADaMIVICS: a software tool to mine satellite data, *in* '87th AMS Annual Meeting'.

Droegemeier, K. K., Gannon, D., Reed, D., Plale, B., Alameda, J., Baltzer, T., Brewster, K., Clark, R., Domenico, B., Graves, S., Joseph, E., Morris, V., Murray, D., Ramachandran, R., Ramamurthy, M., Ramakrishnan, L., Rushing, J., Weber, D., Wilhelmson, R., Wilson, A., Xue, M. and Yalda, S. (2005), 'Service-oriented environments in research and education for dynamically interacting with mesoscale weather', *IEEE Computing in Science and Engineering* **7**, 24–32.

Dunham, M. H. (2003), *Data Mining: Introduction and Advanced Topics*, Pearson Education.

Gannon, D., Plale, B., Christie, M., Huang, Y., Jensen, S., Liu, N., Marru, S., Pallickara, S. L., Perera, S., Shirasuna, S., Simmhan, Y., Slominski, A., Sun, Y. and Vijayakumar, N. (2007), Building grid portals for e-science: a service-oriented architecture, *in* L. Grandinetti, ed., 'High Performance Computing and Grids in Action', IOS Press.

Gu, Y. and Grossman, R. L. (2007), 'UDT: UDP-based data transfer for highspeed wide area networks', *Computer Networks* **51**, 1777–1799.

Hinke, T. and Novotny, J. (2000), Data mining on NASA's Information Power Grid, *in* '9th International Symposium on High-Performance Distributed Computing', pp. 292–293.

Hinke, T., Rushing, J., Kansal, S., Graves, S., Ranganath, H. and Criswell, E. (1997), Eureka phenomena discovery and phenomena mining system, *in* 'AMS 13th International Conference on Interactive Information and Processing Systems (IIPS) for Meteorology, Oceanography and Hydrology', pp. 277–281.

Hinke, T., Rushing, J., Ranganath, H. and Graves, S. (2000), 'Techniques and experience in mining remotely sensed satellite data', *Artificial Intelligence Review: Issues on the Application of Data Mining* **14**, 503–531.

Kandaswamy, G., Fang, L., Huang, Y., Shirasuna, S., Marru, S. and Gannon, D. (2006), 'Building Web services for scientific application', *IBM Journal of Research and Development* **50**, 249–260.

Ludäscher, B., Altintas, I., Berkley, C., Higgins, D., Jaeger-Frank, E., Jones, M., Lee, E., Tao, J. and Zhao, Y. (2005), 'Scientific workflow management and the Kepler system', *Concurrency and Computation: Practice and Experience* **18**, 1039–1065.

Lynnes, C. (2006), The simple, scalable, script-based science processor, *in* J. Qu, W. Gao, M. Kafatos, R. Murphy and V. Salomonson, eds, 'Earth Science Satellite Remote Sensing', Springer, pp. 146–161.

Ramachandran, R., Conover, H., Graves, S. J. and Keiser, K. (2000), Algorithm development and mining (ADaM) system for earth science applications, *in* 'Second Conference on Artificial Intelligence, 80th AMS Annual Meeting'.

Ramachandran, R., Li, X., Movva, S., Graves, S., Nair, U. S. and Lynnes, C. (2007), Investigating data mining techniques to detect dust storms in MODIS imagery, *in* '32nd International Symposium on Remote Sensing of Environment'.

Rushing, J., Ramachandran, R., Nair, U., Graves, S., Welch, R. and Lin, A. (2005), 'ADaM: a data mining toolkit for scientists and engineers', *Computers and Geosciences* **31**, 607–618.

Rushing, J. A., Ranganath, H. S., Hinke, T. and Graves, S. J. (2001), 'Using association rules as texture features', *IEEE Transactions on Pattern Analysis and Machine Intelligence* **23**, 845–858.

Slominski, A. (2007), Adapting BPEL to scientific workflows, *in* I. Taylor, E. Deelman, D. Gannon and M. Shields, eds, 'Workflows for e-Science', Springer, pp. 208–226.

5

Mining for misconfigured machines in grid systems

Noam Palatin, Arie Leizarowitz, Assaf Schuster and **Ran Wolff**

ABSTRACT

This chapter describes the *Grid Monitoring System* (*GMS*) – a system which adopts a distributed data mining approach to detection of misconfigured grid machines. The GMS non-intrusively collects data from sources available throughout the grid system. It converts raw data to semantically meaningful data and stores these data on the machine from which, it was obtained limiting incurred overhead and allowing scalability. When analysis is requested, a distributed outliers detection algorithm is employed to identify misconfigured machines. The algorithm itself is implemented as a recursive workflow of grid jobs and is especially suited to grid systems in which the machines might be unavailable most of the time or often fail altogether.

5.1 Introduction

Grid systems are notoriously difficult to manage. First, they often suffer more faults than typify other large systems: their hardware is typically more heterogeneous, as are the applications they execute, and since they often pool resources that belong to several administrative domains, there is no single authority able to enforce maintenance standards (e.g. with respect to software updates.) Once problems occur, the huge complexity of the system makes it very hard to track them down and explain them. Maintenance, therefore, is often catastrophe driven and rarely preventive.

The scale of a typical grid system and the volume of the data associated with and resulting from its operation clearly invite automated assistance for maintenance. Such automation can usually be done in two ways: an expert system can be built, which detects previously studied patterns of failure. Otherwise, knowledge discovery can be applied to the data, aiming to find ill behaving machines – a more fuzzy concept, which may encompass errors not yet studied by the expert. In the context of maintenance it is crucial that the outcome of knowledge discovery would be interpretable – i.e. that the user would be able to discern why a certain machine was indicated as suspected. Both approaches have been studied in real systems (IBM, 2002; Chen *et al.*, 2004), with no conclusive evidence favoring one over the other.

Data Mining Techniques in Grid Computing Environments Edited by Werner Dubitzky

Since expert systems can be difficult to apply in systems where one installation is very different from another, and are also hard to port from one system to the next, the knowledge discovery approach seems more fit for grid systems. However, applying knowledge discovery to raw data (log files, etc.) would lead to results that are difficult to interpret. It is thus required to first elevate the data to a semantically meaningful representation – an ontological domain. This has the further benefit that by executing knowledge discovery in an abstracted domain this portion of the monitoring system becomes highly portable. Raw data, in contrast, is always system specific and systems that rely on it are hard to port.

A grid system may well have thousands of machines, each producing megabytes of monitoring data daily. If these data are to be centralized for processing, they would suffocate the system, and require specialized storage. For instance, other work (Chen *et al.*, 2004) that used centralization to a dedicated server was forced to limit the number of attributes it considers to just six in order to avoid such suffocation. Furthermore, it is typically the case in knowledge discovery that only a tiny portion of the data available actually matters – i.e. influences the results. In view of this the architecture chosen for the monitoring system is distributed. By and large, data are preprocessed and stored where they originate. It is up to the knowledge discovery process to selectively pull the data it needs from the different machines. In this way, communication and storage loads – both potential chocking points of the system – are, at least potentially, minimized. The actual saving depends, naturally, on the choice of knowledge discovery method and on the data itself.

The distributed knowledge discovery process requires the coordinated execution of multiple queries on all of the machines. Such coordination is very problematic in grid systems for two reasons. One reason is that grid systems intentionally disassociate resources from execution – a job (a query) awaiting execution will be sent to the next suitable resource that becomes available. There is not even a guarantee that any specific resource will ever become available. The chosen algorithm should thus rely on data redundancy and never wait for the input of any specific computer. Another reason is that current communication infrastructure of grid systems has considerable limitations. The only communication mode of two jobs is if the second is executed with the outcome of the first as its input. This type of communication is available in many grid systems through a workflow management mechanism, or can be implemented using pre-execution and post-execution scripts.

We chose to implement a distributed outlier algorithm which was originally presented in the context of wireless sensor networks (Branch *et al.*, 2006). The algorithm looks for job executions that seem odd with respect to similar jobs executing on other machines as measured by 56 different attributes of the state of the job and the system. The data and its pre-processing are described in Section 5.3 below. The algorithm was implemented as a series of recursive workflows – each executing a job at a resource that becomes available, and then feeding its output (if there is any) to other similar workflows. The details of the algorithm and the implementation are given in Sections 5.4 and 5.5 below.

We experimented with the GMS in a Condor pool of 42 heterogeneous machines. Of these, the GMS indicated that four were misconfigured. In the first machine indicated, we found that an incorrect operating system configuration turned on a network service that overloaded the machine CPU. This overload caused Condor to evict jobs after a constant period of time. The GMS pointed out that the user receives very little CPU, which made debugging easy. In the second machine an incorrect BIOS setting turned HyperThreading on and divided each processor into two weaker virtual machines. The GMS identified that the CPU model does not match job launching and execution times. The third misconfigured machine had a disk that was nearly full. The GMS indicated the higher than usual disk usage. The fourth machine had a temporary phenomenon that disappeared when we reran the experiment. We deem this result a false positive. Further experiments verified that the GMS is scalable and suitable for operation in a real-life system. Details of the validation process are provided in Section 5.6. Finally, conclusions are drawn in Section 5.7.

5.2 Preliminaries and related work

5.2.1 System misconfiguration detection

There are two general approaches to automated system management: white-box and black-box. The white-box approach relies on knowledge of the system and its behaviour. A typical white-box management system such as Tivoli's TEC (IBM, 2002) interprets system events according to a set of rules. These rules specify exceptional behaviour patterns and appropriate responses to them. The biggest drawback of the white-box approach is the reliance on domain experts for the definition of patterns. Expert opinion is hard to obtain in the first place, and even harder to maintain when the system develops and updates. Furthermore, an expert system can never exceed the capabilities and knowledge of the experts.

In the black-box approach, problems are detected and diagnosed with limited, if any, knowledge about the system. Black-box management systems are far more economical and simple to deploy and maintain. In the long run, they can yield results that are comparable to those of expert systems – even ones that are well updated and maintained. The black-box approach is implemented today in several systems. For example, eBay (Chen *et al.*, 2004) experimented with a system that diagnoses failures by training a degraded decision tree on the record of successful and failed network transactions. The route leading to a leaf associated with failed transactions identifies feature–value combinations that may explain the reason for the failure. The main disadvantage of eBay's system is that each transaction is described by just six features. This is mainly because the system centralizes the data and processes it on the fly. Obviously, six features offer a very limited description of the variety of possible reasons for failure. A further disadvantage is that the algorithm used focuses on negative examples (failures) and thus performs sub-optimal learning.

Kola, Kosar and Livny (2004) describe a client-centric grid knowledge base that keeps track of the performance and failures of jobs submitted by the client to different machines. The client can use this knowledge base to choose an appropriate machine for future jobs. This system has several limitations. First, it does not allow clients to learn from each other. This would become a problem when the number of machines is large and when some clients have either few jobs or jobs that are very different from each one another. Furthermore, the system takes no advantage of the wealth of information available on the machine side, and this imposes a static view on every machine. To exemplify, consider that a specific machine has a high load during the execution of a client's job. When the client submits a new job this machine would not be used although it may now have a relatively low load.

The PeerPressure system, proposed by Wang *et al.* (2004), automatically troubleshoots misconfigured registry entries of one machine by comparing their configuration with those of many other machines. It is, however, the user's responsibility to identify the misconfigured machine in the first place, and to operate the troubleshooting process.

The system described here – the GMS – is different from previous work in several key factors. First, it is fully distributed. It processes and stores the data where they were created, which means that the overhead it normally creates is negligible. The low overhead permits the GMS to process a large number of attributes (hundreds in this implementation), where other monitoring systems use just a few. Overhead is imposed on the system only during analysis. Then, too, the overhead is moderate compared with that of centralization because the distributed algorithm only collects a portion of the data. Second, unlike previous work, the GMS does not process raw data but rather translates it first into semantically meaningful terms using an ontology. Besides leveraging output interpretability, this translation also rids the data of semantically meaningless terms and constructs semantically meaningful features from combinations of raw features. In this way, the quality of the knowledge discovery process is greatly improved. In addition to these two differences, the GMS is also non-intrusive, relying solely on existing logging capabilities of the monitored system. Finally, for data analysis we chose an unsupervised outlier detection algorithm, removing the need for tagged learning examples, which are often hard to come by.

5.2.2 Outlier detection

Hodge and Austin (2004) define an outlier point as "... one that appears to deviate markedly from other members of the sample in which it occurs". Of course, this definition leaves a lot to interpretation. This is not accidental: in the absence of a model for the data, the definition of an outlier is necessarily heuristic. Of the many proposed heuristics, this paper focuses on unsupervised methods, which are less demanding of the user. Specifically, we choose to focus on distance-based outlier detection methods, in which the main responsibility of the user is to define a distance metric on the domain of the data. In the following, we outline some of the more popular heuristics.

Several distance-based outlier detection methods are described in the literature. One of the first is that of Knorr and Ng (1998). They define a point p as an outlier if no more than k points in the data set are within a d-size neighbourhood of the point p, where the k and d parameters are user defined.

Definitions such as that of Knorr and Ng consider being an outlier a binary property. A more sophisticated approach is to assign a degree of being an outlier to each point. Ramaswamy, Rastogi and Shüm (2000) presented a distance-based outlier detection algorithm, which sorts the points according to their distance from their respective mth nearest neighbour. Hence, the point farthest from its mth nearest neighbour is the most likely outlier, etc. Breunig *et al.* (2000) extend this idea by presenting the *local outlier factor* (*LOF*) – a measure distinguishing outliers proportionally to the density of points in their near environment. This definition handles well data sets in which multiple clusters exist with significantly different densities.

In the GMS, we use the definition proposed by Angiulli and Pizzyti (2002) for the HilOut Algorithm. By this definition the outlier score of a point is its average distance to its m nearest neighbours. This can be seen as a compromise between the definition of Ramaswamy, Rastogi and Shüm (2000) and the LOF. Of further importance to our purposes is that this definition lends itself, to parallelization more easily than LOF. A distributed version of the HilOut algorithm can be found in the literature (Branch *et al.*, 2006), in the context of *wireless sensor networks* (*WSNs*). Grid systems, however, are different from WSNs in that the main challenge they pose is not reducing messaging but rather addressing the limited availability of resources. Thus, we describe a different distributed scheme for HilOut, which is innovative in its own right.

5.3 Acquiring, pre-processing and storing data

The quality of any knowledge discovery process is known to be critically dependent on data acquisition and pre-processing. Furthermore, organization of the data is crucial in determining which algorithms can be used and, consequently, the performance of the process. This section outlines the main points of interest in the GMS approach to data acquisition, pre-processing and organization.

5.3.1 Data sources and acquisition

There are two main approaches to system data acquisition: intrusive and non-intrusive. Intrusive data acquisition requires the manipulation of the system for the extraction of data. The non-intrusive approach, which is the one employed by the GMS, prefers using existing data sources. This is both because these sources include data about high level objects (jobs, etc.) and because the overhead incurred by monitoring is thus minimized. The sources used by the GMS are the following.

- *Log files.* Every component of the grid system is represented by a daemon. Daemons log their actions in dedicated log files, mainly for the purpose of debugging. This file also contains time stamps, utilization statistics, error messages and other important information. Log files are read as event streams.
- *Utilities.* Grid systems supply utilities, which extract useful information about the state of each machine, the overall status of the pool, machine configurations and information about the jobs. Utilities are used through sampling – they are called periodically and can be matched against log file data, primarily by using time stamps.
- *Configuration files.* Each grid system has configuration files, which contain hundreds of attributes. These files are read (once) and the data are stored in reference tables.

5.3.2 Pre-processing

Pre-processing of data in the GMS mainly takes the form of converting raw data into semantically meaningful data. This is done by translating them into ontological terms. The use of ontologies in knowledge discovery has been discussed widely in the literature and has three main benefits. First, it increases the interpretability of the outcome. Second, it enriches the data with background information. For instance, it allows linking multiple executions of a single job, and constructing features such as the cumulative running time of a job across all executions. Third, it allows data reduction by focusing on those parts of the data that the ontology architect deems more important. Additionally, from a system design perspective, the use of an ontology allows the data to be virtualized, resulting in easy porting of the higher levels (the analytical part) of the GMS from one grid system to the another.

The ontology used in the GMS follows the principles suggested by Cannataro, Massara and Veltri (2004). It is hierarchical, with the highest level of the hierarchy containing the most general concepts. Each of these concepts is then broken down into more subtle concepts until, at the lowest level, *ground concepts* or *basic concepts* are considered. The relation between levels is that an upper level concept is defined by the assignment of its values to lower level

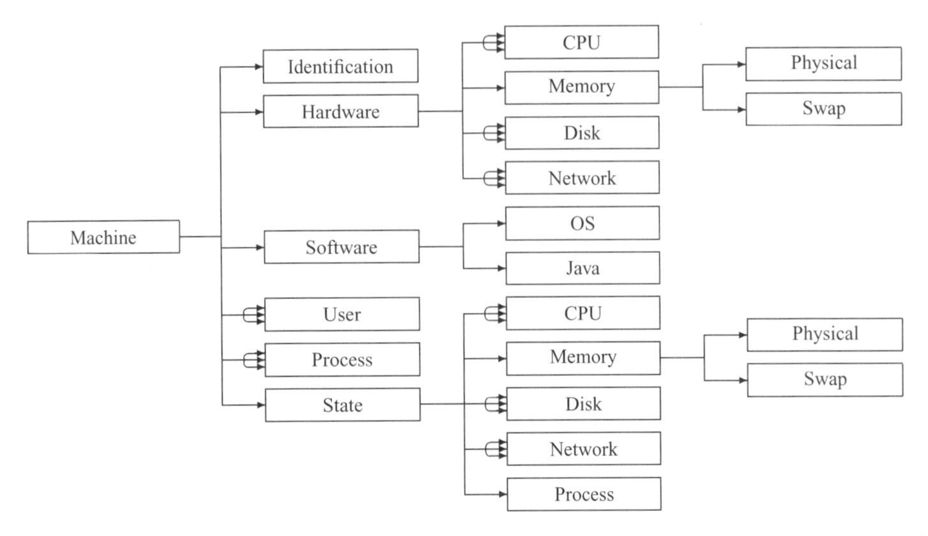

Figure 5.1 A schematic layout of the machine concept. Solid arrows represent sub-concept relationships, dashed arrows represent one-to-one relationships. Ground concepts are removed in order to simplify the chart – e.g. Physical Memory state is further divided into Buffered, Cache, Used and Free Memory

concepts. For instance, Figure 5.1 schematically depicts the concept hierarchy starting at the *machine* general concept. As can be seen, a machine, from a grid system perspective, is defined by its hardware, software, state, configuration and a unique identifier.

The other general concepts of the ontology are *a job*, *a pool* and *matchmaking*. The *job* concept contains the definition of the job in terms of inputs and requirements, and the events that occur with respect to this job's execution. The *matchmaking* concept contains statistics about the status of the machines in the pool, and the result of the matchmaking process. The *pool* concept depicts the deployment of the batch system daemons in the entire pool. The full ontology, containing hundreds of concepts, is too large to include in this context.

Beside translation into ontological terms, the other part of pre-processing in the GMS addresses missing values. Missing values mostly stem from three sources. There are concepts, such as utilization of some of the levels of the memory hierarchy, whose reading is not available in some architectures (Windows). Other concepts may become temporarily unavailable due to the failure of a component (e.g. NFS statistics when the NFS fails). Last, since there are concepts that are combined from both sampled and event-based data, there are cases in which the system is unable to complete the sampled detail (e.g. return code) of event data (respectively, job termination), and thus the higher-level concept becomes missing. Our choice for dealing with missing values is to treat the first kind as a value in its own right, and to purge altogether records of the second and third kinds.

5.3.3 Data organization

The data in the GMS are fully distributed. The reasons for this are the desire to restrict to a minimum the GMS overhead while no analysis is taking place, and the need for scalability with the number of machines. Regarding the latter, it is enough to note that the average data rate at an execution machine can top one kilobyte per second. Accumulating these data over a

24-hour sliding window in a 1000-machine pool would yield more than 100 gigabytes of raw data, an amount requiring specialized resources for management if centralized. However, if the data remain distributed, the amount per execution machine – a mere 100 megabytes – can be supported by off-the-shelf, even free, databases such as MySQL (Axmark *et al.*, 2001).

With respect to this, the GMS stores each piece of data where it originates. For instance, all machine-related statistics (load averages etc.) are stored in the machine where they are measured, all matchmaking data are stored at the machine running the matchmaker daemons and so forth. The resulting database is, therefore, a mix of a horizontally partitioned database – with every execution machine, e.g., storing the same kinds of datum regarding jobs it executed – and a vertically partitioned database – with, e.g., submission machines storing different data from the execution machines.

Motivated by the same desire for overhead limitation, the GMS processes data lazily. That is to say that data is stored in its raw form unless requested. When they are, preprocessed and stored in their ontological form. This is especially important for concepts that require, for their instantiation, raw data that are produced at different machines. For instance, computing the total CPU time spent for a job that ran on several machines (e.g. due to repeated failure to terminate) will cause communication overhead and had better not be carried out unless the concept is requested.

5.4 Data analysis

5.4.1 General approach

Two major assumptions guide our approach to detecting misconfigured machines. First, we assume that the majority of machines in a well maintained pool are properly configured. Second, we assume that misconfigured machines behave differently from other, similar machines. For example, a misconfigured machine might run a job significantly slower, or faster, than that job would run on other, similar, machines. The first of these assumptions limits our approach to systems that are generally operative, and predicts that it would fail if most of the resources are misconfigured. The second assumption limits the usefulness of the GMS to misconfigurations that affect the performance of jobs (and not, e.g., system security).

Our choice of an algorithm is strongly influenced by two computational characteristics of grid systems. First, function shipping in grid systems is, by far, cheaper than data shipping – i.e. it pays to process the data where they reside rather than ship them elsewhere for processing. This, together with the difficulty in storing centralized data (as discussed in the previous section), motivates a distributed outlier detection algorithm. Second, machines in a grid system are expected to have very low availability. Thus, the algorithm needs to be able to proceed asynchronously and produce results based on the input of only some of the machines.

Finally, our approach is influenced by characteristics of the data itself. It takes many features to accurately describe events which occur in grid systems, and these events are very heterogeneous. For instance, Intel NetBatch reportedly serves more than 700 applications and about 20 000 users. Since every application may put a different load on the system (and furthermore this load may greatly vary according to the input) and since the working habits of users can be very different from one another, the log data generated by NetBatch has a very complex distribution. To simplify the data, one can focus on jobs that are associated with a particular application receiving varied input parameters. For instance, in NetBatch a great percentage of the actual executions are by a single application, which takes in the description of a VLSI circuit and a random input vector and validates that the output of the circuit adheres to its specification. In our implementation, we did not have access to this kind of application. Rather we emulated it by executing standard benchmarks with random arguments.

Because the data regarding an execution are always generated and stored on the execution machine the resulting database is horizontally partitioned: a property of which we take advantage in our implementation. On the other hand, sampling, which is popular in knowledge discovery, does not seem to be suited to our particular problem for several reasons. First, sampling is in general less appropriate for sparse distributions, and specifically it is prone to missing outliers. Second, because the distribution of the data is typically highly dependent on the machine from which it was sampled, it seems that uniform sampling would still have to visit each individual machine and would thus not achieve substantial performance gains. Instead, we concentrated on an algorithm that guarantees exact results once all of the machines are available, and which would often yield accurate results even in the absence of many of the well configured machines.

5.4.2 Notation

Let $P = \{P_1, P_2, \ldots\}$ be a set of machines in the algorithm, and let $S_i = \{x_i^1, x_i^2, \ldots\}$ be the input of machine P_i. Each input tuple x_i^j is taken from an arbitrary metric space $\mathbb{D}$, on which the metric $d : \mathbb{D} \to \mathbb{R}^+$ is defined. We denote by S_N the union of the inputs of all machines. Throughout the remainder of this chapter we assume that the distances between points in S_N are unique[1]. Among other things, this means that for each $S \subseteq S_N$ the solution of the HilOut outlier detection algorithm is uniquely defined.

For any arbitrary tuple x we define the *nearest neighbours* of x, $[x|S]_m$, to be the set of m points in S which are the closest to x. For two sets of points $S, R \subseteq \mathbb{D}$ we define the nearest neighbours of R from S to be the union of the nearest neighbours from S for every point in R. We denote by $\hat{d}(x, S)$ the average distance of x from the points in S. Consequently, $\hat{d}(x, [x|S]_m)$ denotes the average distance of x from its m nearest neighbours in S. For any set S of tuples from $\mathbb{D}$, we define $\mathcal{A}_{k,m}(S)$ to be the top k outliers as computed by the (centralized) HilOut algorithm when executed on S. By definition of HilOut, these are k points from S such that for all $x \in \mathcal{A}_{k,m}(S)$, $y \in S \setminus \mathcal{A}_{k,m}(S)$ we have $\hat{d}(x, [x|S]_m) > \hat{d}(y, [y|S]_m)$.

5.4.3 Algorithm

The basic idea of the Distributed HilOut algorithm is to have the machines construct together a set of input points S_G from which the solution can be deduced. S_G would have three important qualities. First, it is eventually shared by all of the machines. Second, the solution of HilOut, when calculated from S_G, is the same one as is calculated from S_N (i.e. $\mathcal{A}_{k,m}(S_G) = \mathcal{A}_{k,m}(S_N)$). Third, the nearest neighbours of the solution on S_G from S_G are the same ones as the nearest neighbours from the entire set of inputs S_N (i.e. $[\mathcal{A}_{k,m}(S_G)|S_G]_m = [\mathcal{A}_{k,m}(S_G)|S_N]_m$).

Since many of the machines are rarely available, the progression of S_G over time may be slow. Every time a machine P_i becomes available (i.e. a grid resource can accept a job related to the analysis), it will receive the latest updates to S_G and will have a chance to contribute to S_G from S_i. By tracking the contributions of machines to S_G, an external observer can compute an

[1]This assumption is easily enforced by adding some randomness to the numeral features of each data point.

ad hoc solution to HilOut at any given time. Besides permitting progress even when resources are temporarily (sometime lastingly) unavailable, the algorithm has two additional benefits. One, the size of S_G is often very small with respect to S_N, and the number of machines contributing to S_G very small with respect to the overall number of machines. Two, $\mathcal{A}_{k,m}(S_G)$ often converges quite quickly, and the rest of the computation deals solely with the convergence of $[\mathcal{A}_{k,m}(S_G)|S_G]_m$. $\mathcal{A}_{k,m}(S_G)$ converges quickly because many of the well configured machines could be used to weed out a nonoutlier that is wrongly suspected to be an outlier.

The details of the Distributed HilOut algorithm are given in Algorithms 1–3. The algorithm is executed by a sequence of recursive workflows. The first algorithm, Algorithm 1, is run by the user. It submits a workflow (Algorithm 2) to every resource in the pool and terminates. Afterwards, each of these workflows submits a job (Algorithm 3) to its designated resource and awaits the job's successful termination. If the job returns with an empty output, the workflow terminates. Otherwise, it adds the output to S_G and recursively submits another workflow – similar to itself – to each resource in the pool.

If there are points from S_i that should be added to S_G, they are removed from S_i and returned as the output of the job. This happens on one of two conditions: (1) when there are points in the solution of HilOut over $S_i \cup S_G$ which come from S_i and not S_G. (2) when there are nearest neighbours from $S_i \cup S_G$ to the solution as calculated over S_G alone, which are part of S_i and not S_G.

Moving points from S_i to S_G may change the outcome of HilOut on S_G. Thus, the second condition needs to be repeatedly evaluated by P_i until no more points are moved from S_i to S_G. Strictly for the sake of efficiency, this repeated evaluation is encapsulated in a while loop; otherwise, the same repeated evaluation would result from the recursive call to Algorithm 3.

ALGORITHM 1. Distributed HilOut - User Side

Input: The number of desired outliers – k and the number of nearest neighbours to be considered for each outlier – m

Initialization:

Set $S_G \leftarrow \emptyset$

For every P_i submit a Distributed HilOut workflow with arguments P_i, k, and m

On request for output: Provide $\mathcal{A}_{k,m}(S_G)$ as the ad hoc output.

ALGORITHM 2. Distributed HilOut Workflow

Arguments: P_i, k, and m

Submit a Distributed HilOut Job to P_i with arguments k, m and S_G

Wait for the job to return successfully with output R

Set $S_G \leftarrow S_G \cup R$

If $R \neq \emptyset$ submit a Distributed HilOut workflow for every $P_j \neq P_i$ with arguments P_j, k, and m.

Optimizations One optimization that we found necessary in our implementation is to store at every execution machine the latest version of S_G it has received. In this way, the argument to every Distributed HilOut job can be the changes in S_G rather than the full set. A second optimization is to purge, at the beginning of a workflow, all of the other workflows intended for the same resource from the Condor queue. This is possible since the purpose of every such

ALGORITHM 3. Distributed HilOut Job

Job parameters: k, m, S_G
Input at P_i: S_i
Job code:
Set $Q \leftarrow \mathcal{A}_{k,m}(S_G \cup S_i)$
Do
– $Q \leftarrow Q \cup [\mathcal{A}_{k,m}(S_G \cup Q)|S_G \cup S_i]_m$
While Q changes
Set $R \leftarrow Q \setminus S_G$
Set $S_i \leftarrow S_i \setminus R$
Return with R as output.

workflow is only to propagate some additional changes to the resource. Since the first workflow transfers all of the accumulated changes, the rest of the workflows are not needed.

5.4.4 Correctness and termination

Termination of Distributed HilOut is guaranteed because points are transferred in one direction only – from S_i to S_G. Assuming static S_i (or, rather, that S_i is a static query even though the data might change), the algorithm has to terminate either when all S_i are empty or some time before this. It is left to show that at termination the outcome of the algorithm – $\mathcal{A}_{k,m}(S_G)$ is correct.

Lemma 1. *Once no jobs are pending or executing* $\mathcal{A}_{k,m}(S_G) = \mathcal{A}_{k,m}(S_N)$

Proof. Since no jobs are pending and none are executing the last job that has returned from each P_i has necessarily returned an empty output. Assume $\mathcal{A}_{k,m}(S_G) \neq \mathcal{A}_{k,m}(S_N)$. Thus, there are two points x, y such that $x \in \mathcal{A}_{k,m}(S_N)$ and $x \notin \mathcal{A}_{k,m}(S_G)$, and $y \notin \mathcal{A}_{k,m}(S_N)$ and $y \in \mathcal{A}_{k,m}(S_G)$.

Since $x \in \mathcal{A}_{k,m}(S_N)$ and not y, and since $y \in \mathcal{A}_{k,m}(S_G)$ and not x, we know that $\hat{d}(y, [y|S_G]_m) > \hat{d}(x, [x|S_G]_m)$ while $\hat{d}(y, [y|S_N]_m) < \hat{d}(x, [x|S_N]_m)$. Furthermore, by definition it is true that for any z, S, and R, $\hat{d}(z, [z|S \cup R]_m) \leq \hat{d}(z, [z|S]_m)$. Specifically, $\hat{d}(x, [x|S_N]_m) \leq \hat{d}(x, [x|S_G]_m)$. We conclude that $\hat{d}(y, [y|S_G]_m) > \hat{d}(x, [x|S_G]_m) \geq \hat{d}(x, [x|S_N]_m) > \hat{d}(y, [y|S_N]_m)$.

Since $\hat{d}(y, [y|S_G]_m) > \hat{d}(y, [y|S_N]_m)$ there must be some P_j such that $\hat{d}(y, [y|S_G]_m) > \hat{d}(y, [y|S_G \cup S_j]_m)$. Hence, there is some $z \in [y|S_G \cup S_j]_m$ such that $z \notin [y|S_G]_m$. When P_j executes the last job y is already part of $\mathcal{A}_{k,m}(S_G)$ – because $\mathcal{A}_{k,m}(S_G)$ does not change following this last job and because y is eventually in $\mathcal{A}_{k,m}(S_G)$. Thus, z is in $[\mathcal{A}_{k,m}(S_G)|S_G \cup S_j]$. Hence, P_j has to return z. However, this stands in contradiction to our premise that on this last round P_j returns an empty output. Thus, our initial assumption has to be wrong and there could be no such y.

5.5 The GMS

Architecturally, the GMS is divided into two parts. One part – the data collector – takes charge of data acquisition while the other part – the data miner – is in charge of analysis. The

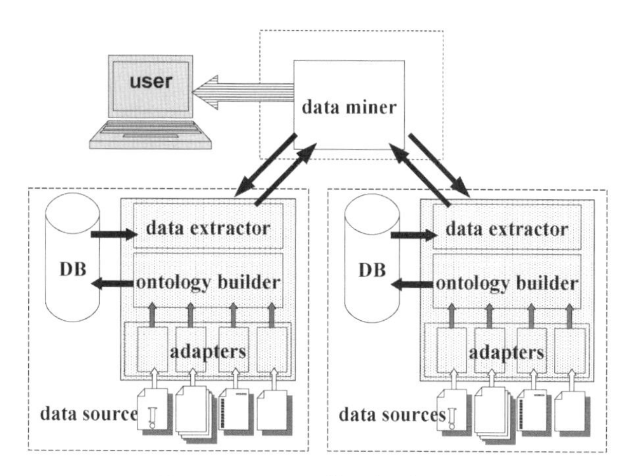

Figure 5.2 Data-flow in the GMS

data collector is a stand-alone software component installed on the resources that are to be monitored by the GMS. The data miner, on the other hand, is a grid workflow, which executes jobs, exchanges data between resources and produces the outcome. A schematic view of the GMS can be seen in Figure 5.2.

Naturally, the data collector is the more complex part. It is made up of three layers. The lowest layer contains a set of *adapters*. An adapter is a module designed to siphon data from a particular data source. The adapters convert raw data, collected from the data sources, into an intermediate representation. In this intermediate form, data can be stored in a MySQL database, but are still not expressed in ontological terms. We implement the adapter as a series of Java programs and Perl script, which, in a Linux system, are periodically executed by the cron daemon. Our current implementation was not ported to other operating systems.

The middle layer is the *ontology builder* which organizes the data acquired by the adapters, into relational tables according to the ontology scheme. This mapping process, from the adapter tables to the ontology tables, involves the integration of data from various data sources to a single ground concept. It is at this stage that complex features are created and that redundant ones are eliminated. The ontology builder is implemented as a Java program that runs a set of SQL queries by the JDBC engine and is therefore machine independent.

The top layer of the data collector is the data extractor. It is a standard SQL interface, which receives queries in ontological terms (e.g. `SELECT* FROM Job WHERE Job.Status=FAILED`) and provides the appropriate data. The analytical part of the GMS uses this interface to execute a knowledge discovery algorithm. In our implementation this part is a set of prepared SQL queries.

The analytical part of the GMS is implemented using Condor DAGs. The Condor DAGman (Condor Team, 2002) is a workflow management engine. It supports execution of mutually dependent Condor jobs, and limited control structures. In our implementation, there are two dependences. The system submits a job to a specified execution machine and then waits for it to terminate successfully. Following this if the job returns non-empty output then the workflow dictates that a number of new workflows be instantiated – one for every execution machine. These new workflows, in turn, repeat the same process recursively. It should be noted that the use of DAGs means that Condor itself takes charge of the server side of the knowledge discovery algorithm.

5.6 Evaluation

To validate the usefulness of the GMS we conducted an experiment in a pool of 42 heterogeneous Linux machines (84 virtual machines in all): 10 dual Intel XEON 1800 MHz machines with 1 GB RAM, six dual Intel XEON 2400 MHz with 2 GB RAM and 26 dual IBM PowerPC 2200 MHz 64-bit machines with 4 GB RAM

We ran two benchmarks independently: BYTEmark[2] – a benchmark that tests CPU, cache, memory, integer and floating-point performance – and Bonnie (Bray, 1996), which focuses on I/O throughput. We sent multiple instances of these benchmarks as Condor jobs to every machine in the pool, varying their arguments randomly across a large range. In all, of the order of 9000 jobs were executed. We then independently ran Distributed HilOut on the two resulting data sets. After pre-processing the data set, we selected 56 attributes, which describe job execution and the properties of the execution machine. Job execution attributes include, for example, runtime, while the machine properties include attributes such as CPU architecture, memory size, disk space and Condor version.

We used the following distance metric d. First, all of the numerical features were linearly normalized to the range $[0, 1]$, so that the weight of different attributes would not be influenced by range differences. Given two points, the distance between numerical features was calculated using a weighted L2 norm – with more weight allocated to job runtime indicators and to configuration features (memory size etc.). This different weighting encourages the algorithm to gather records of jobs that ran on similar machines and have similar runtime. Nominal features contributed zero to the overall distance if their value was the same, and a constant otherwise. This is with the exception of the machine identifier. This field contributed a very large constant if it was the same and zero otherwise. The reason is that we wanted neighbour points to belong to different machines – so that our algorithm would detect exceptional machines rather than exceptional executions in the same machine.

Below, we describe three different experiments. We note that between experiments some of the misconfigured machines were fixed, which explains the differences in the number of outlying machines, data points etc. Although experimentally undesirable, this is an unavoidable outcome of working in a truly operational system. Finally, we note that in all our experiments the number of neighbours (m) was set to five.

5.6.1 *Qualitative results*

We ran Distributed HilOut separately on the records of each benchmark. The outcome of the analysis was a list of suspect machines. Additionally, as shown in Table 5.1, the algorithm ranked for each outlier the main attributes that contributed to the high score. In both tests two machines, i3 and bh10, were indicated as misconfigured. Additionally, the test based on BYTEmark data indicated that bh13 and i4 were also outlying. With the help of a system administrator, we analysed the four machines that had the highest ranking.

The machine bh10 was ranked highest because of excessive swap activity, but we were not able to recreate the phenomenon and so concluded that it was temporary.

The next highest-ranked machine was i4, which contributed five of the top nine outlying points. In all of the outlying executions, the extraordinarily low user CPU was one of the

[2] http://www.byte.com/bmark/bmark.htm

Table 5.1 The output of Distributed HilOut on BYTEmark data

Rank	Machine	Score	Main contributing attributes
1	bh10	0.476	SwapIn, SwapOut, AvgCPU_sys
2	bh10	0.431	SwapOut, SwapIn, AvgCPU_user
3	i4	0.425	DiskUsedRoot, MaxCPU_user, AvgCPU_user
4	i4	0.422	DiskUsedRoot, MaxCPU_user, AvgCPU_user
5	bh13	0.422	LauncherLen, CPUModel, JobLen
6-8	i4	...	...
9	i3	0.421	AvgCPU_user, MaxCPU_user, MaxRootUsed%

outstanding attributes. A quick check revealed that the CPU load of this machine was very high. The source of the high load turned out to be a network daemon (Infiniband manager) that was accidentally installed on the machine. As a result, the user was only allocated a small percentage of the CPU time. After the system administrator shut the daemon down, the machine started to behave normally.

The third-ranked machine was bh13. With this machine the algorithm indicated a mismatch of the CPU model and the time it takes to launch a job. As it turned out, this machine had a wrong BIOS setup: it was configured with active HyperThreading, which meant each CPU (the machine had two) was represented as two CPUs with half the system resources (memory etc.). Consequently, it launched jobs more slowly than other machines with the same CPU model.

The fourth-ranked machine was i3. Here, the GMS indicated a higher than usual use of the root file system. We found that the root file system was nearly full. This led to the failure of the benchmark, and thus to much shorter runtime than usual.

Although qualitative, we consider the validation process highly successful. Of the four highest-ranked machines, three were found to actually have been misconfigured. The GMS contributed to the analysis of all three by pointing out not only which machines to check but also which attributes in the outlying machine differ from those in comparable machines. Further analysis of the misconfiguration, beyond using the GMS, required no access of logs or configuration files. Unlike the BYTEmark benchmark, the Bonnie benchmark missed two out of the three misconfigured machines. We attribute this to the narrow nature of this benchmark, which focuses on I/O.

5.6.2 Quantitative results

The goal of our second experiment was to evaluate the scalability of the Distributed HilOut algorithm. Specifically, we wanted to examine what portion of the entire data set, S_N, is

Table 5.2 The output of Distributed HilOut on Bonnie (Bray, 1996) data

Rank	Machine	Score	Main contributing attributes
1	bh10	0.447	SwapOut, AvgCPU_user, MaxCPU_sys
2	i3	0.444	MaxCPU_sys, MaxCPU_idle, MaxRootUsed%
3-5	i3	...	...
6	bh10	0.417	SwapOut, MaxCPU_user, AvgCPU_user

Table 5.3 Scalability with the number of outliers (k) – BYTEmark benchmark, 30 machines

k	$\|S_N\|$	$\|S_G\|$	Percent
3	2290	304	13%
5	2290	369	16%
7	2290	428	19%

collected into S_G. To be scalable, S_G needs to grow sublinearly with the number of execution machines and at most linearly with k – the number of desired outliers.

Tables 5.3 and 5.4 depict the percentage of points – out of the total points produced by 30 machines – that were collected when the user chose to search for three, five, or seven outliers. This percentage increases linearly.

Tables 5.5 and 5.6 depict the percentage of points – out of the total points produced by 10, 20 or 30 machines – collected into S_G. As the number of machines grows, this percentage declines.

Because we were unable to devote enough machines to check the scalability in a large-scale set-up with hundreds of machines, we were forced to simulate the run of the algorithm in such a set-up on a single machine. The data for the scalability check (taken from metric space $\mathbb{R}^{70}$) were randomly generated according to the following procedure. The set of machines P was randomly partitioned into 10 clusters $\{C_1, \ldots, C_{10}\}$. Seven machines from P were selected to be outlier machines independently of their cluster. For each cluster C_r a point $\Phi_i = \{\phi_i^1, \ldots, \phi_i^{70}\} : \forall l = 1, \ldots, 70\phi_i^l \in [0, 1]$ was selected with a uniform distribution. After the above steps had been performed, we generated the data set $S_i = \{x_i^1, \ldots, x_i^{100}\}$ for every machine $P_i \in C_r$ as follows: $\forall j = 1, \ldots, 100$ and $\forall l = 1, \ldots, 70 x_i^j(l) = \phi_r^l + R_i^j(l)$ Where $x_i^j(l)$ is the l component of the j point of machine P_i and $R_i^j(l)$ is a random value drawn from $N(0, 1/7)$ if P_i is a normative machine, or from the uniform distribution $U[-\frac{1}{2}, \frac{1}{2}]$ if P_i is an outlier machine.

In real life, there is a partial random order between machine executions. Namely, at each time a random selection of the machines execute on the same S_G. To simulate this, we first randomized the order in which machines execute (by going through them in round robin and skipping them at random). Then, to simulate concurrency, we accumulated the outcomes of machines and only added it to S_G at random intervals. The resulting simulation is satisfactorily similar to those experienced in real experiments with fewer machines.

Figure 5.3 depicts the percentage of points collected into S_G out of the total points produced for 100, 200, 500 or 1000 machines. As the number of machines grows, this percentage declines and it drops below two percent for large systems that contains at least 1000 machines. In

Table 5.4 Scalability with the number of outliers (k) – Bonnie benchmark, 30 machines

k	$\|S_N\|$	$\|S_G\|$	Percent
3	2343	287	12%
5	2343	372	16%
7	2343	452	19%

Table 5.5 Scalability with the number of machines and data points – BYTEmark benchmark. $k = 7$

Number of machines	$\|S_N\|$	$\|S_G\|$	Percent
10	751	190	25%
20	1470	336	23%
30	2290	428	19%

Table 5.6 Scalability with the number of machines and data points – Bonnie benchmark. $k = 7$

Number of machines	$\|S_N\|$	$\|S_G\|$	Percent
10	828	186	22%
20	1677	373	22%
30	2343	452	19%

addition, as can be shown from Figure 5.4, the amount of workflow performed by the algorithm grows approximately linearly with the number of machines. That is, on average, the overhead of the algorithm on each machine is fixed regardless of the number of machines. These results provide strong evidence that our approach is indeed scalable.

5.6.3 Interoperability

Our final set of experiments validated the ability of the GMS to operate in a real-life grid system. For this purpose we conducted two experiments. In the first, we ran the Distributed HilOut algorithm while the system was under low working load. We noted the progression of the algorithm in terms of the recall (portion of the outcome correctly computed) and observed both the recall in terms of outlier machines (Recall – M) and in terms of data points (Recall – P). The data points are important because, for the same outlier machines, several indications of its misbehaviour might exist, such that the attributes explaining the problem differ from one point to another.

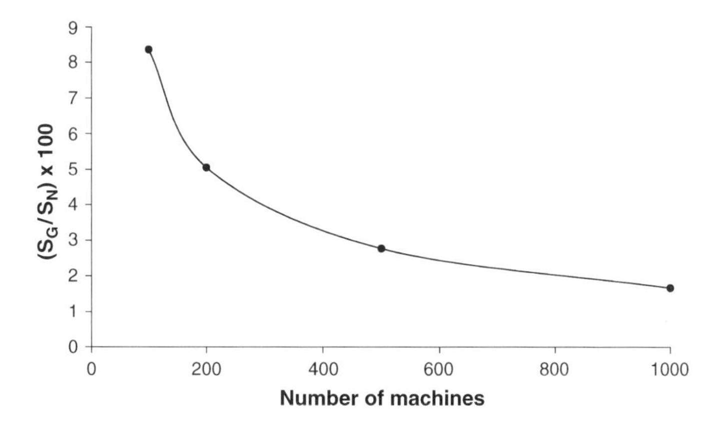

Figure 5.3 S_G as percent of S_N, synthetic data, k = 7

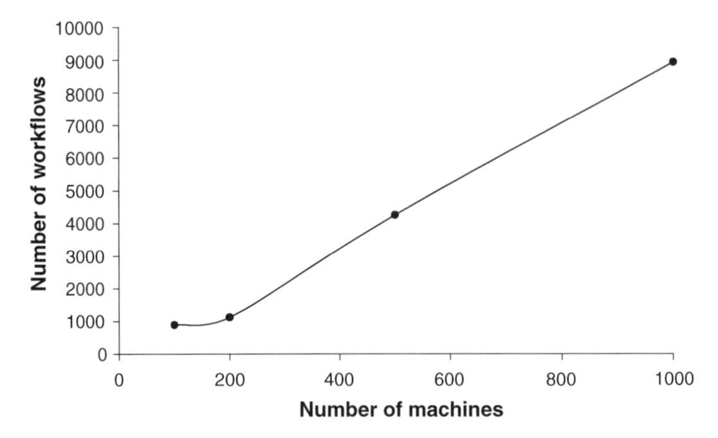

Figure 5.4 Relationship between the amount of running workflows and the number of machines, data points – synthetic data, k = 7

As can be seen in Tables 5.7 and 5.8, the progression of the recall is quite fast: Although the entire execution took 46 minutes for the BYTEmark data (88 for Bonnie), within 10 minutes one or two misconfigured machines were already discovered. This means the administrators were able to start analysing misconfigured machines almost right away. In both benchmarks, a quarter of an hour was sufficient to discover all outlying machines, and most of the patterns indicating the reason for them being outliers. The computational overhead was not very large – in all, about 40 workflows have resulted in additions to S_G. The percentage of S_G in S_N, however, was quite high – about 25 percent. Further analysis shows that many of the points in S_G were contributed in response to the initial message – the one with an empty S_G. We expand on this below.

In the second set, we repeated the same experiment with 20 of the well-configured machines shut down. Then, when no more workflows were pending for the active machines, we turned the rest of the machines on. The purpose of this experiment was to observe the GMS behaviour in the presence of failure.

The results of this second test are somewhat surprising. With respect to the recall our expectations – fast convergence regardless of the missing machines – were met. As Tables 5.9 and 5.10 show, the rate of convergence was about the same as that in the previous experiment. Furthermore, nearly complete recall of the patterns was achieved in both benchmarks, and all of the outlying machines were discovered. What stands out, however, is that all this was achieved with far less overhead in both workflows and data points collected. The overhead remained low even after we turned the missing machines on again (as described below the double line).

Table 5.7 Interoperability – a grid pool in regular working load – BYTEmark benchmark

Time	$\lvert S_G\rvert/\lvert S_N\rvert$	Workflows	Recall – P	Recall – M
0:00	0/4334	0	0/7	0/2
0:09	476/4334	17	1/7	1/2
0:14	708/4334	26	6/7	2/2
0:38	977/4334	37	7/7	2/2
0:46	1007/4334	39	7/7	2/2

Table 5.8 Interoperability – a grid pool in regular working load – Bonnie benchmark

Time	$\vert S_G\vert/\vert S_N\vert$	Workflows	Recall – P	Recall – M
0:00	0/4560	0	0/7	0/3
0:05	295/4560	9	2/7	1/3
0:09	572/4560	18	4/7	2/3
0:14	745/4560	24	5/7	3/3
1:28	1068/4560	44	7/7	3/3

Table 5.9 Interoperability – a grid pool with 20 of the machines disabled – BYTEmark benchmark

Time	$\vert S_G\vert/\vert S_N\vert$	Workflows	Recall – P	Recall – M
0:00	0/4334	0	0/7	0/2
0:03	82/4334	6	4/7	1/2
0:07	460/4334	20	5/7	2/2
0:10	462/4334	22	6/7	2/2
0:23	495/4334	26	6/7	2/2
1:38	617/4334	37	7/7	2/2
2:54	619/4334	39	7/7	2/2

Further analysis of the algorithm reveals the reason for its relative inefficiency in the first set-up. Had all of the computers been available at the launching of the algorithms, and had Condor been able to deliver the jobs to them simultaneously, they would all receive an empty S_G as an argument. In this case, each computer would return its seven most outlying points with their nearest neighbours (five points to each outlier). Because of the possible overlap between outliers and nearest neighbours, this could result in 7–42 points per computer. Multiplied by the number of computers, this gives anything between around 300 and around 1600 points that are sent before any real analysis can begin.

We see three possible solutions to this. One, we could emulate the second set-up by sending the initiation message to small sets of computers. Two, we could add a random delay to the first workflow, increasing the chance of non-empty S_G being delivered with it. Three, we could restrict the return value of the first job to just some of the outliers and their nearest neighbours.

Table 5.10 Interoperability – a grid pool with 20 of the computers disabled – Bonnie benchmark

Time	$\vert S_G\vert/\vert S_N\vert$	Workflows	Recall – P	Recall – M
0:00	0/4560	0	0/7	0/3
0:05	112/4560	4	2/7	1/3
0:07	417/4560	16	4/7	2/3
0:10	475/4560	22	5/7	3/3
0:15	478/4560	24	7/7	3/3
0:17	484/4560	27	6/7	3/3
0:31	494/4560	30	7/7	3/3
2:03	578/4560	40	7/7	3/3

However, we believe none of these solutions is actually needed. In a real system some of the jobs would always be delayed and some of the machines would always be unavailable when the algorithm initiates. Furthermore, under sufficient workload Condor itself delays jobs enough to stop this worst-case scenario from ever happening.

5.7 Conclusions and future work

We have studied the problem of detecting misconfigured machines in large grid systems. Because these systems are heterogeneous, a rich description of their operation is required for more accurate analysis. Moreover, their scale makes centralization of the data as well as its manual analysis inefficient. We therefore proposed a distributed architecture which enables automatic analysis of the data via knowledge discovery algorithms.

We implemented the highly portable Grid Monitoring System, known as the GMS, that relies on an ontology for virtualization of the underlying batch system and for enhancing data quality. We deployed our system on a heterogeneous Condor pool and demonstrated its effectiveness by discovering three misconfigured machines.

Outlier detection is just one of the algorithms that can be implemented on top of the GMS. Correspondingly, misconfiguration detection is just one of the possible applications of knowledge discovery for grid systems. We intend to extend the GMS for various applications, such as resource preservation in workflows, runtime predication etc.

Monitoring data is not static by nature. One of the more interesting aspects of this data, which we have not yet investigated, is its temporal nature. In the future, we intend to look into ways to analyse the temporal behaviour of a grid system via the GMS.

References

Angiulli, F. & Pizzuti, C. (2002), Fast outlier detection in high dimensional spaces, *in* 'Proceedings of the European Conference on Principles of Data Mining and Knowledge Discovery (PKDD'02)', pp. 15–26.

Axmark, D., Widenius, M., Cole, J. & DuBois, P. (2001), 'MySQL Reference Manual', http://www.mysql.com/documentation/mysql

Branch, J. W., Szymanski, B., Giannella, C., Wolff, R. & Kargupta, H. (2006), In-network outlier detection in wireless sensor networks, *in* 'Proceedings of the International Conference on Distributed Computing Systems (ICDCS'06)', p. 51.

Bray, T. (1996), 'Bonnie – File System Benchmarks', http://www.textuality. com/bonnie/.

Breunig, M., Kriegel, H.-P., Ng, R. & Sander, J. (2000), LOF: Identifying density-based local outliers, *in* 'Proceedings of SIGMOD'00', pp. 93–104.

Cannataro, M., Massara, A. & Veltri, P. (2004), The OnBrowser ontology manager: managing ontologies on the grid, *in* 'International Workshop on Semantic Intelligent Middleware for the Web and the Grid'.

Chen, M., Zheng, A., Lloyd, J., Jordan, M. & Brewer, E. (2004), Failure diagnosis using decision trees, *in* 'Proceedings of the International Conference on Autonomic Computing (ICAC'04)', pp. 36–43.

Condor Team (2002), 'The directed acyclic graph manager (DAGman)', http://www.cs.wisc.edu/condor/dagman/.

Hodge V. and Austin J. (2004), 'A survey of outlier detection methodologies', *Artificial Intelligence Review* **22**, 85–126.

IBM (2002), 'IBM Tivoli Monitoring Solutions for Performance and Availability', ftp://ftp.software.ibm.com/software/tivoli/whitepapers/wp-monitor-solutions.pdf

Knorr, E. M. and Ng, R. T. (1998), Algorithms for mining distance-based outliers in large datasets, *in* A. Gupta, O. Shmueli and J. Widom, eds, 'VLDB'98, Proceedings of 24rd International Conference on Very Large Data Bases, August 24–27, 1998, New York City, New York, USA', Morgan Kaufmann, pp. 392–403.

Kola, G., Kosar, T. and Livny, M. (2004), A client-centric grid knowledgebase, *in* 'Proceedings of 2004 IEEE International Conference on Cluster Computing', IEEE, San Diego, CA, pp. 431–438.

Ramaswamy, S., Rastogi, R. and Shim, K. (2000), Efficient algorithms for mining outliers from large data sets, *in* 'Proceedings of the ACM SIGMOD Conference', pp. 427–438.

Wang, H., Platt, J., Chen, Y., Zhang, R. and Wang, Y.-M. (2004), 'Peerpressure for automatic troubleshooting', *ACM Sigmetrics Performance Evaluation Review* **32** (1), 398–399.

6

FAEHIM: Federated Analysis Environment for Heterogeneous Intelligent Mining

Ali Shaikh Ali and **Omer F. Rana**

ABSTRACT

Data mining is a process of mapping large volumes of data onto more compact representations, which may be used to support decision making. At the core of the process is the application of specific data mining methods for pattern discovery and extraction. This process is often structured from interactive and iterative stages within a discovery pipeline and workflow. At these different stages of the discovery pipeline, a user needs to access, integrate and analyse data from disparate sources, to use data patterns and models generated through intermediate stages and to feed these models to further stages in the pipeline.

The availability of Web service standards and their adoption by a number of communities, including the grid community, indicates that development of a data mining toolkit based on Web services is likely to be useful to a significant user community. Providing data mining Web services also enables these to be integrated with other third party services, allowing data mining algorithms to be embedded within existing applications.

We present a data mining toolkit, called FAEHIM, that makes use of Web service composition, with the widely deployed Triana workflow environment. Most of the Web services are derived from the Weka data mining library of algorithms.

6.1 Introduction

The capabilities of generating and collecting data have been growing in recent years. The widespread use of sensors for scientific experiments, the computerization of many business and government transactions, and the advances in data collection tools have provided us with huge amounts of data. Many data collections are being used in business management, scientific and engineering data management and many other applications. It is noted that the number of such data collections keeps growing rapidly because of the availability of powerful and

Data Mining Techniques in Grid Computing Environments Edited by Werner Dubitzky

affordable database systems. This explosive growth in data and databases has generated an urgent need for new techniques and tools that can automatically transform the processed data into useful information and knowledge. Hence, machine learning algorithms and data mining have become research areas with increasing importance.

There is often a distinction being made between machine learning algorithms and data mining. The former is seen as the set of theories and computational methods needed to deal with a variety of different analysis problems, whereas the latter is seen as a means to encode such algorithms in a form that can be efficiently used in real-world applications. Often data mining applications and toolkits contain a variety of machine learning algorithms that can be used alongside a number of other components, such as those needed to sample a data set, read and write output from and to data sources and visualize the outcome of analysis algorithms in some meaningful way. Visualization is also often seen as a key component within many data mining applications, as the results of data mining applications and toolkits are often used by individuals not fully conversant with the details of the algorithm deployed for analysis. Further, users of results of data mining are generally domain experts (and not algorithm experts), and often some (albeit limited) support is needed to allow such a user to choose an algorithm.

The basic problem addressed by the data mining process is one of mapping low-level data (which are typically too voluminous to understand) into other forms that might be more compact (e.g. a short report), more abstract (e.g. a descriptive approximation or model of the process that generated the data) or more useful (e.g. a predictive model for estimating the value of future cases). At the core of the process is the application of specific data mining methods for pattern discovery and extraction. This process is often structured from interactive and iterative stages within a discovery pipeline and workflow. At these different stages of the discovery pipeline, a user needs to access, integrate and analyse data from disparate sources, to use data patterns and models generated through intermediate stages and to feed these models to further stages in the pipeline. Consider, for instance, a breast-cancer data set acquired by a cancer research centre, where a physician carries out a series of experiments on breast-cancer cases and records the results in a database. The data now needs to be analysed to discover knowledge of the possible causes (or trends) of breast cancer. One approach is to use a classification algorithm. However, applying an appropriate classification algorithm requires some preliminary understanding of the approach used in the classification algorithm and, in the instance where the size of data is large, for processing of the data to be carried out on computational resources suitable to handle the large volume of data.

The availability of Web service standards (such as WSDL, SOAP), and their adoption by a number of communities, including the grid community as part of the *Web services Resource Framework* (*WSRF*) indicates that development of a data mining toolkit based on Web services is likely to be useful to a significant user community. Providing data mining Web services also enables these to be integrated with other third party services, allowing data mining algorithms to be embedded within existing applications.

We present a data mining toolkit, called FAEHIM, that makes use of Web service composition, with the widely deployed Triana workflow environment.[1] Most of the Web services are derived from the Weka data mining library of algorithms (Witten and Frank, 2005), and contain approximately 75 different algorithms (primarily classifier, clustering and association rule algorithms). Additional capability is provided to support attribute search and selection within a numeric data set, and 20 different approaches are provided to achieve this (such as a genetic

[1]The Triana workflow system: http://www.trianacode.org/

search operator). Visualization capability is provided by wrapping the GNUPlot software; additional capability may be supported through the deployment of a Mathematica Web service (developed using the MathLink software). However, this is constrained by the availability of Mathematica licences to render the output using Mathematica. Other visualization routines include a decision tree and a cluster visualizer.

6.2 Requirements of a distributed knowledge discovery framework

We define a *distributed knowledge discovery framework (DKDF)* as *a computational system that provides a complete and convenient set of high-level tools for knowledge discovery*. A DKDF should allow users to define and modify problems, choose solution strategies, interact with and manage appropriate distributed resources, visualize and analyse results and record and coordinate extended knowledge discovery tasks. Furthermore, interaction with the user needs to be in a way that is intuitive to the way a user solves problems, and must not necessitate the understanding of a particular operating system, programming language or network protocol. We identify a list of requirements that influenced our framework design, and divide these into two broad categories.

6.2.1 Category 1: knowledge discovery specific requirements

The first set of requirements relates to requirements specific to knowledge discovery. These requirements are generally also valid for data mining tools that do not make use of grid resources.

- *Handling different types of datum.* As data sources can vary, we may expect that the framework should offer a set of tools to manipulate different data types. For instance, the framework should have a tool for converting data formats, i.e. comma-separated values to attribute-relation file format and vice versa.
- *Choosing a data mining algorithm.* A variety of algorithms exist, from machine learning algorithms that operate on data (such as neural networks) to statistical algorithms (such as regression). Although an extremely challenging task, we should require the toolkit to provide some support in algorithm choice (based on the characteristics of the problem being investigated). The framework should enable the discovered knowledge to be analysed from different perspectives. For instance, the framework should enable expressing the result from a classification algorithm in the form of a decision tree, as well as a textual summary.
- *Mining information from different sources of data.* The framework should enable data mining from different sources of formatted or unformatted data, supporting a diverse set of data semantics. For example, the framework should allow the streaming of data from a remote machine along with the capability to process the data locally. Data streaming is particularly important when large volumes of data cannot be easily migrated to a remote location.
- *Utilize users' experience.* The wide range of available data mining tools introduces a new challenge for selecting the right tool for solving a specific problem. The framework should assist the users to utilize previous experience to select the appropriate tool.

- *Testing the discovered knowledge.* The discovered knowledge may not always be accurate. Therefore, the framework should have a set of tools to test the discovered knowledge with real data and validate the accuracy of the knowledge.

6.2.2 Category 2: distributed framework specific requirements

The second set of requirements concerns issues relating to distributed problem-solving and computing frameworks.

- *Problem oriented.* The framework should allow specialists to concentrate on their discipline, without having to become experts in computer science issues, such as networks, or grid computing.
- *Collaborative.* An increasing number of science and engineering projects are performed in collaborative mode with physically distributed participants. It is therefore necessary to support interaction between such participants in a seamless way.
- *Visualization.* The use of graphics can enhance the usability of the framework, for example through animated tables and directed graphs to visualize the state of the application.
- *Fault tolerance.* The use of distributed resources can provide additional capability. The framework must therefore include the ability to complete the task if a fault occurs by moving the job to another available resource.
- *Service monitoring.* The framework should allow users to monitor the progress of their jobs, as they are executed on distributed resources. Such feedback is dependent, however, on the availability of monitoring tools on the resources being used.
- *Data privacy.* Moving data to a remote resource causes some problems with privacy. For example, a centre for disease control may want to use remote data mining to identify trends and patterns in spread of a disease, but are unwilling to disclose the data for the remote analysis due to patient privacy concerns.

6.3 Workflow-based knowledge discovery

Although workflows play an important role in data mining (as specified in the stages below), user interaction between stages is important, as it is often necessary for a user to make decisions during the process (depending on partial results of each stage). Generally, the following steps are involved.

- The first stage involves the selection of a data set. The data set may be in a variety of different formats (such as comma-separated value and attribute-relation file formats), and often converters may be necessary to convert to a format required by the data mining algorithm. A library of such converters may be necessary.
- In the second stage, a data mining algorithm is selected. The selection of an algorithm depends on the nature of the data set and the required knowledge to be extracted from the data set. The selection process could be automated through the use of pre-defined rules. It could also be based on the past experience of a user. This stage can be skipped if the user

knows which algorithm to select. This is often a difficult decision to make, and generally little support is provided in existing toolkits for this. The user is often presented with available algorithms, and has to make a choice manually.

- The third stage involves the selection of the resources on which the data mining algorithm needs to be executed. The choice may be automated by the underlying resource management system, if multiple instances of the same algorithm can be found, or the user needs to manually request access to a particular algorithm on a specific resource.
- Once the resources are selected, the algorithm is executed – by checking security constraints on the remote resource, and often the data set is also migrated to the remote resource.
- The generated model from the data mining algorithm is now presented textually or graphically to the user. The model may now be verified through the use of a test set.

6.4 Data mining toolkit

The presented data mining toolkit is designed to address the requirements outlined above. It consists of a set of data mining services exposed via an *application programming interface (API)*, a set of tools to interact with these services and a workflow system that can be used to assemble these services and tools to solve specific data mining problems. The workflow system that we use to assemble the components is Triana. Triana is an open source problem solving environment, developed as part of the European gridLab project.[2] Triana has a user base in the astrophysics, biodiversity and healthcare communities.

Figure 6.1 illustrates the composition tool provided for the data mining workspace. On the left hand side, the user is provided with a collection of pre-defined services that can be composed together. This composition takes place within a window on the right hand side. Each service illustrated is a Web service with a WSDL interface. The ports of the Web service are exposed as input and output elements that can be connected with other components. The stages as illustrated in Figure 6.1 involve the following:

A user first selects the *get classifier* Web *service*. This allows the selection of a number of different classification algorithms available within the Weka Toolkit. The *get classifier service* therefore acts as a discovery tool to create a list of all the classification services that are known to it via a registry service.

Once a classification algorithm is picked, this is instantiated as a *classifier selector service*. This Web service now contains a single classification algorithm that has been picked by the user. Therefore, although this is a workflow operation, the user has to make a decision on which algorithm should be used. This demonstrates the user interaction within a workflow session.

Once a classification algorithm has been selected and instantiated, the user needs to know the parameters that need to be passed to this algorithm. These are generally run-time parameters associated with the algorithm, and are often contained within a configuration file. The user may reference a configuration file as a URL, or specify these parameters directly within a text box. For instance, in the case of a neural network (back-propagation algorithm), such run-time options include the number of neurons in the hidden layer, the momentum and the learning

[2]The European GridLab project: http://www.gridlab.org/

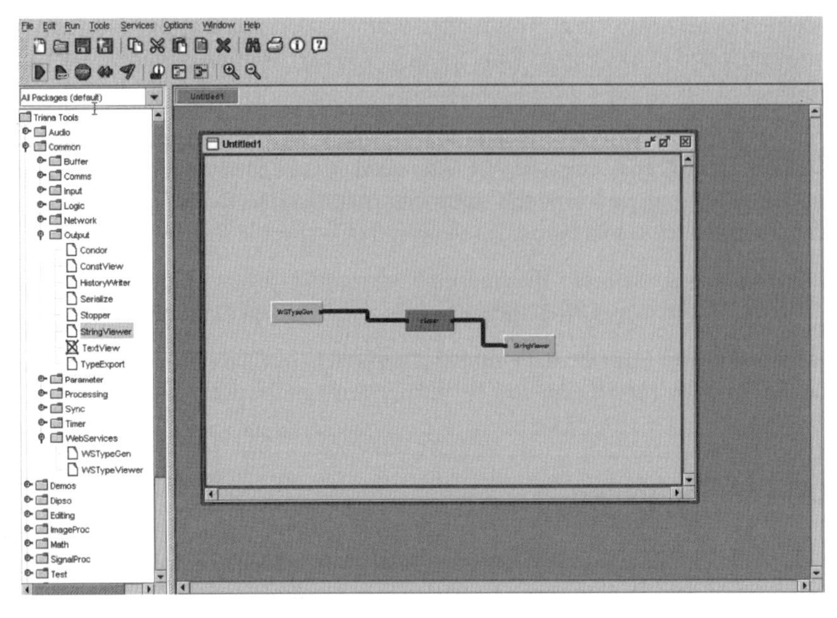

Figure 6.1 Data mining toolkit

rate. Such parameter evaluation is achieved by the *get options* and the *option selector service* demonstrated in Figure 6.1.

The `ClassifyInstance` contains the actual algorithm, and also the parameters that are passed to it at run time. It also contains as input the data set on which the classification algorithm needs to be applied. The data set may be contained locally (on the local file system), or it may be transferred from a remote location via a URL. A third option of streaming the data set is also provided, as long as the classification algorithm supports this.

Finally, the last stage – the `stringViewer` in Figure 6.1 – displays the output to the user. Here the user is presented with the option of either plotting the output using a GNUPlot Web service, using a specialist tree viewer (for decision trees) or generating the output in a textual form. Figure 6.2 illustrates the output displayed in textual form within a window on the right hand side of the composition space.

6.5 Data mining service framework

Several data mining services are implemented to support the data mining process. The implemented data mining services are categorized into classification and clustering algorithms as follows.

Classification algorithms Each classification algorithm available in the Weka Toolkit was converted into a Web service, for example, a J4.8 Web service that implements a decision tree classifier, based on the C4.5 algorithm. The J4.8 service has two key options: (i) classify, and (ii) classify_graph. The classify option is used to apply the J4.8 algorithm to a data set specified by the user. The data set must be in the attribute-relation file format, which essentially involves a description of a list of data instances sharing a set of attributes. The result of invoking the classify

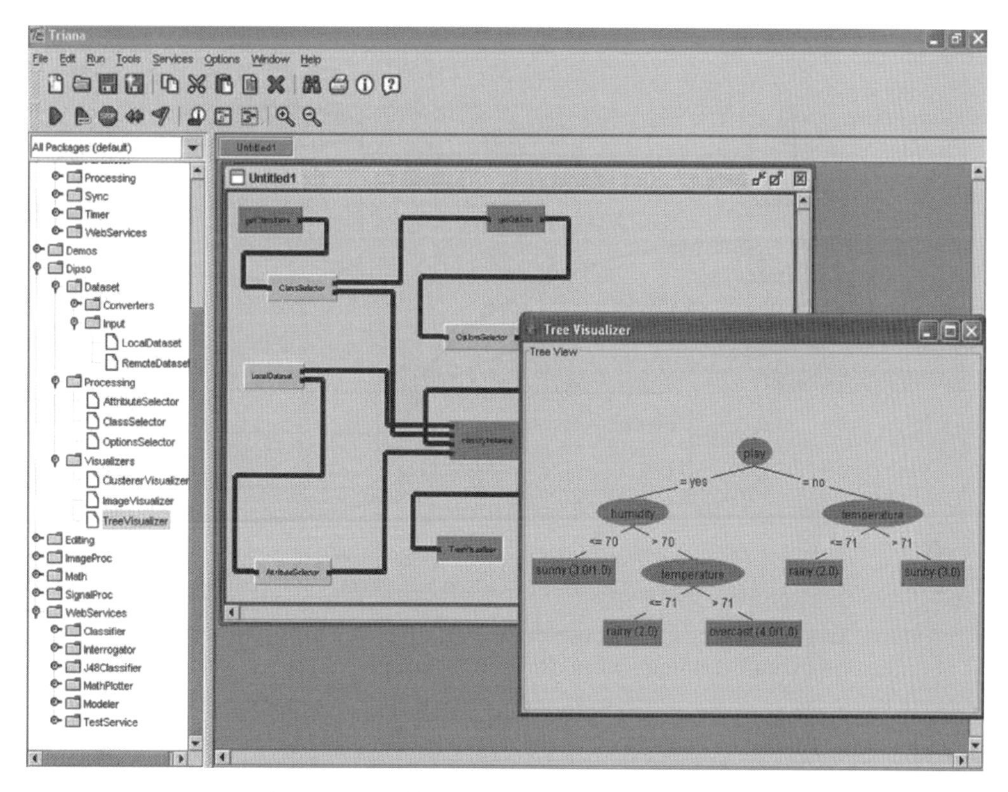

Figure 6.2 Decision tree displayed in graphical form

operation is a textual output specifying the classification decision tree. The classify_graph option is similar to the classify option, but the result is a graphical representation of the decision tree created by the J4.8 algorithm. The graph can then be plotted using an appropriate tree plotter (a service to achieve this is also provided).

Clustering algorithms Similar to classification algorithms, Web services have been developed and deployed for a variety of different clustering algorithms. For example, a Web service was implemented for the Cobweb clustering algorithm. The Web service has the following operations: (i) cluster, (ii) `getCobwebGraph()`. The cluster operation is used to apply the algorithm on a specified data file. The data set must be in the attribute-relation file format format. The result of invocation is a textual output describing the clustering results. The `getCobwebGraph()` operation is similar to the equivalent operation for the clustering algorithm, but the result is a tree created by the Cobweb algorithm.

This is a useful approach if only a single algorithm is to be used, such as one used extensively by a scientist for a given type of data. As there are many available classification algorithms, it is necessary to implement a Web service for each algorithm. This is a time consuming process, and requires the scientist to have some understanding of Web services. Therefore, we have opted to implement a single Web service that acts as a wrapper for a set of classification and clustering algorithms. We also define an interface that algorithm developers must implement if they want their algorithms to be used with the wrapper. The interface defines two request-response operations.

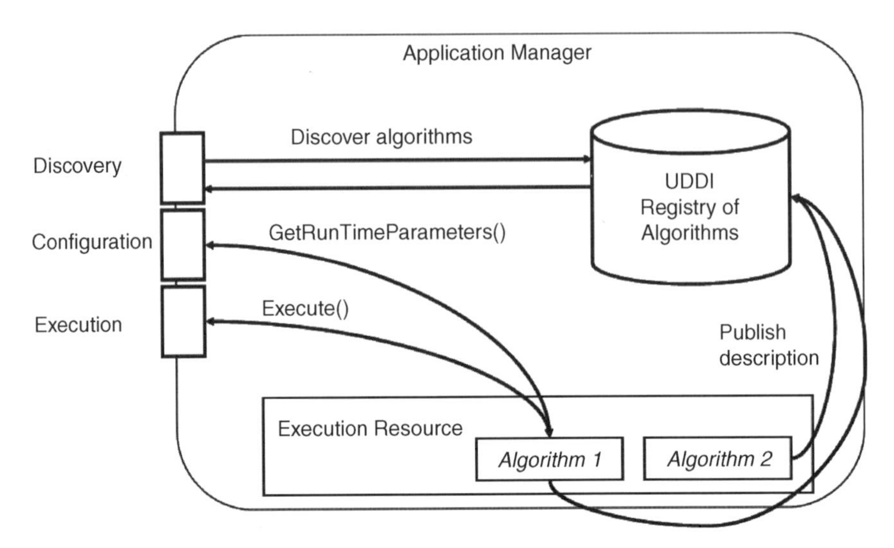

Figure 6.3 Data mining Web service architecture

- `getRunTimeParameters()` must return a list of the run time parameters that the user of the algorithm must configure when using the algorithm.
- `execute(dataSet f, param p)` must execute the algorithm for the data set file and the run time parameters passed in the operation input parameters. The data set file must be in the attribute-relation file format.

All implemented algorithms are placed in a library on the machine where the Web service is hosted. A description of each algorithm is published in a local UDDI registry.

The architecture of this Web service is shown in Figure 6.3. The architecture consists of two main components: *Application Manager* and *Service Registry* (*UDDI*). The Application Manager is responsible for receiving requests from users and executing these request. Three types of requests are accepted by the Application Manager: discovery, configuration and execution requests.

Discovery requests These are concerned with discovering and selecting data mining algorithms. Examples of such requests include `getClassifiers()`, which returns the names of the classifier Web services available, and `getClusterers()`, which returns the names of the clustering Web services available. The Application Manager on receiving a discovery request, calls UDDI to retrieve a list of available data mining algorithms.

Configuration requests These retrieve the run-time parameters associated with a data mining algorithm. This set of requests includes the following:

- `getOptions(dataMiningAlgorithm a)` returns the run-time parameters that must be specified for the data mining algorithm *a*. The algorithm *a* could be either a classification or clustering algorithm. The name of the algorithm is obtained by invoking `getClassfiers()` or `getClusterers()` requests.

- `getRunTimeParameters()`: the Application Manager on receiving a configuration request for a particular data mining algorithm creates an object for the data mining algorithm and executes the `getRunTimeParameters()` method on the created object. The method returns a list of run-time parameters, which are sent to the user by the Application Manager.

Execution requests These execute a data mining algorithm. Four execution requests are provided:

- `classify(dataSet f, class a, param p)` requests the Application Manager to run the specified classifier on the data set using the run-time parameters specified in the argument. The data set is sent to the Application Manager as a SOAP attachment.
- `classifyRemoteDataset(dataSet URL, class a, param p)` requests the Application Manager to perform the same operation as the `classify()` request. However, instead of sending the data set in a SOAP attachment, the URI of the data set is sent. The Application Manager will stream the file from the specified URI.
- `cluster(dataSet file, clust a, param p)` requests the Application Manager to run the specified clustering algorithm on the data set using the run time parameters specified in the argument. The data set is sent to the Application Manager as a SOAP attachment.
- `clusterRemoteDataset(dataSet URL, clust a, param p)` requests the Application Manager to perform the same operation as the `cluster()` request. However, instead of sending the data set file in a SOAP attachment, the URI of the data set is sent. The Application Manager will stream the file from the specified URI.

The Application Manager handles an execution request by creating an object for the requested data mining algorithm and calls the `execute(dataSet f, param)` method on the created object. If the user passes a URI for a data set, the Application Manager streams the data set file to the machine where the algorithm resides and then passes the downloaded data set to the algorithm object.

In order to use the Web service for classifying a data set, the following operations need to be performed by the user: the user needs first to obtain the list of the available classifiers by invoking the `getClassifier()` operation. The Web service will return a list of the available classifiers known to it. This is achieved by searching a pre-defined list of UDDI registries.

Once a classifier has been selected, the user should obtain the run-time parameters that need to be specified for that classifier. This is achieved by using the `getOption()` operation. The Web Service will return a list of the run-time parameters that the user should pass to the Web service. The last step will involve invoking the `classify()` operation on a specified data set instance.

6.6 Distributed data mining services

In the previous section we presented our implementation of the data mining algorithm as Web services where each data mining request is executed on a single machine. In this section, we present a distributed architecture where a data mining request is executed on distributed machines via distributed Web services.

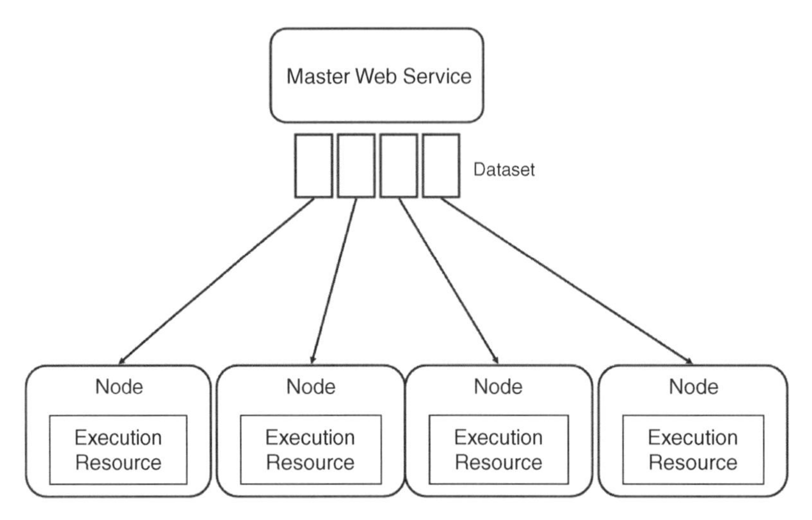

Figure 6.4 Distributed Web service architecture

Figure 6.4 shows the architecture of our implementation of the distributed data mining services. The key elements of our distributed architecture are the master and the nodes (they have the same architecture as shown in Figure 6.3). However, the main difference between the server and node is that the master Web service accepts a data mining request and distributes the data mining tasks on the node Web services.

The master Web service is responsible for accepting a data mining request and input data set from a user and distributing the work to these nodes. The master Web service implements the necessary functionality for load balancing and fault monitoring and recovery. The master Web service distributes the work between the nodes by splitting the input data set and sending each part to a node that runs an instance of the requested data mining Web service. When all nodes return the result of executing the data mining function on their input data set part, the master Web service merges the results and returns it to the user. The system uses SOAP interfaces for communication between the master and nodes.

In the distributed architecture, we use data mining algorithms from the GridWeka Toolkit (Khoussainov, Zuo and Kushmerick, 2004), a modification of the Weka Toolkit, which enables Weka to utilize resources of several computers. The toolkit provides the necessary capabilities of splitting the input data set and merging the data mining results coming from the nodes into a single result.

6.7 Data manipulation tools

In order create a workflow and use the developed Web services, additional support for data mining has been implemented by extending Triana. Three types of tools have been developed-these are illustrated in Figure 6.5(a) and Figure 6.5(b).

Data set manipulation tools These tools are used to convert data sets from one format to another, and include a tool for loading a data set into Triana and sending it to a Web service and a tool for converting comma-separated values into attribute-relation file format (particularly useful for using data sets obtained from commercial software such as Microsoft Excel).

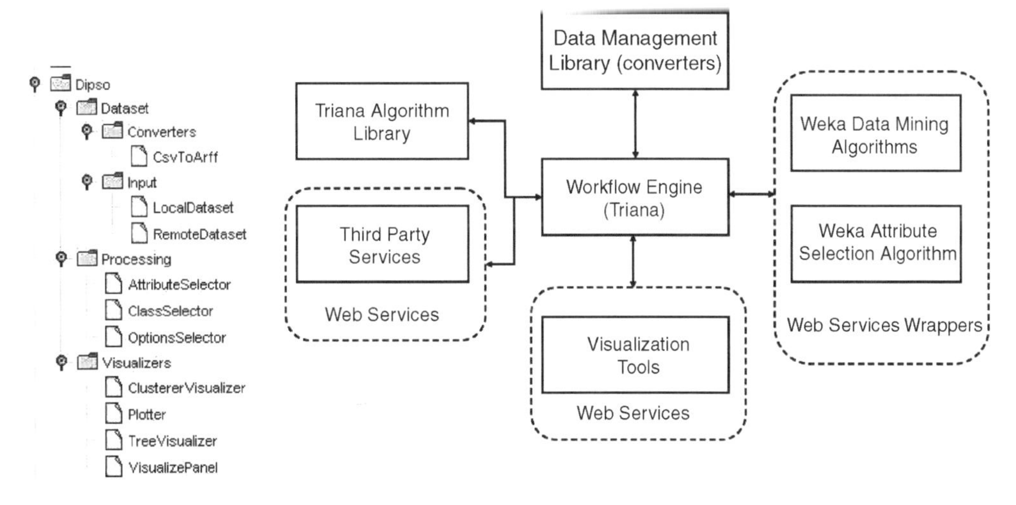

Figure 6.5 Elements of the data mining toolkit

Processing tools These tools are used to process data generated as output from a Web service. The processing tools include a tool to visualize the classifiers list from the Classifier Web service as a tree according to their types, a tool to assist the user in selecting the run time option parameters and a tool to visualize the attributes embedded in a data set.

Visualization tools These tools implement various visualization services based on the output generated by a particular Web service. These visualization tools include Tree Plotter, Image Plotter and Cluster Visualizer. Additional Web services have been implemented that communicate with GNUPlot.

Figure 6.5(b) illustrates the components in the data mining toolkit. Triana provides a centralized enactor that coordinates the execution of a number of different Web services. It is also possible to integrate services provided within a signal processing toolbox (containing a fast Fourier transform, StringViewer, StringReader etc.) with inputs of Web services.

6.8 Availability

The toolkit can be downloaded under the GNU Public License. To use the toolkit, the Triana workflow engine needs to be downloaded and installed. The data mining toolkit can then be installed as a folder within Triana, and can be downloaded from the FAEHIM download site.[3] Installation instructions are also provided. A user can also add additional Web services to the toolkit within the data mining toolbox.

6.9 Empirical experiments

To evaluate the effectiveness of the distributed analysis approach, it is necessary to identify metrics that are quantifiable and can be easily measured. These metrics help determine the value of the framework and its strengths and weaknesses. We use accuracy and running time as evaluation criteria.

[3]http://users.cs.cf.ac.uk/Ali.Shaikhali/faehim/Downloads/index.htm

```
J48 pruned tree
---------------
node-caps = yes
| deg-malig = 1: recurrence-events(1.01/0.4)
| deg-malig = 2: recurrence-events(26.2/8.0)
| deg-malig = 3: recurrence-events(30.4/7.4)
node-caps = no: no-recurrence-events(228.39/53.4)
Number of Leaves: 4
Size of the Tree: 6
```

Figure 6.6 The output of the J4.8 classifier

- *Accuracy.* The data mining Web service must produce the same result as the monolithic data mining algorithm.
- *Running time.* The amount of time that the Web services require to perform the data mining process compared to the stand-alone algorithm. We are also interested in comparing the running time difference of processing a data mining request by the single Web service framework and the distributed Web service framework.

6.9.1 Evaluating the framework accuracy

The data used in this experiment are obtained from the UCI Machine Learning Repository.[4] The data set contains information about a number of different breast-cancer cases – with patient data being anonymized. It includes 201 instances of one class and 85 instances of another class. The instances are described by nine attributes, some of which are linear and others nominal. The data set is already in the *attribute-relation file format* (ARFF).

Using the Weka Toolkit to extract knowledge from the data set The Weka Toolkit is installed on the machine where the data set resides. The J4.8 classifier is used to perform classification on the data set. In this instance, the attribute 'node-caps' has been chosen to lie at the root of the tree. The result of the J4.8-based classification is shown in Figure 6.6.

Using the Classification Web service to extract knowledge from the data set In order to use the Classifier Web service, we first need to obtain the available classifiers that the Web service supports, and the options for the selected classification algorithm. The supported classifier algorithms are obtained by invoking the `getClassifiers()` operation. For our case study, we select J4.8. The run-time options that the J4.8 algorithm requires are obtained by invoking the `getOptions()` operation. Once the classifier algorithm and the options have been identified, we can invoke the `classifyInstance()` operation. The result of the classification in this instance is viewed in a text viewer. As expected, the result of the J4.8 Web service classifier is exactly like the stand alone version as shown above (J4.8 pruned tree).

This example involved the use of a four Web services: (i) a call to read the data file from a URI and convert this into a format suitable for analysis, (ii) a call to perform the classification (i.e. one that wraps the J4.8 classifier), (iii) a call to analyse the output generated from the decision tree, and (iv) a call to visualize the output.

[4]http://archive.ics.uci.edu/ml/

Through the two experiments above we confirmed that the Web services produce the same results as the stand-alone Weka Toolkit.

6.9.2 Evaluating the running time of the framework

In this set of experiments, we evaluated the running time required by a single Web service to respond to a classification request compared to the time that a stand-alone Weka Toolkit required to classify a data set that resides on the same machine. We also compared the running time required by a single Web service to respond to a classification request compared with the time that distributed Web services require to respond to the same classification request.

Comparing the running time of a Web service and a stand-alone Weka Toolkit The data used in this experiment are stored in a 518 KB file, in ARFF, containing information about weather. The data set is derived from the work of Zupan *et al.* (1997) and available at the UCI Machine Learning Repository. The data set contains information about weather conditions. It includes 1850 instances and each instance is described by five attributes (outlook, temperature, humidity, windy and play), some of which are ordinal and others nominal.

We chose the J4.8 classifier to carry out the classification and we kept the default run-time parameters for the J4.8 classifier. The execution time for the Weka Toolkit to process a classification request for the data set, which resided on the same machine, for 20 requests was measured. We also measured the running time that the Web service required to respond to the same classification requests. Figure 6.7 shows a comparison between the running time of the Weka Toolkit and the Web service.

The above experiment demonstrated that the running time required to perform a classification request, on the same dataset, by a Web service was, on average, twice the time required by the stand-alone algorithm from the Weka toolkit. This additional time can primarily be attributed to messaging delays and time to migrate the dataset to the remote location.

Comparing the running time of a single Web service and distributed Web services In this experiment, we used a 10 MB data set containing information on weather obtained from the UCI Machine Learning Repository. We also chose the J4.8 classifier and kept the default run-time parameters for the J4.8 classifier.

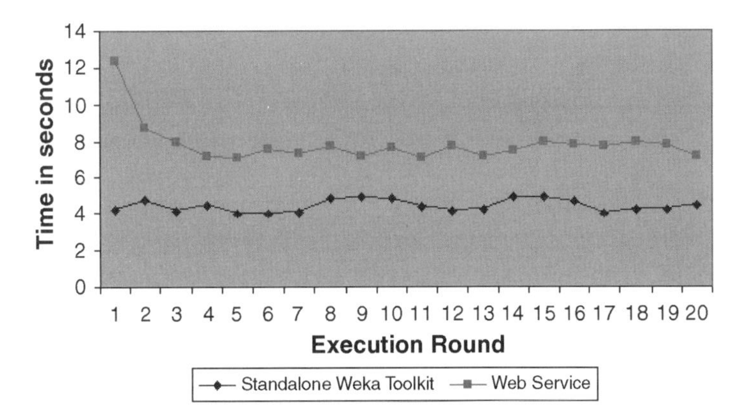

Figure 6.7 Running time of the Weka Toolkit versus Web service

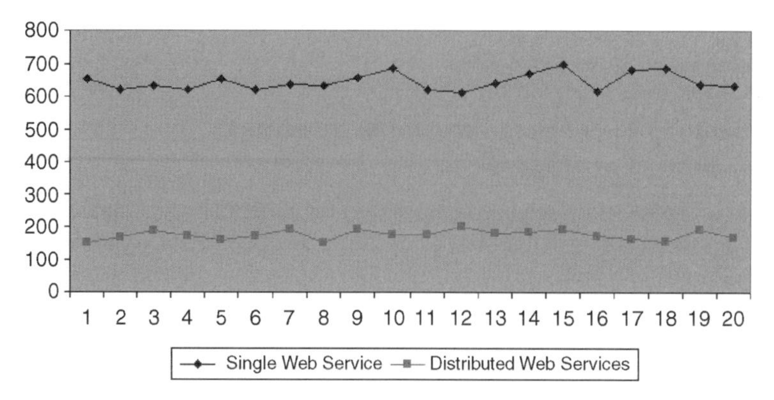

Figure 6.8 Running time of a single Web service versus distributed Web service

We measured the running time of a single Web service for 20 classification requests on the data set and compared it with the running time distributed Web services (consisting of a master and two nodes) required to process the same requests. Figure 6.8 shows a comparison between the running time of the single and distributed Web services.

The experiment showed that the total running time of the Web services was shorter than a single Web service. This arises as a consequence of splitting the data set across multiple Web services, enabling a "high throughput" mode of execution. Figure 6.8 demonstrates the benefits of using multiple distributed services for processing large data sets – under the assumption that a large data set can be split into multiple independent sets.

6.10 Conclusions

This chapter describes the FAEHIM Toolkit. FAEHIM enables composition of Web services from a pre-defined toolbox. Web services have been developed from the Java-based templates provided by the Weka library of algorithms. Using the Triana workflow system, data analysis can be performed on both local and remote data sets. A variety of additional services to facilitate the entire data mining process are also supported, for data translation, visualization and session management. A set of empirical experiments are reported, which demonstrate the use of the Toolkit on publicly available data sets (at the UCI Machine Learning Repository).

References

Khoussainov, R., Zuo, X. and Kushmerick, N. (2004), 'Grid-enabled Weka: a toolkit for machine learning on the grid', *ERCIM News*.

Witten, I. H. and Frank, E. (2005), *Data Mining: Practical Machine Learning Tools and Techniques*, Morgan Kaufmann.

Zupan, B., Bohanec, M., Bratko, I. and Demsar, J. (1997), Machine learning by function decomposition, *in* 'ICML-97'.

7

Scalable and privacy preserving distributed data analysis over a service-oriented platform

William K. Cheung

ABSTRACT

Scalability and data privacy are two main challenges hindering distributed data analysis from being widely applied in many collaborative projects. In this chapter, we first review a recently proposed scalable and privacy-preserving distributed data analysis approach. The approach computes abstractions of distributed data which are then used for mining global data patterns. Then, we describe a service-oriented realization of the approach for data clustering and explain in detail how the analysis process is deployed in a BPEL platform for execution. In addition, lessons learned in the implementation exercise and future research directions regarding how distributed data analysis platforms can be built with even higher scalability and improved support for privacy preservation is also discussed.

7.1 Introduction

With the advent of the Web and grid computing, distributed data are now much easier to gain access to and distributed computing in a heterogeneous environment is becoming much more feasible. In the past few years, there have been a number of large-scale cross-disciplinary and cross-institution research collaboration projects launched in different application domains including e-science, e-business and e-government, to name just a few. Among them, it is common to see distributed data analysis as an important part of the project. The underlying scalability and collaboration requirements cause a unique set of data analysis challenges to be faced.

For example, systems that provide cross-institution remote access to individuals' clinical data for supporting medical research (e.g. the CABIG Project[1]) and disease diagnosis (e.g. Organization for Economic Co-Operation and Development) need to ensure the processes are properly governed so as to comply with privacy policies required by, say, the government,

[1] http://cabig.nci.nih.gov

Data Mining Techniques in Grid Computing Environments Edited by Werner Dubitzky

institutions and individuals (Organization for Economic Co-Operation and Development, 1980). Also, the advantage of sharing data or resources on a large scale to speed up science discovery has recently been advocated,[2] which however also triggered scientists' concern on how to share their data (typically terabyte in volume) only on the *need-to-know* basis to avoid competing research groups taking any advantage from them. For homeland security, extensive use of data mining tools in the past few years for analysing data which are widespread in different companies and organizations has also highlighted the need to properly protect personal data at the same time (Weitzner *et al.*, 2005, 2006). Such an analysis versus privacy dilemma also exists in the commercial field, as it is not uncommon for customers to expect better personalized services while at the same time they expect their privacy rights to be well respected (Kobsa, 2007). The issue becomes further complicated for collaborative business. For example, sharing distributed data collected from different regional branches of different business partners can support more personalized worldwide cross-selling recommendation services. However, this may cause offence to customers who do not expect that third parties can gain access to their personal information.

All the data analysis examples just illustrated share three key characteristics: (i) the data are of large scale, (ii) the data are distributed and (iii) the privacy concern works against the goal of collaborative analysis. To develop data analysis platforms to cater for these applications, new challenges are faced.

To contrast with conventional data analysis systems hosted in a single machine running centralized data analysis algorithms, the data analysis systems which are of primary concern in this chapter require (i) *data analysis techniques* that are scalable, distributed and privacy-preserving, (ii) *software infrastructures* that allow data analysis modules distributed across different institutions to be highly reusable and extensible and (iii) *execution environments* that support the infrastructure and help manage the distributed data analysis processes on behalf of the user.

7.2 A service-oriented solution

To address the data analysis challenges as explained in the previous section, a data analysis prototype has recently been developed for realizing a new data analysis approach called learning-by-abstraction (Cheung *et al.*, 2006). The system is characterized by the following features.

A scalable and privacy preserving data mining paradigm This approach is adopted where data analysis is performed on *abstracted data* that are computed by first pre-grouping data items and then retaining only the first and second order statistics for each group. It is privacy preserving in the sense that individual data items after being grouped can no longer be identified. It is scalable and distributed, as data mining can now be applied directly to a compact representation of the data, instead of the raw data. The feasibility of such a paradigm has been demonstrated for applications such as clustering and manifold discovery.

The service-oriented architecture This approach provides seamless integration of remote self-contained data analysis modules implemented as *Web services* so that they can be coordinated

[2]http://grants.nih.gov/grants/ policy/data_sharing/data_sharing_workbook.pdf

to realize distributed data analysis algorithms. With the recent advent of Web-service-related standards and technologies, including *Web Service Description Language* (*WSDL*), *Universal Description, Discovery and Integration* (*UDDI*) and *Simple Object Access Protocol* (*SOAP*), which take care of the basic interfacing and messaging details, developing data analysis systems essentially becomes an exercise of *discovering* and *composing* relevant data analysis components.

A BPEL system BPEL is employed for creating and executing distributed data mining processes as composite services. *BPEL*[3] stands for *Business Process Execution Language* and it is a markup language originally proposed for describing service-oriented business processes. With the help of a graphical user interface provided by the BPEL system, a distributed data mining process can be schematically created and then described in BPEL. After deploying the described process onto a BPEL execution engine (middleware), the data mining process execution can then be readily taken care of.

7.3 Background

7.3.1 Types of distributed data analysis

The literature discusses a wide collection of distributed data analysis. There are many ways to decompose and distribute a particular data analysis task over different machines.

Decomposition with global analysis The data analysis process is first decomposed into steps. Data are aggregated in a server and *all* the data flow through all the steps for computing the final analysis result, each being taken care of by a distributed resource. For example, a data mining process may involve a data cleansing step, followed by a data pre-processing step, a data mining one and so on (Cannataro and Talia, 2003). Specific distributed resources can be delegated to the corresponding mining steps for executing the complete data mining process. For this type of data analysis, no special attention is paid to addressing the data scalability and privacy issues. The complete set of data is accessed and processed by all the involved computational resources.

Decomposition with local and global analysis The data analysis process involves data that are distributed and intermediate analysis steps are applied to each set of local data at their own source. Only these intermediate results are then combined for further global analysis to obtain the final analysis result. In the literature, most of the work on distributed data mining refers to processes of this type. As an example, a global classifier can take input from the outputs of a set of classifiers, each being created based on a local data set (Prodromidis and Chan, 2000). The scalability of this data analysis paradigm rests on the fact that some intermediate steps are performed only on local data. Also, for the global analysis part, one gains computational saving as the intermediate analysis results are normally more compact than the raw data and thus take less time to process.

Decomposition with peer-to-peer analysis The data analysis process involves data that are distributed and only local analysis steps are applied to local data. Without a global mediator,

[3]http://docs.oasis-open.org/wsbpel/2.0/OS/wsbpel-v2.0-OS.html

global analysis is achieved via exchange of intermediate local analysis with the *peers* of each local data source. As an example, it has been shown that frequent counts (which is the basis of association rule mining) over distributed data sources can be computed when the distributed sources follow a certain peer-to-peer messaging protocol (Wolff and Schuster, 2004). The advantages of such a peer-to-peer paradigm include fault tolerance and better privacy protection. So far, only some restricted types of analysis can be shown to be computed in a peer-to-peer manner (Datta *et al.*, 2006).

7.3.2 A brief review of distributed data analysis

Regarding the type of distributed data analysis that involves both local and global analysis, a meta-learning process has been proposed for combining locally learned decision tree classifiers based on horizontally partitioned data (Prodromidis and Chan, 2000). Beside, collective data mining works is another approach that works on vertically partitioned data and combines immediate results obtained from local data sources as if they are orthogonal bases (Kargupta *et al.*, 2000). Data from the distributed sources form an orthogonal basis and are combined to obtain the global analysis result. This method was later on applied to learning Bayesian networks for Web log analysis (Chen and Sivakumar, 2002). Also, a number of research groups have been working on distributed association rule mining with privacy preserving capability (Agrawal and Aggarwal, 2001; Kantarcioglu and Clifton, 2004; Gilburd, Schuster and Wolff, 2004). With the heterogeneity in privacy concern for different local data sources as previously mentioned, there also exist works studying the trade off between the two conflicting requirements – data privacy and mining accuracy (Merugu and Ghosh, 2003). As uncontrolled local analysis could result in losing information that is salient to the subsequent global analysis, sharing and fusion of intermediate local analysis results to enhance the global analysis accuracy has also been studied (Zhang, Lam and Cheung, 2004).

7.3.3 Data mining services and data analysis management systems

In this section, we describe a number of systems implemented to support data analysis service provisioning and management. The viability of remote data analysis service provisioning has been demonstrated in a number of projects, e.g. DAMS[4] and XMLA,[5] where data mining requests described in XML format can be sent to the services for fulfillment. Also, there are systems with data mining services on top of the grid middleware, e.g. Weka4WS[6] and GridMiner.[7] The use of the grid middleware can free the data analysts from taking care of management issues related to distributed resources, which is especially important if the complexity and the scalability of the analysis are high.

The recent advent of e-science also triggered the implementation of management systems for supporting data analysis processes (also called scientific workflows). Related systems were designed with the requirement of hiding details from users, i.e. the scientists who care more about the data analysis processes than the system for handling them. Most of them contain three major parts: (i) a workbench environment for the user to create

[4]http://www.csse.monash.edu.au/projects/MobileComponents/projects/dame/

[5]http://www.xmlforanalysis.com/

[6]http://grid.deis.unical.it/weka4ws/

[7]http://www.gridminer.org/

and modify workflows, (ii) an engine to orchestrate the execution of the deployed workflows and (iii) a repository of computational/data components that are properly designed regarding their application interfacing and communication protocol so as to be effectively composed and orchestrated. Examples include Wings (Gil *et al.*, 2007*b*) and Pegasus (Deelman *et al.*, 2005), Kepler (Ludascher *et al.*, 2006), Taverna (Oinn *et al.*, 2006), Sedna (Emmerich *et al.*, 2005) and ActiveBPEL.[8] Among them, some have followed Web service standards for better reusability and extensibility. For example, WSDL has been used for specifying the component interfaces and BPEL has been used for specifying the analysis processes.

7.4 Model-based scalable, privacy preserving, distributed data analysis

In this section, we review a distributed data analysis approach that can perform analysis directly on a compact statistical representation of local data sets. By computing the compact local representations of the distributed data sets, scalability and privacy preserving can both be satisfied at the same time. With this approach, data privacy can be protected by not only the conventional type of access control policy but also a new type of privacy policy can control the level of data granularity to be shared (see Section 7.7).

7.4.1 Hierarchical local data abstractions

To represent data in a distributed data source in a compact manner so as to achieve scalability and privacy preservation, a *Gaussian mixture model* (*GMM*) is used. GMMs have long been known to be effective for approximating multimodal data distributions (McLachlan and Basford, 1988). Let $t_i \in \Re^d$ denote a feature vector (also a data item) and

$$p_{local}(t_i|\theta_l) = \sum_{j=1}^{K_l} \alpha_{lj} p_j(t_i|\theta_{lj})$$
$$p_j(t_i|\theta_{lj}) = (2\pi)^{-\frac{d}{2}} |\Sigma_{lj}|^{-\frac{1}{2}} \exp\{-\tfrac{1}{2}(t_i - \mu_{lj})^T \Sigma_{lj}^{-1}(t_i - \mu_{lj})\}.$$

denote its probability distribution where K_l is the number of Gaussian components, α_{lj} is the mixing proportion such that $\sum_{j=1}^{K_l} \alpha_{lj} = 1$, and $\theta_{lj} = \{\mu_{lj}, \Sigma_{lj}\}$ parameterizes the jth Gaussian component of the abstraction at the lth source with a mean vector μ_{lj} and a covariance matrix Σ_{lj}. Given the lth local data set, its abstraction will be parameterized as $\{K_l, \{\alpha_{lj}\}, \{\mu_{lj}, \Sigma_{lj}\}$. This abstraction essentially provides only the first and second order statistics of the data clusters that the data set contains. By varying the value of K_l, each local data source can be represented as a hierarchy of GMMs. The GMM with $K_l = 1$ corresponds to the abstraction with the coarse granularity level where the whole local data set is summarized by only a mean vector and a covariance matrix. The other extreme is the GMM that has the granularity level (the number of Gaussian components) such that one data item is represented by one component.

Depending on the privacy preservation requirement of the local source, each local data source observed from outside is only a GMM abstraction at a particular granularity level out of its abstraction hierarchy. See Figure 7.1 for an illustration.

[8] http://www.active-endpoints.com

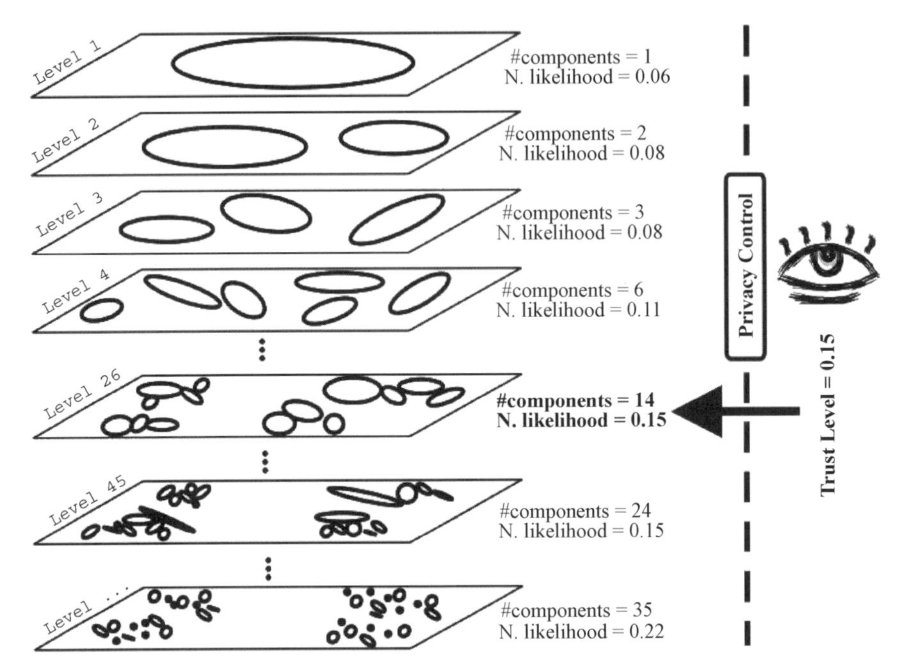

Figure 7.1 A hierarchy of data abstractions acquired by merging and computing the statistics of the two nearest data subgroups at the next granularity level. Given the user's trust level, the matched abstraction level will be the output (N. likelihood = Normalized likelihood)

To compute the hierarchical local data abstraction, agglomerative hierarchical clustering is applied to each local data subset. Given the dendrogram (i.e. the clustering hierarchy) obtained, the mean vector and covariance matrix of each cluster are computed at each level of the hierarchy in an iterative manner and thus the abstraction can be computed accordingly. For data with a more complex distribution, better abstraction techniques are essential (Zhang, Cheung and Li, 2006).

7.4.2 Learning global models from local abstractions

After aggregating the data abstractions (the sets of GMM parameters, i.e. $\{\theta_1, \theta_2, ..., \theta_L\}$) obtained from the local data sources, the model-based approach is adopted for the global data analysis. In general, an arbitrary global latent variable model $p_{\text{global}}(t_i|\Phi)$ (Bishop, 1999) can be specified according to the need of analysis, where Φ is the model parameters characterizing the probabilistic structure of the latent variables and the mapping between the latent space and the data space. In the literature, commonly used latent-variable models include GMM, generative topographical mapping and hidden Markov models. They all model local data clusters as Gaussians given a particular Φ.

The conventional *expectation-maximization* (*EM*) algorithm (Dempster, Laird and Rubin, 1977) is effective for learning different latent variable models given the raw data. The closer is the raw data to one of the Gaussians of the latent variable model's data model, the better will be the quality of the model learned and thus the analysis result. Given only the data abstractions, the distance between a local Gaussian component (belonging to one of the local sources) and a global Gaussian component (belonging to the global latent variable model) is measured. The Kullback–Leibler divergence (also information divergence, information gain or relative entropy) is an information metric commonly used for measuring the distance between two probability distributions and fits well here, as there exists a closed-form solution for computing the divergence between two Gaussian components. An EM algorithm derived for learning models from data abstractions has been shown to be effective for distributed data clustering (Zhang and Cheung, 2005*a*) and distributed manifold unfolding (Zhang and Cheung, 2005*b*). Two key characteristics of this learning-from-abstraction approach are the following.

- The analysis quality degrades only gracefully as the level of details of the data abstraction decreases (or the privacy requirement increases).
- The optimality of the local data abstraction determines the scalability, the degree of privacy protection and the accuracy of the underlying distributed data analysis processes.

Recently, the issue of determining the degree of abstraction for each local source has been formulated as a cooperative game. Under the game formulation, the learning-from-abstraction process becomes active in exploring the right level of details for each source in terms of overall cost-effectiveness (Zhang and Cheung, 2007).

7.5 Modelling distributed data mining and workflow processes

To realize the scalable and privacy preserving distributed data analysis approach described in the previous section with the reusability and extensibility objective in mind, we adopted a service-oriented architecture. Through the *Web Services Description Language* (*WSDL*), it is possible to standardize the way in which *distributed data mining* (*DDM*) components can import and export their functionalities for ease of interoperation over the Internet, where their ports (locations to invoke them), the corresponding port types (abstractions of messages and methods) and bindings (protocols for message transportation) are precisely specified. In addition, a UDDI service repository can be set up so that high-level descriptions of the available DDM component services' functionalities can readily be described, categorized and thus later discovered.

7.5.1 DDM processes in BPEL4WS

Web services support only *stateless* and *synchronous* interactions of computing processes and are not designed to handle *stateful* long-running interactions among a group of service providers and consumers, which however is typically required in distributed data analysis. The need to model stateful Web services was recognized years ago. Related mark-up languages proposed early on (WSCI, WSFL and XLANG) were later combined and replaced by the *Business Process Execution Language for Web Services* (*BPEL4WS*), which is now the *de facto* standard for specifying business process behaviours.

BPEL4WS defines a model and a grammar for specifying interactions between the business process and its partners through Web service interfaces. Relationships between the Web services appear as *partner links* in BPEL4WS. The BPEL4WS *process* primarily defines how multiple service interactions with the partners are coordinated to achieve a business goal. The interactions can be *sequential*, *concurrent* or *conditional*. BPEL4WS also introduces mechanisms for dealing with business exceptions and supports compensation for undoing steps in the process when needed.

While a service-oriented data analysis application can be implemented by developing a system that interfaces with different data analysis component services, adopting BPEL4WS facilitates data analysis processes to be specified in a declarative manner. The concept of a business goal in BPEL4WS becomes that of a data analysis request. A process in BPEL4WS corresponds to the plan for executing the data analysis component services.

Regarding the distributed data analysis approach described in this chapter, the service relationships are essentially the links between the local abstraction and global analysis component services. Once the BPEL4WS description is properly specified, the subsequent orchestration of service executions, exception handling processes and quality-of-service performance optimization can be handled by the BPEL4WS middleware. In other words, the data analysts can focus on the design of the data mining part. In addition, they can easily reuse and modify previously composed DDM processes for other related analysis tasks.

7.5.2 *Implementation details*

The first implementation issue is to ensure the compatibility of the versions of the publicly available SOA tools, which are still in the development stage. In our implementation, we used Java SDK Version 1.4.2 and Apache Axis Web services library Version 1.2 for developing the data analysis services, and hosted them at servers running Apache Tomcat version 5.5. Also, we used the Oracle BPEL4WS execution engine Version 10.1.2 to orchestrate the data analysis services according to the data analysis process specified in BPEL4WS using the Oracle BPEL4WS Designer, which is an Eclipse plug-in that supports visual modelling for BPEL4WS processes.

7.6 Lessons learned

7.6.1 *Performance of running distributed data analysis on BPEL*

To evaluate the performance of running the proposed distributed clustering over the Oracle BPEL execution engine, several data sets of different sizes were created for testing. Each data set was partitioned, with each partition being hosted at a particular Web server. While the main purpose of partitioning is to simulate the scenarios with data distributed in different machines, the partitioning would also lead to higher parallelism for the local abstraction step and thus improve computation performance. The experimental results are shown in Table 7.1. The computational time needed for each step is based on the execution audit log reported by the engine.

As revealed in Table 7.1, the server we used for hosting local data and computing their abstraction reached its limit when handling the abstraction of data sets of size close to 1500 data items. This is mainly because agglomerative hierarchical clustering used for computing the abstraction requires a distance matrix of dimension $N \times N$, where N is the number of data items. In our implementation, a complete version of agglomerative hierarchical clustering was implemented and as a result the memory requirement was high. In fact, a major portion of the computational time reported is due to that needed for the local data abstraction step. Fortunately, the abstraction computation needs only to be computed once. Different abstraction levels computed for a source can then be reused. Also, as the desired number of global clusters is

Table 7.1 Time needed for clustering data sets of different sizes and of different numbers of partitions

No. of partitions/ No. of data	Time for clustering in seconds					
	500	1000	1500	2000	2500	3000
1	13.8	37.3	–	–	–	–
2	17.0	23.7	44.5	94.0	221.6	–
3	15.3	19.3	27.0	49.0	90.8	166.3
4	15.6	18.4	23.5	36.9	60.2	98.8
5	15.4	17.7	22.1	28.9	40.2	62.0

normally much smaller than that of the local data items, local abstractions that have the number of Gaussian components close to the number of data items are often not required. Therefore, one possibility to further speed up the local abstraction step is to apply some efficient data pre-grouping to the data first and then the hierarchical clustering to reduce the computational complexity needed for abstracting each source.

Regarding the overhead cost for communication between services and analysis process handling, we found that it is significant when the size of the data set is small. According to Table 7.1, when the number of data is 500, the computational gain due to partitioning the data into two parts cannot compensate the overhead cost involved. The corresponding time needed for clustering increases from 13.8 seconds to 17.0 seconds. When the number of partitions continues to increase, no speed-up can be observed. When the number of data increases to 1000, the computational gain due to the partitioning starts to outweigh the overhead cost and thus speed-up starts to be observed (from 37.3 seconds for one partition to 17.7 seconds for five partitions). As the number of data continues to increase (from 1500 to 3000), the computational speed-up due to the partitioning becomes even more significant. For example, when the number of data is 2500, it takes 221.6 seconds for clustering data with two partitions, but only 40.2 seconds for that with five partitions.

7.6.2 Issues specific to service-oriented distributed data analysis

Data passing and interoperability In data analysis processes, the need to pass intermediate data analysis results is common. For example, each local abstraction service in the DMM process we implemented needs to pass to the global service its abstraction parameters, which include mean vectors, covariance matrices and their corresponding mixing coefficients. XML-formatted representations for vectors and matrices are used as it is typical for service-oriented systems to exchange messages in XML. In general, analysis tasks of different domains involve data and intermediate analysis results taking different structures. XML-formatted representations for the respective domains will be needed. Via SOAP messaging, such well formed data objects in XML format can then be passed along a BPEL4WS process. However, the data interoperability issue could be further complicated by the possibility that there exist more than one XML-based languages for describing the same type of datum. Data transformers implemented based on *Extensible Stylesheet Language Transformation* (*XSLT*) will then be needed too.

Another issue related to data interoperability is that it is common for data analysts to make use of existing data analysis tools instead of developing their own analysis programs. Regarding our distributed clustering implementation, the local abstraction service was coded from scratch in Java for efficiency reasons. For the global clustering service, we made use of some previously implemented and optimized programs coded in Matlab. Interoperating with the Matlab code was achieved by building a Web service wrapper and deploying it in the server that had Matlab installed. The data received via the Web service wrapper was first saved as a local file on the server and the wrapper then triggered the Matlab code for the executing the global clustering.

Reusability versus performance Implementing a particular step of a distributed data analysis process as a service essentially implies that the step can later be reused via the standardized protocol for Web services. From the perspective of software design, one can consider a service as a remotely invocable module or component. In general, a modular data analysis process that requires frequent transfer of large numbers of data over large distances is problematic, even though the modularization seems to involve highly reusable steps. For example, to realize the proposed distributed clustering, we need to perform local data abstraction whose performance relies on the distance metric used to measure distance between data items. Different distance metrics will lead to different abstraction results. In principle, we can decompose the data abstraction step into two services, where the first one computes the distances among the data items and the subsequent one performs the abstraction based on the computed distances. While such a design can improve the reusability of the implemented services, it will result in the transfer of a large data matrix of order $O(N^2)$, where N is the number of data. Thus, we implemented the local abstraction step as one service hosted at one server. In general, for service-oriented implementations, a data analysis step that is time critical should be designed and implemented as a single service instead of a composition of several services.

7.6.3 Compatibility of Web services development tools

While we expect that the adoption of Web services and BPEL4WS can increase the reusability and reconfigurability of data mining modules, we found that special attention is still needed to ensure the compatibility among versions of the Java SDK, Apache Axis Web services library, Apache Tomcat and BPEL4WS execution engine. Unless the related standards converge and provide backward compatibility, considerable effort is needed to make such analysis processes work.

7.7 Further research directions

7.7.1 Optimizing BPEL4WS process execution

Most existing BPEL4WS process execution engines are designed with the primary objective of providing a reliable execution environment for processes described in BPEL4WS. As the scale of the BPEL4WS applications increases, the execution engine's performance enhancement will soon be an issue. Optimizing the engine's system architecture design at the operating system level (say, with improved memory management or more effective job queuing strategy) will no doubt become more important as BPEL4WS becomes more widely used, e.g. in products such as ActiveBPEL, Oracle BPEL Manager and IBM WebSphere Process Server.

Another direction for optimizing BPEL4WS process execution is to model BPEL4WS processes as workflows and then perform workflow analysis before submitting jobs to the BPEL execution engine with the aim to improve resource allocation and thus a higher throughput rate (Sion and Tatemura, 2005). In large-scale, grid-enabled data mining applications, it has been shown that significant performance improvement can be achieved by properly controlling the complexity of the workflow analysis performed at the middleware to ensure a high enough job submission rate (Singh, Kesselman and Deelman, 2005). Also, the complexity of a data analysis workflow could be reduced if some immediate analysis results were cached (Deelman *et al.*, 2005). In addition, dynamically allocating resources (binding services) during the BPEL process execution is another feature discussed by Deelman *et al.* (2005), but without a proof-of-concept demonstration so far. We envision that the need for executing a large number of data analysis workflows in a highly distributed environment will eventually blossom and such a dynamic property will be essential.

7.7.2 Improved support of data analysis process management

Other than ensuring reliable and efficient execution, another important concern of managing large-scale distributed data analysis processes is how the user can be assisted to create and validate workflows as the correct annotations of the analysis processes expected before the process execution can take place. Most of the existing BPEL process creation tools such as Oracle BPEL designer or ActiveBPEL provide the capability to help users create BPEL processes visually only at the abstraction where all the execution details of a workflow have to be specified for deployment. For creating a large-scale data analysis application, much implementation and execution related knowledge and experience is normally expected for the user, let alone the manual effort required for creating a workflow with, say, thousands of nodes. As simple as the distributed data clustering described in this chapter is, the corresponding process management cost could just be too high for an ordinary user. In our case, the issue becomes important when the number of data sources involved is large.

Fortunately, many distributed data analysis processes can be modelled as workflows with some *repeated* structures. For example, for the distributed data clustering we described, the local abstraction steps taken at the local data sources are in fact functionally identical and thus can be described in a much more compact manner if the iteration structure can be specified in the workflow description. The idea of providing a higher level of process abstraction has been proposed in the literature (Gil *et al.*, 2007*b*). A workflow can first be represented as a workflow template, and then bound with data sources to form a workflow instance, and finally optimized based on some pre-computed intermediate results to give an executable workflow. The user needs only to take care of the workflow template creation step and the data binding step with the help of the workflow creation tool called Wings. Another feature of the tool is that it is enabled for the Semantic Web. Given the semantics of the data sources and the workflow components, various types of constraint for workflow validation can be specified and checked. For example, it is common for a user to expect that the workflow system can ensure reliability of the output results of the workflow (e.g. file consistency constraints (Kim, Gil and Ratnakar, 2006)) or can check the compliance of the workflow against some process related policies (e.g. data privacy policies (Gil *et al.*, 2007*a*)).

We believe that further development along this direction can further help BPEL details to be hidden from the user and the user need only focus on the conceptual design of the processes for supporting their research or applications.

7.7.3 Improved support of data privacy preservation

In many distributed systems, data privacy refers to whether the data transmitted via an open network are properly encrypted so that only the data recipients can read the data. This sense of data privacy is important and a number of related protection mechanisms have been included in the design of some SOA and grid computing environments. However, in many cases, data privacy in distributed data analysis also refers to whether the data being released are properly 'anonymized' so that no one, including the data recipients, can read the data, while some meaningful data analysis should still be possible (as explained in Section 7.3.2). The proposed learning-from-abstraction approach for scalable and privacy preserving data analysis is essentially addressing the privacy issue.

The distributed data analysis processes being considered in this chapter can be modelled as direct acyclic graphs and they fit especially well to most of the workflow and BPEL systems. However, to provide better data privacy preservation support, one may want to explore more privacy preserving data analysis approaches and to incorporate some autonomy in the data analysis services to gain further adaptation and robustness in privacy protection. Then, supporting only simple direct acyclic graph structures will be insufficient. For example, security multiparty computation is another common approach adopted for privacy preserving data analysis (Clifton *et al.*, 2002), where a special secure computation protocol is needed to circulate some privacy protected intermediate computing results through a set of involved hosts. Also, for the learning-from-abstraction approach described in this chapter, a negotiation protocol can be incorporated so that the trade-off between the global analysis accuracy and the degree of local data abstraction can be dynamically computed on a need-to-know basis via negotiation between the services under a game theoretic framework (Zhang and Cheung, 2007).

Unlike some existing workflow systems (Gil *et al.*, 2007*b*), BPEL, and thus the related execution systems, allows the processes that cannot be represented as acyclic directed graphs to be modelled. However, to what extent issues such as confidentially, reliability and flexibility can be supported by the existing BPEL platforms remains to be investigated.

7.8 Conclusions

Learning-from-abstraction is a recently proposed approach for scalable and privacy preserving distributed data analysis applications (Cheung *et al.*, 2006). We discussed the issues related to the design of a service-oriented implementation of the approach. Also we presented the evaluation results regarding the performance of a distributed data clustering process and the lesson learned. We believe that modelling distributed data analysis processes using BPEL is an effective means to enable researchers and data analysts to be able to perform large-scale experimental studies without the need to develop customized analysis software systems.

For future work, incorporating more intelligent execution optimization, providing better process creation and validation support and supporting workflows with less structural constraints are believed to be the important research problems to be addressed before large-scale privacy preservation distributed data analysis can eventually be widely applied.

Acknowledgements

The author would like to say thank to Yolanda Gil for her comments and suggestions for the section on future research directions. This work is partially supported by RGC Central Allocation HKBU 2/03C, RGC Grant HKBU 2102/06E and HKBU Faculty Research Grant FRG/05-06/I-16.

References

Agrawal, D. and Aggarwal, C. (2001), On the design and quantification of privacy preserving data mining algorithms, *in* 'Proceedings of the Twentieth ACM–SIGACT–SIGMOD–SIGART Symposium on Principles of Database Systems', Santa Barbara, CA, pp. 247–255.

Bishop, C. (1999), Latent variable models, *in* 'Learning in Graphical Models', MIT Press, pp. 371–403.

Cannataro, M. and Talia, D. (2003), 'The knowledge grid', *Communications of the ACM* **46** (1), 89–93.

Chen, R. and Sivakumar, K. (2002), A new algorithm for learning parameters of a Bayesian network from distributed data, *in* 'Proceedings of the 2002 IEEE International Conference on Data Mining', Maebashi City, Japan, pp. 585–588.

Cheung, W., Zhang, X., Wong, H., Liu, J., Luo, Z. and Tong, F. (2006), 'Service-oriented distributed data mining', *IEEE Internet Computing* **10** (4), 44–54.

Clifton, C., Kantarcioglu, M., Vaidya, J., Lin, X. and Zhu, M. (2002), 'Tools for privacy-preserving distributed data mining', *ACM SIGKDD Explorations Newsletter* **4** (2), 28–34.

Datta, S., Bhaduri, K., Giannella, C., Wolff, R. and Kargupta, H. (2006), 'Distributed data mining in peer-to-peer networks', *IEEE Internet Computing* **10**, 18–26.

Deelman, E., Singh, G., Su, M., Blythe, J., Gil, Y., Kesselman, C., Mehta, G., Vahi, K., Berriman, G., Good, J., Laity, A., Jacob, J. and Katz, D. (2005), 'Pegasus: a framework for mapping complex scientific workflows onto distributed systems', *Scientific Programming Journal* **13** (3), 219–237.

Dempster, A. P., Laird, N. M. and Rubin, D. B. (1977), 'Maximum likelihood from incomplete data via the EM algorithm', *Journal of the Royal Statistical Society. Series B (Methodological)* **39** (1), 1–38.

Emmerich, W., Butchart, B., Chen, L., Wassermann, B. and Price, S. L. (2005), 'Grid service orchestration using the business process execution language (BPEL)', *Journal of Grid Computing* **3** (3–4), 283–304.

Gil, Y., Cheung, W., Ratnakar, V. and Chan, K. (2007*a*), Privacy enforcement through workflow system in e-Science and beyond, *in* 'Proceedings of Workshop on Privacy Enforcement and Acccountability with Semantics, Held in Conjunction with the Sixth International Semantic Web Conference', Busan, Korea.

Gil, Y., Ratnakar, V., Deelman, E., Mehta, G. and Kim, J. (2007*b*), Wings for Pegasus: creating large-scale scientific applications using semantic representations of computational workflows, *in* 'Proceedings of the 19th Annual Conference on Innovative Applications of Artificial Intelligence (IAAI)', Vancouver.

Gilburd, B., Schuster, A. and Wolff, R. (2004), k-TTP: a new privacy model for large-scale distributed environments, *in* 'Proceedings of the Tenth ACM SIGKDD International Conference on Knowledge Discovery and Data Mining', Seattle, WA, pp. 563–568.

Kantarcioglu, M. and Clifton, C. (2004), 'Privacy-preserving distributed mining of association rules on horizontally partitioned data', *IEEE Transactions on Knowledge and Data Engineering* **16** (9), 1026–1037.

Kargupta, H., Park, B., Hershberger, D. and Johnson, E. (2000), Collective data mining: a new perspective towards distributed data mining, *in* 'Advances in Distributed and Parallel Knowledge Discovery', MIT/AAAI Press, pp. 133–184.

Kim, J., Gil, Y. and Ratnakar, V. (2006), Semantic metadata generation for large scientific workflows, *in* 'Proceedings of the 5th International Semantic Web Conference', Athens, GA.

Kobsa, A. (2007), 'Privacy-enhanced personalization', *Communication of ACM* **50** (8), 24–33.

Ludascher, B., Altintas, I., Berkley, C., Higgins, D., Jaeger, E., Jones, M., Lee, E., Tao, J. and Zhao, Y. (2006), 'Scientific workflow management and the Kepler system', *Concurrency Computation: Practice and Experience* **18** (10), 1039–1065.

McLachlan, G. J. and Basford, K. E. (1988), *Mixture Models – Inference and Applications to Clustering*, Dekker, New York.

Merugu, S. and Ghosh, J. (2003), Privacy-preserving distributed clustering using generative models, *in* 'Proceedings of the Third IEEE International Conference on Data Mining', Melbourne, FL, pp. 211–218.

Oinn, T., Greenwood, M., Addis, M., Alpdemir, M., Ferris, J., Glover, K., Goble, C., Goderis, A., Hull, D., Marvin, D., Li, P., Lord, P., Pocock, M., Senger, M., Stevens, R., Wipat, A. and Wroe, C. (2006), 'Taverna: lessons in creating a workflow environment for the life sciences', *Concurrency Computation: Practice and Experience* **18** (10), 1067–1100.

Organization for Economic Co-Operation and Development (1980), 'Guidelines on the protection of privacy and transborder flow of personal data', http://www.oecd.org/document/18/0,2340,en_2649_34255_1815186_1_1 _1_1,00.html

Prodromidis, A. and Chan, P. (2000), Meta-learning in distributed data mining systems: issues and approaches, *in* 'Advances of Distributed Data Mining', MIT/AAAI Press.

Singh, G., Kesselman, C. and Deelman, E. (2005), 'Optimizing grid-based workflow execution', *Journal of Grid Computing* **3** (3–4), 201–219.

Sion, R. and Tatemura, J. (2005), Dynamic stochastic models for workflow response optimization, *in* 'Proceedings of the IEEE International Conference on Web Services', IEEE Computer Society, Washington, DC, pp. 657–664.

Weitzner, D., Abelson, H., Berners-Lee, T., Hanson, C., Hendler, J., Kagal, L., McGuinness, D., Sussman, G. and Waterman, K. (2006), Transparent accountable data mining: new strategies for privacy protection, Technical Report MIT-CSAIL-TR-2006-007, MIT.

Weitzner, D., Hendler, J., Berners-Lee, T. and Connolly, D. (2005), Creating a policy-aware Web: discretionary, rule-based access for the World Wide Web., *in* E. Ferrari and B. Thuraisingham, eds, 'Web and Information Security', IRM Press.

Wolff, R. and Schuster, A. (2004), 'Association rule mining in peer-to-peer systems.', *IEEE Transactions on Systems, Man, and Cybernetics, Part B* **34** (6), 2426–2438.

Zhang, X. and Cheung, W. (2005*a*), Learning global models based on distributed data abstractions, *in* 'Proceedings of the International Joint Conference on Artificial Intelligence', Edinburgh, pp. 1645–1646.

Zhang, X. and Cheung, W. (2005*b*), Visualizing global manifold based on distributed local data abstraction, *in* 'Proceedings of the 5th IEEE International Conference on Data Mining', Houston, pp. 821–824.

Zhang, X. and Cheung, W. (2007), A game theoretic approach to active distributed data mining, *in* 'Proceedings of the IEEE/WIC/ACM International Conference on Web Intelligence and Intelligent Agent Technology', IEEE Press, Silicon Valley, USA.

Zhang, X., Cheung, W. and Li, C. H. (2006), Graph-based abstraction for privacy-preserving manifold visualization, *in* 'Proceedings of the IEEE/WIC/ACM International Conference on Web Intelligence and Intelligent Agent Technology', Hong Kong, pp. 94–97.

Zhang, X., Lam, C. and Cheung, W. (2004), 'Mining local data sources for learning global cluster models via local model exchange', *IEEE Intelligent Informatics Bulletin* **4** (2), 16–22.

8

Building and using analytical workflows in Discovery Net

Moustafa Ghanem, Vasa Curcin, Patrick Wendel and **Yike Guo**

ABSTRACT

The Discovery Net platform is built around a workflow model for integrating distributed data sources and analytical tools. The platform was originally designed to support the design and execution of distributed data mining tasks within a grid-based environment. However, over the years it has evolved into a generic data analysis platform with applications in such diverse areas as bioinformatics, cheminformatics, text mining and business intelligence. In this work we present our experiences in designing the platform. We also map the evolution of its workflow server architecture to meet the demands of these different applications.

8.1 Introduction

The recent interest of the scientific and business communities in data mining has been primarily driven by our increasing ability to generate, capture and share more data than could be analysed using traditional methods.

Motivating examples abound in the domain of bioinformatics where routine scientific experiments can generate thousands of gene measurements from a single biological sample. The samples themselves are typically collected from large numbers of subjects under different conditions. With a large number of data points generated, the analysis can proceed only with the use of high-performance computing resources. Furthermore, the interpretation of the experimental data and analysis results typically requires integrating information from other data sources such as online genomic and proteomic data banks.

Motivating examples in the context of business applications also abound. Consider data mining activities conducted to understand the factors leading to customer churn in the mobile telecommunication industry. The operator typically holds information on the service usage patterns of its several million customers, going back a number of years. It also has access to previous marketing campaign data, results of previous questionnaires and sociodemographic customer data, either its own or from third-party providers, as well as tariff information, its

Data Mining Techniques in Grid Computing Environments Edited by Werner Dubitzky

own and the competitors'. The data sets reside in multiple databases and an effective data mining exercise would demand their integration, bringing in the need for high-performance computing resources.

The above two examples share a number of common features that highlight the key challenges faced in complex data mining projects. The analysis itself is an interactive process where a domain expert (e.g. researcher or business analyst) needs to explore the patterns available in the data to generate new hypotheses and then select the best analytical techniques to test them. Analysis techniques cannot be decided on beforehand, since the choice may depend on the problem at hand. The suitability of a particular data cleaning technique, for example, will depend on the distribution and nature of outliers in the data set. Similarly, the choice between predictive modelling techniques (e.g. logistic regression, decision trees or support vector machines) to be used will also be determined by the properties of cleaned sampled data. Each technique may require the use of specific data pre-processing to put the data in the right form (e.g. binning numerical data into categories, or alternatively representing categorical variables in numeric form). Furthermore, and depending on the nature of the problem, users may decide to explore using new data from other sources in their analysis (e.g. gene properties from an online data source for scientific experiments, or demographic data on customers from a third-party database for marketing applications). Analysing the new data may in turn require using different analytical tools.

Any data mining environment must be able to support users in conducting such interactive and iterative processes effectively and also support tracking of the different options explored throughout the analysis. When large quantities of data are to be processed and analysed, the environment must support the execution of the relevant tasks on high-performance computing resources. When data are integrated from different sources at different organizations, the environment must also support the user in accessing and integrating such data in their analysis as needed. Grid computing technologies offer possible solutions for addressing these issues effectively.

8.1.1 *Workflows on the grid*

Over the past few years, the service-oriented computing paradigm has been advocated as a new model for developing grid-based applications (Foster *et al.*, 2002; Talia, 2002; Atkinson *et al.*, 2005; Surridge *et al.*, 2005; Foster, 2005). The computational paradigm builds on the notion of a remote service that provides a uniform and standardized representation of any remotely accessible resource, e.g. software tool, computational resource, storage resource, sensor device etc. Such services can be advertised by service providers, discovered by service requestor and accessed remotely via well defined, standards-base interfaces.

Informally, a workflow, Figure 8.1, is an abstract description of steps required for executing a particular real-world process, and the flow of information between them. Each step is defined by a set of activities that need to be conducted. Within a workflow, work (e.g. data or jobs) passes through the different steps in the specified order from start to finish, and the activities at each step are executed either by people or by system functions (e.g. computer programs).

Workflow systems play an important role both within a service-oriented paradigm and in data mining applications. For service-oriented computing, they provide the languages and execution mechanisms that enable users to orchestrate the execution of available services, compose them to build applications and also use them in developing new aggregated services for use in further applications. For data mining applications, the workflow paradigm provides a natural

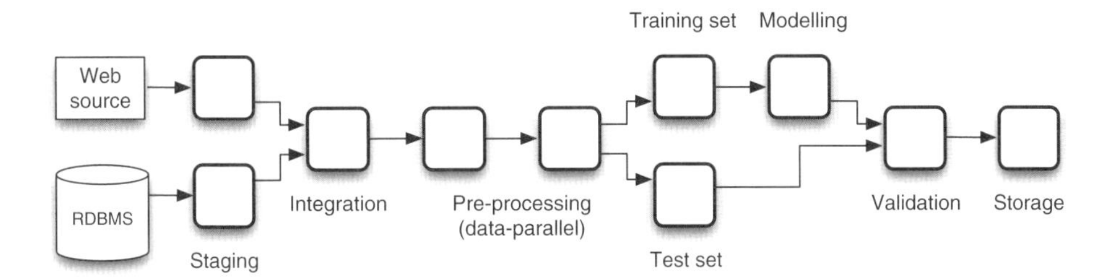

Figure 8.1 Simple workflow

way of describing and managing distributed data mining tasks – data integration, processing and analysis. Moreover, each of the different steps in the workflow can be implemented as a remote service using grid resources. The advantage is that users of the workflow system are shielded from the complexities of dealing with the underlying grid infrastructure and can focus on the logic of their data mining tasks. Access to the remote services and managing data transfers between them becomes the responsibility of the workflow system. We refer to workflows handled by such system as *analytical workflows* since their key feature is managing data analysis tasks.

8.2 Discovery Net system

The Discovery Net system has been designed around an analytical workflow model for integrating distributed data sources and analytical tools within a grid computing framework. The system was originally developed as part of the UK e-science project Discovery Net (2001–2005) (Rowe *et al.*, 2003) with the aim of producing a high-level application-oriented platform, focused on enabling the end-user scientists to derive new knowledge from devices, sensors, databases, analysis components and computational resources that reside across the Internet or grid.

Over the years, the system has been used in a large number of scientific data mining projects in both academia and industry. These include life sciences applications (Ghanem *et al.*, 2002, 2005; Lu *et al.*, 2006), environmental monitoring (Richards *et al.*, 2006) and geo-hazard modeling (Guo *et al.*, 2005). Many of the research ideas developed within the system have also been incorporated within the InforSense[1] KDE system, a commercial workflow management and data mining system that has been widely used for business-oriented applications. A number of extensions have also been based on the research outputs of the EU-funded SIMDAT[2] project.

8.2.1 System overview

Figure 8.2 provides a high-level overview of the Discovery Net system. The system is based on a multi-tier architecture, with a workflow server providing a number of supporting functions needed for workflow authoring and execution, such as integration and access

[1] http://www.inforsense.com/
[2] http://www.simdat.org/

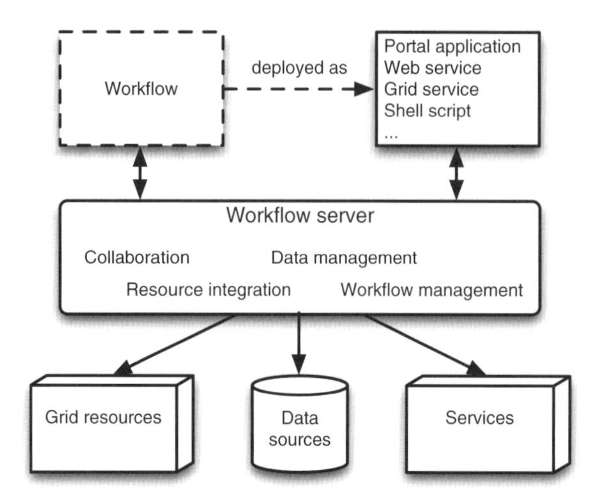

Figure 8.2 Discovery Net concept

to remote computational and data resources, collaboration tools, visualizers and publishing mechanisms.

A key feature of the system is that it is targeted at domain experts, i.e. scientific and business end users, rather than at distributed and grid computing developers. These domain experts can develop and execute their distributed data mining workflows through a workflow authoring client, where users can drag and drop icons representing the task nodes and connect them together. The workflows can also be executed from specialized Web-based client interfaces.

Although generic, the three-tier architecture model presented in Figure 8.2 serves as a consistent and simple abstraction throughout the different versions of the Discovery Net system. The implementation of the system itself has evolved over the past few years from a prototype targetted to specific projects to an industrial strength system widely used by commercial and academic organizations.

8.2.2 Workflow representation in DPML

Within Discovery Net, workflows are represented and stored using *Discovery Process Markup Language* (*DPML*) (Syed, Ghanem and Guo, 2002), an XML-based representation language for workflow graphs supporting both a data flow model of computation (for analytical workflows) and a control flow model (for orchestrating multiple disjoint workflows).

Within DPML, each node in a workflow graph represents an executable component (e.g. a computational tool or a wrapper that can extract data from a particular data source). Each component has a number of parameters that can be set by the user and also a number of input and output ports for receiving and transmitting data, as shown in Figure 8.3. Each directed edge in the graph represents a connection from an output port, namely the *tail* of the edge, to an input port, namely the *head* of the edge. A port is connected if there is one or more connections from/to that port.

In addition, each node in the graph provides metadata describing the input and output ports of the component, including the type of data that can be passed to the component and parameters

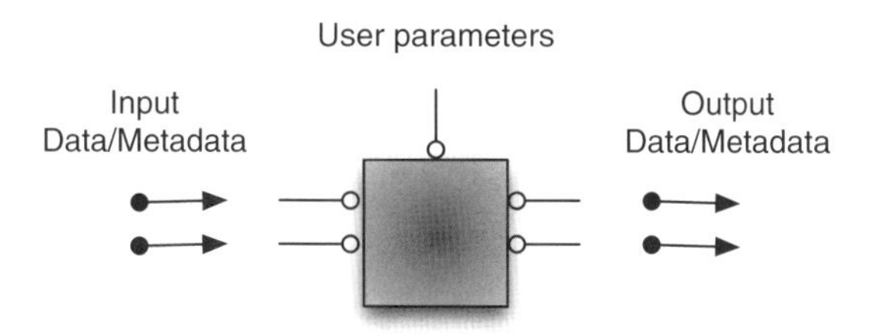

Figure 8.3 Structure of a component in Discovery Net

of the service that a user might want to change. Such information is used for the verification of workflows and to ensure meaningful chaining of components. A connection between an input and an output port is valid only if the types are compatible, which is strictly enforced.

8.2.3 Multiple data models

One important aspect of a workflow system is which data models it supports. Discovery Net is based on a typed workflow language and therefore supports arbitrary data types. However for the purpose of supporting data mining and other scientific applications, we included support for a relational data model, a bioinformatics data model for representing gene sequences and a stand-off markup model for text mining based on the Tipster architecture (Grishman, 1997). Each model has an associated set of data import and export components, as well as specific visualizers, which integrate with the generic import, export and visualization tools already present in the system. As an example, chemical compounds represented in the widely used SMILES format can be imported inside data tables, where they can be rendered adequately using either a three-dimensional representation or its structural formula. The relational model also serves as the base data model for data integration, and is used for the majority of generic data cleaning and transformation tasks. Having a typed workflow language not only ensures that workflows can be easily verified during their construction, but can also help in optimizing the data management.

8.2.4 Workflow-based services

To simplify the process of deploying workflow-based applications, Discovery Net supports the concept of workflow services. A workflow-based service is a service, i.e. either a Web/grid service or a Web application front-end to that service, derived from the workflow definition. Deploying a workflow as service requires selecting which parameters of the components used within a workflow and which input and output data elements are to be exposed to the user of the service. It also requires specifying the properties of the user interface, if any, to be associated with the service. Such services are then created and registered dynamically, from the workflow building client, and do not necessitate any re-initialization of the various servers.

8.2.5 Multiple execution models

Workflow execution within Discovery Net is delegated to the execution module on the workflow server that executes the data flow and control flow logic of the workflow definition, invoking the execution of the different components and also handling and data transfer between them. Each component receives input objects (data in data flow or tokens in control flow) on its input

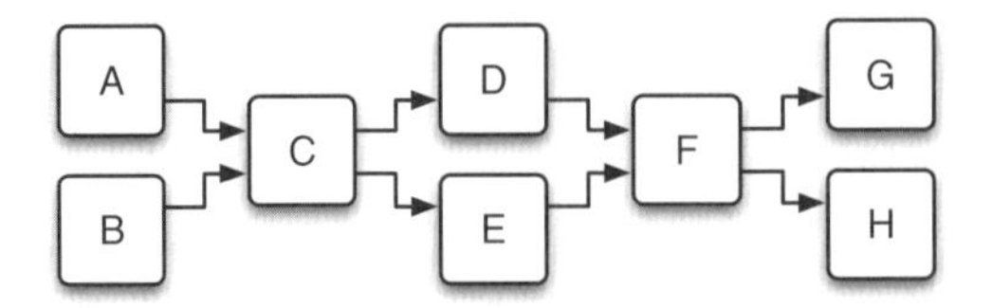

Figure 8.4 Abstract workflow graph

ports, manipulates such objects based on its own predefined implementation and then passes output objects along its output port. In order to understand how a workflow graph is interpreted and executed by the server, we now describe briefly the execution semantics for Discovery Net workflows.

It should be noted that, although different workflow systems all work on graph definitions similar to those of Discovery Net, the meaning of these workflows and how they are executed may vary greatly. A comparison between the execution semantics of Discovery Net and other systems including Taverna (Hull *et al.*, 2006), Triana (Taylor *et al.*, 2005) and Kepler (Ludäscher *et al.*, 2006) is described by Curcin *et al.* (2007). In our discussion below we make use of the simple workflow graph shown in Figure 8.4 to highlight the key concepts behind the workflow execution semantics for Discovery Net. The graph is generic, without assumptions about the nodes' functionality, and we shall use it to explain the data flow and control flow semantics of the system.

8.2.6 Data flow pull model

Discovery Net supports a data *pull* model for data flow graphs with a workflow acting as an acyclic dependence graph. Within this model a user requests execution of one the end-point nodes, e.g. G. This node can only execute if its input is available and it requests an input from the preceding node F, which in turn requests inputs from its preceding nodes, etc. Within this example both D and E are eligible for parallel execution if they represent computations executing on different resources; however, H will not be executed unless its output is requested explicitly by the user at a later time. Note also that the system supports caching of intermediate data outputs at node outputs. This means that if at a later time the user requests the execution of H, it would be able to re-use the pre-cached output of F. This model is typically suitable for data mining workflows.

8.2.7 Streaming and batch transfer of data elements

Discovery Net supports both streaming and batch transfers of data elements between components connected in a data flow graph. Note that, although metadata can be associated with the different input/output ports, specifying which data will be placed on the port, e.g. a list of protein sequences, the definition of the streaming behaviour for each component is not specified. There is no indication in the graph definition whether a component will process list elements on its input or output ports one by one or whether it should wait for the whole list to arrive before it is processed. This means that the same graph can be executed in either mode by the server depending on the nature of the environment and the other components in the workflow. Whether streaming or batch mode is to be used is specified by the user when a workflow is submitted for execution.

8.2.8 Control flow push model

As opposed to data flow graphs, nodes within Discovery Net control flow graphs represent special constructs for manipulating execution tokens and controlling iteration and check point behaviour. These control flow graphs can be cyclic in their definition, and communication between them is based on passing tokens rather than on passing data elements. The execution of Discovery Net control flow graphs is based on a data *push* paradigm, where workflow components are invoked left to right and nodes with multiple output ports can decide on which port(s) they will place the tokens, hence determining the direction of further execution.

8.2.9 Embedding

Since the strength of the workflow approach for data mining lies primarily in the use of data flows for structured application design, control flows in Discovery Net are used mainly to coordinate disconnected workflows whose ordering is driven by business logic, rather than an explicit data dependence. Therefore, control flow nodes contain within them data flows (deployed as services) that are invoked every time the node executes. This concept is shown in Figure 8.5.

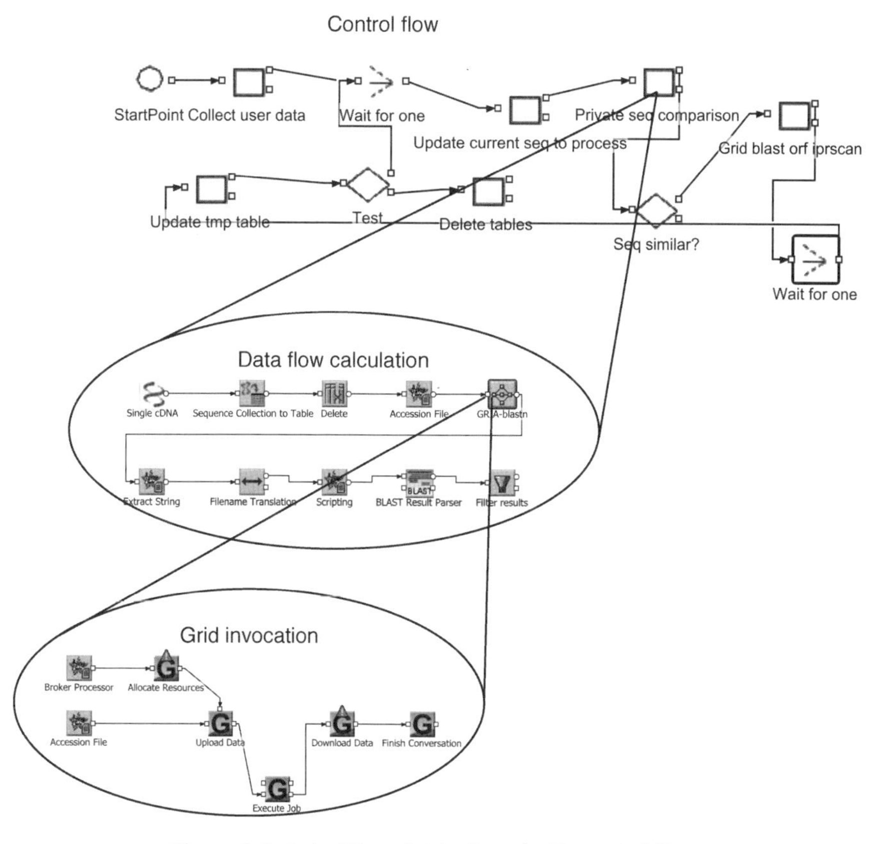

Figure 8.5 Embedding of data flows inside control flows

8.3 Architecture for Discovery Net

In this section, we provide a detailed overview of the Discovery Net 4.0 server architecture. It is however important to note that the architecture of the server and its implementation have gone through a number of iterations to support more features including different workflow types as well as different data models. Compatibility of workflow definitions across different implementations has been maintained through the use of DPML and open standards for accessing remote computational and data resources, such as WSDL/SOAP and OGSA-DAI.

The simple three-tier model itself, presented in the previous section, traces its roots to a predecessor of the Discovery Net system, namely the Kensington Enterprise Data Mining system (Chattratichat *et al.*, 1999). The Kensington system was based on a simple workflow server and the system itself did not support execution within a grid computing environment but rather only enabled users to access and integrate distributed enterprise resources through CORBA (Object Management Group, 2006) and proprietary interfaces.

The architecture of the first grid-enabled version of the Discovery Net system was presented by Rowe *et al.* (2003). Specifically, in that paper we described how the generic three-tier model can be easily extended to support the composition of Web and grid services within data mining workflows. The advantage of using Web/grid services is that they provide a standardized mechanism for accessing remote applications and resources. Alsairafi *et al.* (2003) presented an enhanced version of the architecture, describing how various grid computing protocols can be supported, and also where support for interaction between multiple Discovery Net servers could be supported. The system described in these two papers only supported a data flow model over grid resources without control flow constructs, which were introduced in Discovery Net 3.0. A general description of how different types of workflow (e.g. data flow, control flow or grid flow) interact within the system was described by Curcin *et al.* (2007).

In other papers we have focused on how specialized functionality can be supported within the server itself. For example, Richards *et al.* (2006) describe how the architecture could be further extended to support collection of data from online sensor data sources. Ghanem *et al.* (2002) discuss how the system was extended to support text mining workflows, and Lu *et al.* (2006) describe the extension of the system that supports a data model for bioinformatics data analysis. Furthermore, Wendel, Ghanem and Guo (2006) describe the implementation of workflow scheduling sub-components of the server, and Ghanem *et al.* (2006) describe how integration with the GRIA (Taylor, Surridge and Marvin, 2004) middleware can be supported.

8.3.1 Motivation for a new server architecture

The new Discovery Net architecture has been designed to cope with the ever increasing complexity of the system, in terms of data models being added for particular application domains, new integration paradigms and the need to customize generic components and expand portal design capabilities while retaining the operational simplicity.

The new architecture was also driven by observing how the system was being used by real users to solve their large-scale problems based on the workflow paradigm. The analytical workflows developed within the system were seldom used to execute one-off analytical tasks. Rather, in the first instance they would be developed by domain experts (e.g. bioinformaticians, or business data analysts) who would create the workflows in terms of defining the overall structure, choosing the types of data being integrated and analysed and selecting the data mining and analysis methods to be used. Then, they would configure the workflows to use specific computing resources and make them available to their end users (e.g. biologists or marketing experts) to start using for processing, analysing and interpreting different data runs. The end users would also use the interactivity features from the system in terms of exploring the data itself and controlling how a workflow would proceed, but seldom changed the components used in the workflow itself. To better support such a mode of operation, several key aspects had to be introduced within the system, as discussed below.

8.3.2 *Management of hosting environments*

A hosting environment is one of the different environments in which a workflow service can be deployed, e.g. the authoring, execution, configuration or Web environments. These environments differ in the sense that they may require different resources or libraries to be available in order to perform a particular function. Discovery Net 4.0 supports the definition and use of multiple hosting environments for the purpose of workflow authoring (verification code, complex parameterization and visualization user interfaces), production use (Web components associated with data types and parameters), runtime or configuration. Essentially, a hosting environment enables a workflow application to be packaged with only the resources needed for its target user base, be it embedded inside a device or in an interactive process editor.

8.3.3 *Activity management*

While we used to consider types of activities available to build workflows as a limited but extensible set of functionalities, we have now introduced frameworks to improve the management of these activities. First, we needed a mechanism to add new sets of activities (i.e. their description, implementation classes, libraries and dependences) to a live system, which is why we introduced *activity services* that provide access to the description, definition, resources and deployment information of the set of activities required for workflow execution, verification or configuration and can be registered at runtime. Second, we extended the workflow definition language to support activity *decorators*, i.e. additional pieces of processing, verification or configuration logic, defined using a dynamic programming language that allows us to specialize the behaviour of an activity usually for a purpose specific to a workflow.

8.3.4 *Collaborative workflow platform*

In order to support different types of user with their own skill set, expertise or responsibilities, e.g. workflow definition, system administration, grid computing, data mining or application domain expert, the system needs to support the definition of roles assigned to each users. Being assigned a particular role may or may not allow access to data, activities or services hosted by the server, i.e. it enables role-based access and authorization within and across teams of users collaborating to build a scientific workflow. It also allows activity implementations to specialize their runtime behaviour or define a different set of parameters based on this information (e.g. hiding parameters that are not relevant to particular applications).

8.3.5 *Architecture overview*

An overview of the Discovery Net 4.0 architecture supporting the above features is shown in Figure 8.6. It provides three main types of service.

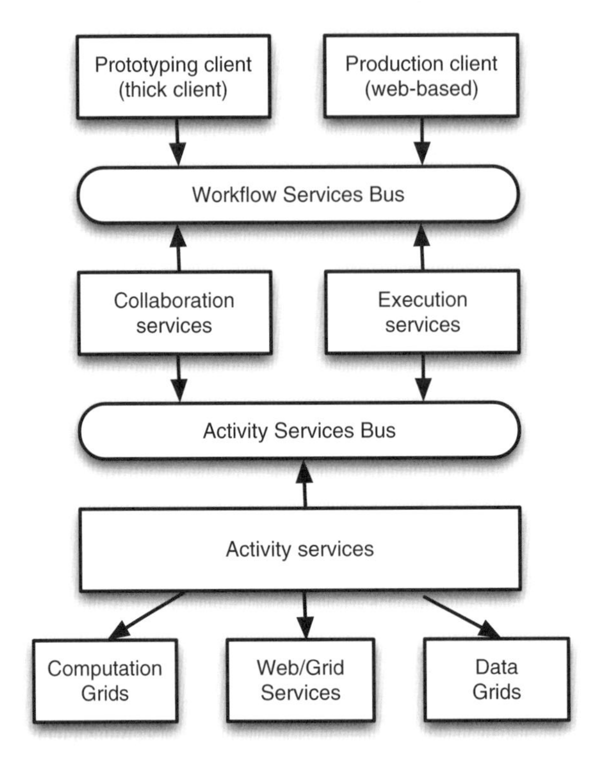

Figure 8.6 Simple architecture

- *Activity services.* These provide the definition, description and implementations of sets of activities that in turn access grids and other third party resources or libraries.
- *Execution services.* These support the workflow execution life cycle (submission, scheduling, enactment, interaction requests, intermediate data product caching, data table management and processing).
- *Collaboration services.* These provide support to share temporary intermediate data products or persistent results between groups of users.

These different Discovery Net services are connected by two buses.

- *Activity services bus.* This acts mainly as a registry for activity services and supports authorization.
- *Workflow services bus.* This acts as a registry for collaboration and execution services as well as authentication and protocol mapping if required by the different types of client.

There are two types of client that can access these buses.

- *Prototyping (thick) client.* This is used for creating and modifying workflows and inspecting intermediate data products.

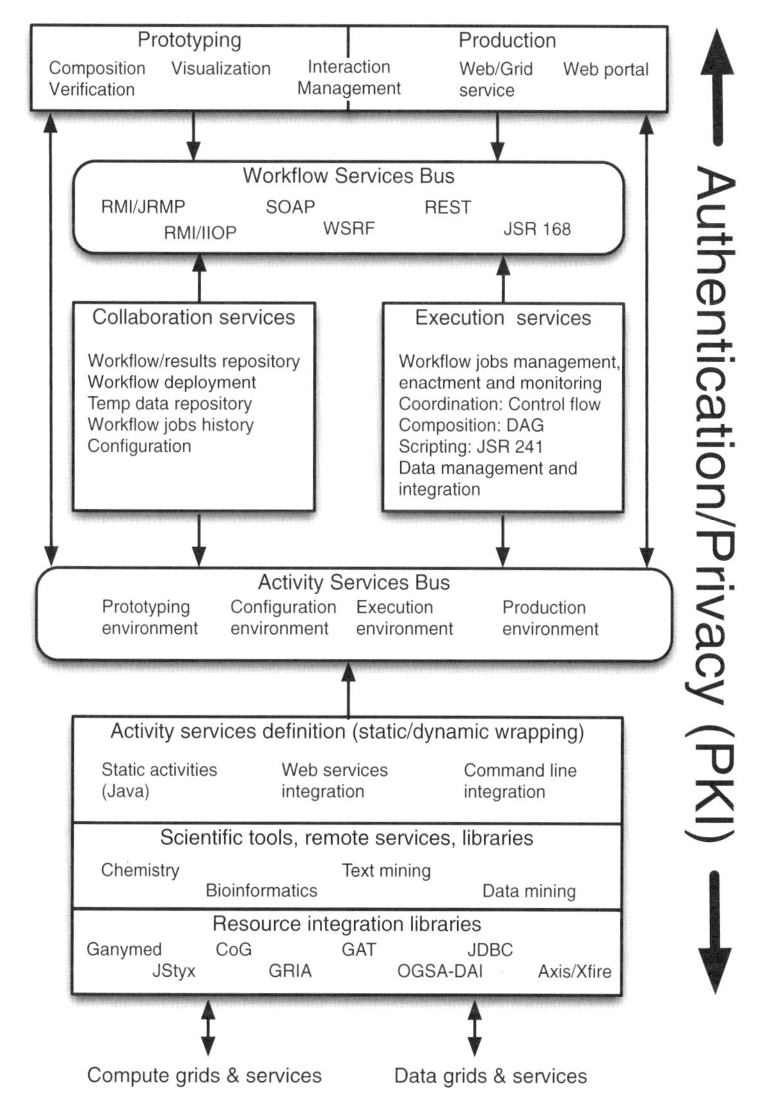

Figure 8.7 Full architecture

- *Production (Web) client.* This is used for consuming production-ready services.

A full schematic view of the architecture is presented in Figure 8.7, with key functions and technologies of each component depicted in text boxes inside the component.

8.3.6 Activity service definition layer

This layer integrates resources and applications. Individual activities can either be statically defined as Java wrappers or can use a number of dynamic mechanisms to connect to implementations (Web service integration, command-line integration, etc). Each activity can access a number of application-specific libraries – BioJava (Pocock, Down and Hubbard, 2000) for bioinformatic applications, or specialized data/text mining libraries such as those provided by InforSense or by Weka (Witten *et al.*, 1999). The activities can also access specialized resource integration libraries such as SSH access (Ganymed), grid access (GAT, CoG, GRIA), data access (OGSA-DAI, JDBC) or Web services (Axis, XFire, HttpClient).

8.3.7 Activity services bus

Activities are accessible through an Activity Services Bus, which handles role-based authorization and provides the information and resources necessary for the instantiation of different environments to compose, consume, configure or enact the activities. An environment in this context represents the libraries and their dependences that are necessary to load the classes and instantiate the objects needed for a particular purpose (i.e. libraries required at run-time or only at design-time, GUI required for configuring the activities).

8.3.8 Collaboration and execution services

Workflow executions are scheduled and managed within the system, though particular activities will delegate their execution to other grid management systems. Task management and farming to execution servers is based on a set of Java commodity services – this is fully described by Wendel, Ghanem and Guo (2006). A set of collaboration services allows sharing data results, workflow and activity definitions among groups of users, using role-based authorization. The authorization information can be retrieved from either local files, LDAP servers or a Web-based authorization server using SAML. The collaboration services include a workflow repository mechanism where DPML workflow definitions can be stored and retrieved and changes tracked. They also include data storage mechanisms for storage of intermediate data results and tables, as well as information on workflow job history, system configuration and deployment parameters.

8.3.9 Workflow Services Bus

The Workflow Services Bus provides access to the collaboration services, handles authorization, hosts the necessary messaging protocols required to communicate control and monitoring information during workflow execution and finally provides access to the workflow-based services.

8.3.10 Prototyping and production clients

The user (workflow builder) is presented with a client tool for prototyping. The interface allows her to compose services, execute them, cache temporary results, visualize intermediate and final data products and modify particular parameters of the workflow either at design time or while its execution is suspended (i.e. when user interaction is requested during the workflow execution). Once the workflow is completed, the client interface also allows the user to define how the workflow should be published, i.e. turned into either a Web-based application for domain experts or into a Web/grid service for integration into a service-oriented architecture. The workflows are then transformed into workflow-based services. These published services can also be reused from the workflow client as particular activities.

8.4 Data management

Figure 8.8 shows the data management system of Discovery Net. It designed to support persistence and caching of intermediate data products. It also supports scalable workflow execution over potentially large data sets using remote compute resources. The processing of data is often performed outside the system. Apart from some libraries such as Weka (Witten *et al.*, 1999) or Colt[3], which process data directly in Java and can therefore be used for tighter integrations, data are processed outside, using local execution with streams or temporary files, socket communication such as the communication with an R[4] server or over SSH protocol for secure communication.

An analytical workflow management system cannot keep storing any data product generated in the workflow, given that during prototyping only a fraction of data is relevant and deserves to be stored back for longer term. The process of storing back the result is a one-off event that occurs when the user is satisfied with the results. Thus these repositories cannot always be used as a place to cache these intermediate data sets. The means and location to stage them must be part of the workflow system itself.

During workflow prototyping we cannot assume a particular, fixed data model, since the workflow is not fixed yet. Re-parameterization or changes to the graph structure may trigger a change in the schemas of these tables. This means that storing these intermediate data products does not follow the same pattern as storing tables in RDBMS, and instead has the following characteristics.

- Many temporary and unrelated tables need to be stored and there is no notion of database schema or referential integrity between these tables.

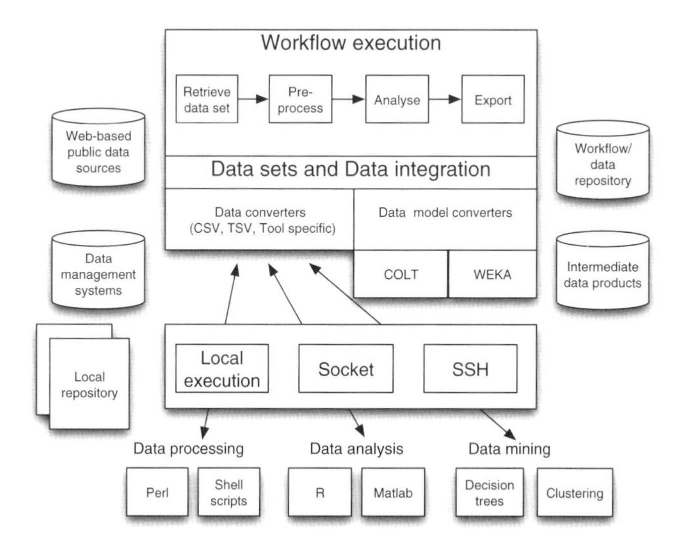

Figure 8.8 Data management in execution

[3]http://dsd.lbl.gov/~hoschek/colt/

[4]http://www.r-project.org/

- For the purpose of caching, any data generated need to be preserved for the duration of the cache. Thus no intermediate data point is modified, but only created and populated once.
- The pattern of access is mainly based on scanning the entire data set at least once for importation, exportation, loading in memory or deriving new attributes. Indexing and optimized querying are not essential, given that this model does not aim to replace a database.
- The ordering of attributes and instances may be relevant and needs to be preserved. For instance, time series may assume a range of attributes in a particular order, or that data instances themselves are ordered. In general, data mining algorithms can be affected by the ordering of the data and it is important to preserve it for the process to be deterministic.

The processing thus bears more resemblance to the way files and documents are usually processed by analysis executables in workflows, by creating new output files instead of modifying the input ones, than to the way data are modified in a relational database through declarative queries. However, relying on text files to contain the data is not practical either, as they cannot handle binary data or documents easily and the cost of parsing the data in and out would take too much time.

The other issue is the execution model over these data sets. Data sets may need to be used in the workflow not only for caching and storage but also for streaming data instances in order to model pipeline parallelism in the workflow, in particular when the processing of these data sets is performed in a remote location. In other words, to the end-user it is natural that the data sets should act simultaneously as a data type and a communication channel between activities.

While this is partly a usability requirement, in the context of data-flow modelling of analysis workflows, having a common model introduces two important changes. First, activities can output new relations derived from their input, by adding, removing or reordering attributes, with only the modifications being stored, for efficiency purposes. This pattern is preferred to the simpler pattern of creation of new output, as it allows easier maintenance of data record provenance through activities. Second, table-based operations such as relational operators, aggregation functions or data mining algorithms, can be used in the workflow regardless of whether the preceding components are streaming or not. Hence streaming and batch components can be freely combined and the component semantics determines the behavior at runtime.

Therefore the main features of Discovery Net data sets are the following.

(a) Temporary and short-term persistence for exploring and caching data products during the workflow prototyping phase.

(b) Streaming of data instances, to enable pipeline parallelism either in memory with no additional I/O cost or over a file-based representation.

(c) Ad hoc integration of information from each activity by creating temporary relations without requiring pre-defined schemas in external databases.

It is important to note that the Discovery Net data set is not a replacement for traditional data management systems, relational database management systems, data warehouses and other long-term data storage mechanisms. It however acts as a data structure for staging the data to be processed in the workflow.

8.5 Example of a workflow study

In this section we briefly present an analytical workflow application developed within Discovery Net to highlight some of the key features of the system. The chosen example is based on developing workflow-based data analysis services for use in studying *adverse drug reaction* (*ADR*) studies. The aim of the analysis is to build a classification model to determine what factors contribute to patients having an adverse reaction to a particular drug.

Although the workflow makes use of many features of the Discovery Net system, including integration of remote data mining tools and underlying data management features of the system, we primarily focus on demonstrating one of the key advantages of the system, its ability to develop and re-use workflow-based services that can be re-used in other studies.

The workflow itself uses multiple data mining tools made accessible as remote grid services executing on high-performance computing resources at the Imperial College London Internet Centre. These include logistic regression, bootstrapping, text mining and others. Previous to use, these tools were integrated as activities within the Activity Service Definition Layer, thereby making the access to them, as well as the associated data management, transparent to the user.

8.5.1 ADR studies

ADR studies are a common type of analysis in pharmacovigilance, a branch of pharmacology concerned with the detection, assessment, understanding and prevention of adverse effects, and particularly long-term and short-term side effects of medicines (World Health Organization, 2002). ADR studies can be contrasted to drug safety trials that are required before the drug enters the market, in that the safety trials typically operate on a limited set of healthy test subjects, while ADR studies focus on patients with specific conditions (diabetes, liver damage etc.) not directly connected to the drug being tested as well as on patients on a combination of drugs. Also, ADR studies typically cover a longer follow-up periods in observations.

However, the mechanisms for conducting ADR studies are still limited, with typical source data consisting of voluntary patient and GP reporting, and lacking in quantity needed for meaningful analysis as well as length coverage. One data source in the UK containing suitable follow-up and all relevant information is the primary care data, available from a number of providers, such as the *General Practitioners Research Database* (*GPRD*) and *The Health Integration Network* (*THIN*), containing full prescription data and observations for a large body of patients over an extended period of time (10–20 years). In this way, researchers can perform an *in silico* study by creating their own control sets and running analyses on them. Furthermore, in addition to traditional statistics techniques, data mining can be utilized to answer more ambitious questions.

8.5.2 Analysis overview

Non-steroidal anti-inflammatory drugs (*NSAIDs*) are one of the most popular classes of drugs in the world, including aspirin, ibuprofen and others. One common problem with them is that they can cause gastro-intestinal side-effects, and significant research effort has gone into producing NSAIDs that did not demonstrate this behaviour. With this goal in mind, selective cyclo-oxygenase-2 (COX-2) inhibitors were developed and marketed in celecoxib and rofecoxib (Vioxx) drugs. However, as demonstrated in numerous research papers, these drugs can still,

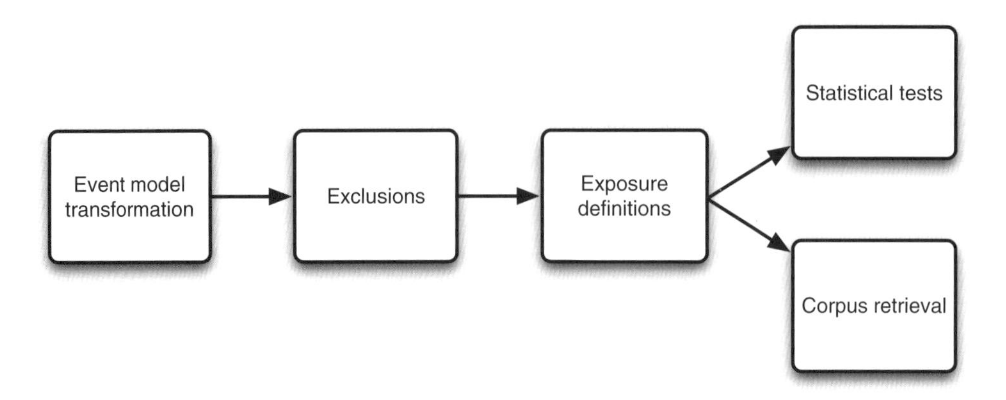

Figure 8.9 Abstract workflow for adverse drug reaction study

in some cases, produce side-effects ranging from gastrointestinal, via skin rashes to respiratory effects (Silverstein *et al.*, 2000).

The study presented here aims to build a classification model to determine what factors contribute to patients having an adverse reaction to COX-2. The list of tasks involves determining side-effects from patients who have been known to previously receive COX-2 inhibitors, defining exposure window, excluding conditions and drugs that may produce similar effects and testing the produced model on a sample of patients on COX-2, both with and without ADRs. The data sources used in this study were derived from GPRD, and consist of a patient table, prescription table and clinical table, containing, respectively, information on the patient, all the prescriptions the patient has been given with full dosage and duration information and all diagnosis and measurements related to the patient. The overall workflow is given in Figure 8.9.

8.5.3 Service for transforming event data into patient annotations

As a first step, the event data in the database, organized by event identifiers and annotated with dates, is transformed into patient samples. For each patient we define a workflow query to establish the first instance of any of the nominated conditions. In order to publish this workflow as a service, several parameters are defined. First, the conditions themselves are made into a parameter. The GPRD selector interface for the conditions utilizes medical dictionaries, and is integrated into the hosting environment, as shown in Figure 8.10. Second, the aggregation function for the condition is generalized, so the researcher can specify first condition, last condition or count of condition occurrences.

8.5.4 Service for defining exclusions

Next step in the analysis is to pick known medical conditions and drugs that may cause gastrointestinal conditions. Event databases are searched for relevant diagnosis and prescriptions and these patients are removed from the data set. The selection dialogue for drug and condition querying, used in the published workflow, is the same as in the previous service.

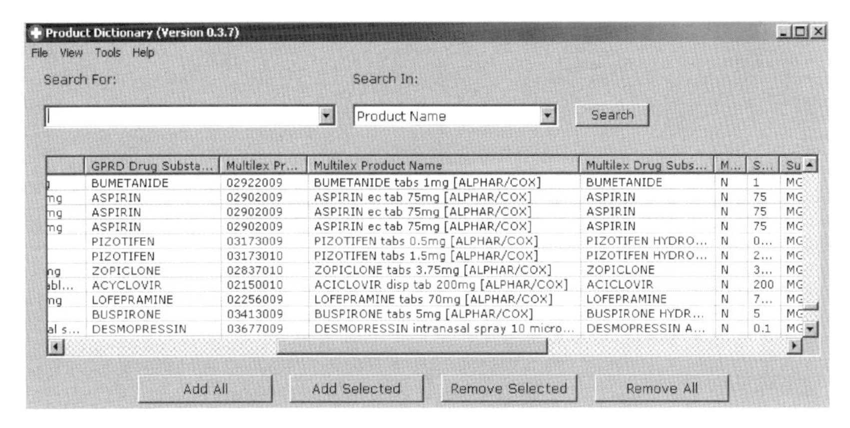

Figure 8.10 Medical term selection widget

8.5.5 *Service for defining exposures*

The next step is to provide a framework for establishing exposure, i.e. duration of time that the patient is on a drug. One valid approach is to only take the first prescription, and the first occurrence of the event, assuming that ADR will happen the first time patient takes the drug. One potential problem with this scenario is that if dual medication needs to be taken into account, it needs to be treated as a separate drug. Taking this into account, the next iteration was to use the dosage information from prescription table to establish an event timeline for each patient and calculate, for each event, a set of events with which it overlaps, including other conditions and other prescriptions. In this way, it is relatively efficient to find valid drug exposures for each event on the timeline. Publishing this workflow as a service includes parameterizing the definition of events that need to be tracked, using a similar selection dialogue as in the service above.

8.5.6 *Service for building the classification model*

Once the data have been filtered and annotated with event information, the training set needs to be determined. Since the cleaned data are likely to contain a relatively low percentage of patients with adverse reactions, over-sampling is utilized so that the sample contains 50 per cent of adverse effects. Default separation then involves 40 per cent for training the logistic regression algorithm, and a further 30 per cent for both the validation and evaluation steps. The outputs from this service are the profit matrix and the lift diagram determining the prediction accuracy of the classifier, as well as the list of variables found to contribute to the adverse reaction to COX-2. Parameterization involves allowing the analyst to specify oversampling, training–test–validation sample ratios and various parameters of logistic regression.

8.5.7 *Validation service*

The last step, after the classification model is calculated, is to create a support corpus for the study. One way of doing this is to query Pubmed with the COX-2 drug names, and retrieve abstracts describing existing scientific knowledge about them. The resulting collection of abstracts can then either be manually checked or passed to a text-mining procedure.

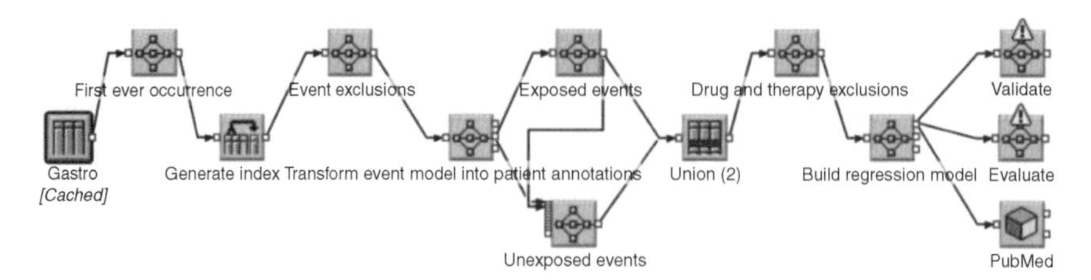

Figure 8.11 ADR workflow

8.5.8 Summary

In this section we have presented how a scientific application for investigating adverse drug reactions can be built using Discovery Net architecture. An example study for finding contributing factors to adverse reaction to COX-2 is presented, with separate workflow-based services designed for each of the main tasks involved in the process. Such services are then published inside a portal environment, with the addition of dictionary lookup widgets for selection of medical terms. The full workflow is presented in Figure 8.11.

8.6 Future directions

Our vision has always been to enable end users, i.e. non-programmers, to build and use their grid-based data mining applications and also to be able to deliver them easily as executable applications to other end users. The overarching theme is to enable a wide range of users to reap the benefits of grid computing technologies while shielding them from the complexities of the underlying protocols. Our experience so far indicates that achieving the balance is indeed possible.

The analytical workflow paradigm is intuitive in this respect since it provides a high-level abstract view of the data analysis steps. Although intuitively simple in terms of usage, designing and implementing an analytical workflow system raises a number of challenges. One issue is deciding the execution semantics of the workflow language used. The next one is enabling extensibility, i.e. enabling users to easily integrate access to new local or remote data sources and data mining tools. The third issue relates to data management. All of these issues have been addressed in the Discovery Net system and its architecture over the years.

The system and architecture presented here constitute a solid foundation for addressing forthcoming research challenges in the area. Here we shall briefly discuss the issues we foresee as being relevant over the next several years.

We have introduced decorators as dynamic modifications to individual components that can manipulate global resources or serve as a way of customizing a predefined workflow. Further use of decorators will ultimately lead to the separation of activities into core definitions and *applicators*, sets of predefined decorators that can adapt an activity to a particular data type or computational resource and are shared among several activities. This separation will, in part, be motivated by the increased component exchange between workflow systems and the need to move it beyond mutual Web service invocation and into the realm of dynamic code exchange. These ideas found their first incarnation in the EU project SIMDAT.

Security and credential management is also an issue that needs to be handled more consistently and safely in all workflow systems. Currently it is up to each activity to deal with the credentials to access a secured service and to provide the necessary options. Furthermore, once the workflow is disseminated, it is no longer possible to control who will be using it. Hence, restricting both modifications and access levels is difficult once the workflow leaves a particular hosting environment, which limits the commercial applicability of workflows as a medium for application creation. The definition of policies to manage remote retrieval of credentials needs to be separated from the workflow definition itself to support a full range of licensing options.

Finally, the increased modularization of workflow servers, resulting in a range of hosting environments, such as embedded workflow-based processing in devices or specialized applications (e.g. visualizer tools or desktop applications), will lead to the next generation of deployment, in which workflows are published together with a custom-made hosting environment created at the deployment stage. Such an environment contains only the resources and the functionality necessary for the application execution, thereby eliminating code bloat and removing dependence on external servers for standalone applications.

The accessibility of data and computational resources to data mining users enables them to conduct more complex analyses over larger data sets in less time. It allows them to become more efficient in discovering new knowledge. The availability of the end-user-oriented frameworks that provide these scientists with full access to the benefits of grid computing technologies, while shielding them from the complexities of the underlying protocols, is essential. Our experience indicates that achieving the balance is indeed possible by providing the users with tools at the appropriate level of abstraction that suits their problem solving methods, and that suits their modes of knowledge discovery.

Acknowledgements

The work described in this chapter has been funded, in part, under different grants including the Discovery Net Project (2001–2005) funded by the EPSRC under the UK e-Science Programme and the SIMDAT project funded by the European Union under the FP6 IST programme. The authors would also like to acknowledge the work conducted by their many colleagues at Imperial College London and InforSense Ltd. in implementing various parts of the Discovery Net system, and Dr. Mariam Molokhia from London School of Hygiene and Tropical Medicine for the ADR case-control workflow.

References

Alsairafi, S., Emmanouil, F.-S., Ghanem, M., Giannadakis, N., Guo, Y., Kalaitzopolous, D., Osmond, M., Rowe, A. and Wendel, P. (2003), 'The design of Discovery Net: towards open grid services for knowledge discovery', *International Journal of High Performance Computing* **17** (3), 297–315.

Atkinson, M., DeRoure, D., Dunlop, A., Fox, G., Henderson, P., Hey, T., Paton, N., Newhouse, S., Parastatidis, S., Trefethen, A., Watson, P. and Webber, J. (2005), 'Web service grids: an evolutionary approach', *Concurrency and Computation: Practice and Experience* **17** (2–4), 377–389.

Chattratichat, J., Darlington, J., Guo, Y., Hedvall, S., Köler, M. and Syed, J. (1999), An architecture for distributed enterprise data mining, *in* 'Proceedings of the 7th International Conference on High-Performance Computing and Networking (HPCN Europe'99)', Springer, London, pp. 573–582.

Curcin, V., Ghanem, M., Wendel, P. and Guo, Y. (2007), Heterogeneous workflows in scientific workflow systems, *in* 'International Conference on Computational Science (3)', pp. 204–211.

Foster, I. (2005), 'Service-oriented science', *Science* **308** (5723), 814–817.

Foster, I., Kesselman, C., Nick, J. M. and Tuecke, S. (2002), 'The Physiology of the Grid: An Open Grid Services Architecture for Distributed Systems Integration', http://www.globus.org/research/papers/ogsa.pdf

Ghanem, M., Azam, N., Boniface, M. and Ferris, J. (2006), Grid-enabled workflows for industrial product design, *in* '2nd International Conference on e-Science and Grid Technologies (e-Science 2006)', p. 96.

Ghanem, M. M., Guo, Y., Lodhi, H. and Zhang, Y. (2002), 'Automatic scientific text classification using local patterns: KDD Cup 2002 (Task 1)', *SIGKDD Explorer Newsletter* **4** (2), 95–96.

Ghanem, M., Ratcliffe, J., Curcin, V., Li, X., Tattoud, R., Scott, J. and Guo, Y. (2005), Using text mining for understanding insulin signalling, *in* '4th UK e-Science All Hands Meeting 2005'.

Grishman, R. (1997), 'Tipster Architecture Design Document, Version 2.3', http://www.itl.nist.gov/iaui/894.02/related_projects/tipster/

Guo, Y., Liu, J. G., Ghanem, M., Mish, K., Curcin, V., Haselwimmer, C., Sotiriou, D., Muraleetharan, K. K. and Taylor, L. (2005), Bridging the macro and micro: a computing intensive earthquake study using Discovery Net, *in* 'Proceedings of the 2005 ACM/IEEE conference on Supercomputing (SC'05)', IEEE Computer Society, Washington, DC, p. 68.

Hull, D., Wolstencroft, K., Stevens, R., Goble, C., Pocock, M. R., Li, P. & Oinn, T. (2006), 'Taverna: a tool for building and running workflows of services', *Nucleic Acids Research* **34**, W729–W732.

Lu, Q., Hao, P., Curcin, V., He, W., Li, Y.-Y., Luo, Q.-M., Guo, Y.-K. and Li, Y.-X. (2006), 'KDE Bioscience: platform for bioinformatics analysis workflows', *Journal of Biomedical Informatics* **39** (4), 440–450.

Ludäscher, B., Altintas, I., Berkley, C., Higgins, D., Jaeger, E., Jones, M., Lee, E. A., Tao, J. and Zhao, Y. (2006), 'Scientific workflow management and the Kepler system', *Concurrency and Computation: Practice and Experience* **18** (10), 1039–1065.

Object Management Group (2006), CORBA component model 4.0 specification, Specification Version 4.0, Object Management Group.

Pocock, M., Down, T. and Hubbard, T. (2000), 'BioJava: open source components for bioinformatics', *ACM SIGBIO Newsletter* **20** (2), 10–12.

Richards, M., Ghanem, M., Osmond, M., Guo, Y. and Hassard, J. (2006), 'Grid-based analysis of air pollution data', *Ecological Modelling* **194**, 274–286.

Rowe, A., Kalaitzopoulos, D., Osmond, M., Ghanem, M. and Guo, Y. (2003), 'The Discovery Net system for high throughput bioinformatics', *Bioinformatics* **19** (1), i225–i231.

Silverstein, F., Faich, G., Goldstein, J., Simon, L., Pincus, T., Whelton, A., Makuch, R., Eisen, G., Agrawal, N., Stenson, W., Burr, A., Zhao, W., Kent, J., Lefkowith, J., Verburg, K. and Geis, G. (2000), 'Gastrointestinal toxicity with Celecoxib vs nonsteroidal anti-inflammatory drugs for osteoarthritis and rheumatoid arthritis. The CLASS study: a randomized controlled trial. Celecoxib long-term arthritis safety study.', *The Journal of the American Medical Association* **284** (10), 1247–1255.

Surridge, M., Taylor, S., Roure, D. D. and Zaluska, E. (2005), Experiences with GRIA – industrial applications on a Web services grid, *in* 'First International Conference on e-Science and Grid Computing', pp. 98–105.

Syed, J., Ghanem, M. and Guo, Y. (2002), 'Discovery Processes: Representation and Reuse', citeseer.ist.psu.edu/syed02discovery.html

Talia, D. (2002), 'The open grid services architecture: where the grid meets the Web', *IEEE Internet Computing* **6** (6), 67–71.

Taylor, I., Shields, M., Wang, I. and Harrison, A. (2005), 'Visual grid workflow in Triana', *Journal of Grid Computing* **3** (3-4), 153–169.

Taylor, S., Surridge, M. and Marvin, D. (2004), Grid resources for industrial applications, *in* 'Proceedings of the IEEE International Conference on Web Services (ICWS'04)', IEEE Computer Society, Washington, DC, p. 402.

Wendel, P., Ghanem, M. and Guo, Y. (2006), Designing a Java-based grid scheduler using commodity services, *in* 'UK e-Science All Hands Meeting'.

Witten, I., Frank, E., Trigg, L., Hall, M., Holmes, G. and Cunningham, S. (1999), Weka: practical machine learning tools and techniques with Java implementations, *in* 'Proceedings of ICONIP/ANZIIS/ANNES'99 International Workshop: Emerging Knowledge Engineering and Connectionist-Based Information Systems', pp. 192–196.

World Health Organization (2002), 'Importance of Pharmacovigilance', http://whqlibdoc.who.int/hq/2002/a75646.pdf

9

Building workflows that traverse the bioinformatics data landscape

Robert Stevens, Paul Fisher, Jun Zhao, Carole Goble and **Andy Brass**

ABSTRACT

The bioinformatics data landscape confronts scientists with significant problems when performing data analyses. The nature of these analyses is, in part, driven by the data landscape. This raises issues in managing the scientific process of *in silico* experimentation in bioinformatics. The myGrid project has addressed these issues through workflows. Although raising some issues of their own, workflows have allowed scientists to effectively traverse the bioinformatics landscape. The high-throughput nature of workflows, however, has forced us to move from a task of data gathering to data gathering and management. Utilizing workflows in this manner has enabled a systematic, unbiased, and explicit approach that is less susceptible to premature triage. This has profoundly changed the nature of bioinformatics analysis. Taverna is illustrated through an example from the study of trypanosomiasis resistance in the mouse model. In this study novel biological results were obtained from traversing the bioinformatics landscape with workflow.

9.1 Introduction

This chapter describes the Taverna workflow workbench (Oinn *et al.*, 2006), developed under the myGrid project[1] and discusses its role in building workflows that are used in bioinformatics.[2] analyses. Taverna is an application that allows a bioinformatician to describe the flow of data between a series of services. These are, in effect, the materials and methods for a bioinformatics *in silico* experiment or analysis. In order to achieve this goal, Taverna has had to

[1] http://www.mygrid.org.uk

[2] The use of mathematical and computational methods to store, manage and analyse biological data to answer biologists' questions. Such a definition would include functional genomics and the other 'omics, as well as systems biology.

Data Mining Techniques in Grid Computing Environments Edited by Werner Dubitzky

navigate through the difficult terrain of the *bioinformatics landscape* – the nature and character of bioinformatics tools and data. Taverna workflows have changed the nature of how bioinformatics analyses are performed and has moved such analyses to a more industrial footing. In this new scenario, ideas of how the bioinformatics landscape is traversed have changed markedly.

Bioinformatics can be seen as a data mining activity. Taking a broad definition of data mining as either 'the analysis of data to establish patterns and relationships within those data' (Frawley, Piatetsky-Shapiro and Matheus, 1992) or 'the science of extracting useful information from large data sets or databases' (Hand, Mannila and Smyth, 2001), then it is easy to see that this is the objective of much of bioinformatics. Such patterns and relationships can be viewed as hypotheses to be tested in the laboratory. Data-driven research, and hypothesis generation, is a direction in which bioinformatics has been moving (Kell and Oliver, 2004), and there is now a need to make such analyses as rigorous, robust and easy to set up as possible.

Bioinformaticians perform analyses where data are gathered from many resources according to some criteria and these data are co-ordinated to establish patterns. For example, a bioinformatician will explore microarray results to find patterns of differentially modulated gene expressions, and co-ordinate this differential modulation with pathways that are connected with some observed characteristic or phenotype. The outcome of such an analysis might be a pattern of relationships between a gene, the perturbing of some pathway and the consequent effect on a phenotype.

Any attempt in bioinformatics to perform such analyses reveals a plethora of issues (Oinn *et al.*, 2006; Stevens *et al.*, 2004). The nature of the bioinformatics data landscape with its heterogeneities (Karp, 1995; Davidson, Overton and Buneman, 1995), both semantic and syntactic, together with the distributed nature of these resources, make such data mining activities problematic. As well as data resources, the bioinformatician is required to use a wide range of tools to further process these data in order to establish the relationships and patterns that are the goals of the analyses.

This landscape, which is described in Sections 9.2 and 9.3, presents a set of issues that bioinformaticians are required to overcome in order to mine their data. Traditionally, bioinformaticians have addressed these problems through large-scale, tedious and error-prone cutting and pasting data between webpages, or through the development of bespoke scripts. Whilst the latter achieve the analytical goal, such scripts tend to be inflexible non-trivial to write and afford the developer and user no easy means of knowing the protocol being used. The re-use and sharing of such scripts is often limited, further increasing the need for individual researchers to write their own scripts, and therefore re-invent the wheel.

The nature of bioinformatics data is too vast, complex and heterogeneous to hope to use traditional data mining techniques efficiently and effectively. Rather, a different approach is needed. The myGrid project has developed a workflow environment named Taverna (Oinn *et al.*, 2006) that addresses the issues of the bioinformatics landscape.

To illustrate the data mining capabilities of Taverna, a case study of a bioinformatics analysis performed with Taverna workflows is presented. In this study, Taverna workflows were used to investigate the tolerance to African tryoanosomiasis infection in the mouse model organism. This case study not only illustrates the power of Taverna workflows in data mining, but also shows how the understanding of how best to use Taverna workflows to traverse the bioinformatics landscape has developed. This has been a move from using workflow to simply gather data efficiently to workflows that gather data efficiently, but also manage it effectively. This case study demonstrates how workflow can be used to overcome a variety of, as yet unaddressed, biases present in many bioinformatics analyses, particularly in microarray and quantitative trait locus studies.

9.2 The bioinformatics data landscape

The number of publicly available bioinformatics data sources is increasing significantly each year. For example, in 2006, the Molecular Biology Database Collection increased by 30 per cent since the previous year and 351 per cent since the first compiled list in 1999 (Galperin, 2006). Bioinformatics data are also becoming more complex. Twenty years ago these data were primarily simple nucleic acid and protein sequences. They are now highly inter-related networks of data that, along with their experimental conditions, provide many axes along which these data can be organized. As well as increasing rapidly in number and complexity, each publisher acts in a highly autonomous manner, designing differing, yet often overlapping, views of the same data. In addition, these data and the views over these data are highly volatile. The same data objects are published by multiple data providers or replicated locally by different organizations (Galperin, 2006). New data entries are submitted on a daily basis, with updated databases being released monthly or even daily (Benson *et al.*, 2005). Also, massive heterogeneity resides in these data sources.

- *The same data object may be given a different identity by each data provider.* For example, the protein 'dual specificity DE phosphatase Cdc25C' is published by different data providers and identified as 'aaa35666' in GenBank (Benson *et al.*, 2005) and as 'p30307' in UniProt (Apweiler *et al.*, 2004).
- *Each data provider publishes its data resources in its own representation.* The *representation format* specifies the entities contained in a data object and their relationship. For example, a sequence record could contain a sequence itself, as well as some metadata about the record (such as its accession number, identifier etc.) or knowledge about the sequence (its species origin or related features etc.). The representations of bioinformatics data are highly heterogeneous, both semantically and syntactically (Karp, 1995; Davidson, Overton and Buneman, 1995). For a DNA sequence record, there are at least 20 different flat file formats for representing such a record (Lord *et al.*, 2005). Most bioinformatics data sources are accessible as files (such as the sequence records) and their internal data structure remain implicit.

Such data resources are published by autonomous groups that make their own decisions on representation. There is no perceived value in common identity or representation.

These heterogeneities are the consequence of the autonomy of data providers. It is these heterogeneous aspects that make it hard to integrate data sources from different publishers as they contain overlapping contents and are not uniquely identified. The volatility forces regular repetition of data analysis to gather updated data products. The heterogeneity diverts scientists into adjusting to the different identities and representation, rather than analysing the data. Much of a bioinformatician's craft is in dealing with the heterogeneities of data.

9.3 The bioinformatics experiment landscape

Bioinformatics experiments are typically dataflow pipelines. Bioinformaticians need to take the data from one resource and deliver them to another resource to retrieve related data, or to be processed for deriving new data. Bioinformaticians need to cope with not only the heterogeneous, autonomous and volatile data sources described in the previous section, but also the heterogeneous, autonomous and volatile tools and applications (Stein, 2002) that work upon these data. The heterogeneity, autonomy and volatility of the tool landscape has the following consequences.

- *No explicit type system.* The data landscape described above means that there is no trans-domain type system used by bioinformatics tools.
- *Heterogeneous interfaces.* Interfaces ranging through program-centred interfaces, database queries, Web-based interfaces including webpages, etc.
- *Heterogeneous cardinalities.* Despite the heterogeneous data representation format, tool providers expose their data at different cardinalities. Some either accept or return a collection instead of a single item (Lord *et al.*, 2005).
- *Volatility.* Similar to the data sources, bioinformatics applications are frequently updated. For example, a BLAST service changes every few months (Altschul *et al.*, 1990). This update is either due to the change of the algorithm used by the tool or the databases queried by the application. The reason for the update, however, often remains implicit and undocumented. Also, in bioinformatics, where many tools and applications are not professionally supported, service failure is a daily issue for the scientists (Oinn *et al.*, 2006).

The autonomy and heterogeneity of the data sources and user interfaces require many *house-keeping* steps during an experiment.

- Many of the experiment data products are transformed and archived in a local customized database. A local identity may be allocated for the data object to replace its identity, published by its original data provider. The metadata associated with the data object may be scrapped during this process (Stein, 2002). The information about where the locally archived data originated is lost.
- Due to the heterogeneous data identities, representation format and domain tools, the scientists have to re-format the data in order to process them further in another interface, map the identities from one source to another, convert between a collection data product and a single item etc.

The volatility of bioinformatics tools and data sources causes repetitive experimental effort. As previously discussed, scientists need to frequently repeat their workflows in case an updated experiment data product could be gathered or computed. Bioinformaticians generally have strong opinions about the particular tool or application that they wish to use (Oinn *et al.*, 2006). The fragility, however, forces them to look for tools or applications with alternative functionalities, which may not always be available (Huang *et al.*, 2006) and often drive bioinformaticians into coping with these heterogeneities.

As data are passed between resources, there is often amplification of such data. For example, one query to a database leads to a collection of candidate similarity entries; each DNA sequence can contain a collection of possible genes; a projection over each sequence record will transform one record into a collection comprising many parts of that record (species, sequence features, etc.). The amplified intermediate data products and their relationships are not always completely captured in an analysis, potentially leading to problems of scientific record (Zhao *et al.*, 2004).

In a 'human-driven' approach, bioinformaticians perform the data pipeline by manually inputting data between different sources. They implement the housekeeping tasks as a set of 'purpose-built scripts' (Stein, 2002). These purpose-built scripts are brittle as the data publishers often update their data representation format. The functionality of these scripts is not always documented, and the tasks performed by the scripts are obscure. Finally, there is massive duplication of effort across the community in order to cope with the volatility (Stein, 2002).

Ideally, the materials and methods used in an experiment are recorded in a scientist's laboratory-book. However, some implicit steps, such as the housekeeping scripts, may be lost in the documentation. Also, as the repetitive nature of experiments and the amplified data results, manually collecting these provenance information is time consuming, error prone and not scalable.

In a 'workflow' approach, the applications from different providers are deployed as (Web) services in order to achieve a unified interface. The workflows make explicit the data passed between the experiment steps (i.e. the services) including any hidden steps, such as the housekeeping steps (Stevens *et al.*, 2004). The experiment protocol documented in a workflow can be reused and repeatedly executed without duplicate efforts (Stevens *et al.*, 2004).

9.4 Taverna for bioinformatics experiments

Taverna (Oinn *et al.*, 2006) is a workflow workbench that is targeted at bioinformaticians who need to deal with the data sources and services described above, including the issues associated with the bioinformatics landscape and the repetitive nature of bioinformatics experiments. To realize these goals, it provides a multi-tiered open architecture and an open-typing and data-centric workflow language, in addition to the following technologies.

- *Shim services.* Implement the housekeeping scripts that reconcile the type mismatches between the heterogeneous bioinformatics services (Hull *et al.*, 2006).
- *Iterations.* Handle the collection data products passed between services.
- *Nested workflows.* Enable modularized workflows. A nested workflow is a workflow that is treated as an experiment step and contained within another higher-level workflow.
- *Fault tolerance.* To help bioinformaticians live with the fragile domain services with dynamic service substitution and re-try (Oinn *et al.*, 2006).
- *Multiple service styles.* Taverna can accommodate a variety of services styles. For example, as well as WSDL services, Taverna can use local Java services, beanshell scripts or R processes. Taverna also provides mechanisms for users to develop their own services. The SoapLab software package allows command-line programmes to be wrapped as Web services, and Taverna is also able to collect and use (scavenge) Java API from a local computer. The extensible plugin framework of Taverna also enables new *processor* types to be developed, bringing an even wider range of services to bear.
- *Provenance and data services.* To collect both final and intermediate experiment data products and their provenance information (Stevens *et al.*, 2004; Stevens, Zhao and Goble, 2007).

Table 9.1 How Taverna addresses issues of the bioinformatics landscape

Issues	Taverna
Heterogeneous data	Use of open typing system
Heterogeneous interface for data access	Use of open typing system
Collection data products	Iterations
Housekeeping scripts	Shim services
Volatile resources and repetitive experiments	Explicit and repeatable workflows; Provenance and data services
Volatile and fragile services	Fault tolerance
Amplified data products and their relationship	Provenance and data services

Table 9.1 summarizes how the issues caused by the bioinformatics data sources and experiments are addressed by Taverna's design considerations and technologies. The autonomy of data sources and services is not managed in Taverna in order to leave bioinformaticians with open choices. These designs and technologies are detailed in Table 9.1.

9.4.1 Three-tiered enactment in Taverna

Taverna provides a three-tiered abstraction architecture, shown in Figure 9.1, in order to separate user concerns from operational concerns:

- The data-centric workflow language, *Simplified Conceptual Unified Workflow Language (SCUFL)* (Oinn *et al.*, 2006), is at the abstraction layer that enables the users to build a workflow without writing a large, bespoke application.
- The run of a workflow is managed at the execution layer by the Freefluo engine following the execution semantics (Turi, 2006), and the experiment data products are also handled at this layer by the Baclava data model. The Freefluo enactor dispatches executions details to

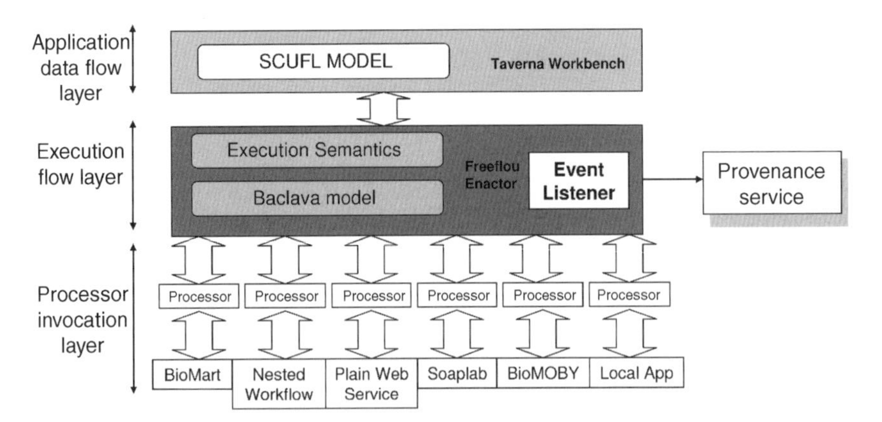

Figure 9.1 The three-tiered abstraction architecture of Taverna. The processor invocation layer provides access to a variety of third-party services. the middle execution flow layer co-ordinates use of these services. The application data-flow layer (top layer) allows the user to describe how these services are to be used

the *Event Listener* plug-in during critical state changes, such as WorkflowRunCompletion, ProcessRunCompletion etc. This enables provenance to be automatically collected on behalf of users by an enactor-centric provenance collection architecture.

- The invocation of different families of services is managed by the Freefluo workflow enactor at the invocation layer. A set of Processor plug-ins is defined in this layer in order to invoke the domain services of heterogeneous styles (Oinn *et al.*, 2006).

This abstraction architecture enables bioinformaticians to design workflows without worrying about the underlying service executions. Taverna's openness to heterogeneous domain data and services is underpinned by both the flexible Processor plug-ins and the open-typing Taverna data models.

9.4.2 The open-typing data models

Neither the SCUFL model for defining a Taverna workflow nor the Baclava model for managing the data products constrains the types of datum passed between services or their invocations.

The SCUFL model The SCUFL model contains the following components.

- A set of input ports (source ports) and output ports (sink ports) for a workflow and for a processor. Each port is associated with free text descriptions and its MIME type. The MIME types describe the data format for transporting between services, such as `text/plain`, and they say little about the type of the data. This is very much a reflection of the bioinformatics data landscape described earlier.
- A set of processors, each representing an individual step within a workflow. The processor specification that specifies the 'style' of the processor, such as a SoapLab service or as a NestedWorkflow.
- A set of data links that link a source port with a sink port. This reflects SCUFL as a data-flow-centric workflow language.
- A set of co-ordination links that link two processors together and enable additional constraints on the behaviour of these linked processors. For instance, a processor B will not process its data until processor A completes.
- A set of optional fault tolerance policies, which either provide alternative services to replace a failed service or specify the number of times a failed service invocation should be re-tried before finally reporting failure.
- A set of optional iteration configurations that specify the behaviour of a processor when it receives an actual data product that mismatches the data cardinality of its port. For example, if a port expects the input as a 'string' but actually receives a 'list', the implicit iteration in Taverna handles this mismatch (Oinn *et al.*, 2006).

Data typing constraints are not associated with a port in the SCUFL model or not in the Baclava data model.

The Baclava data model The Baclava model manages the data products, or data entity, generated during workflow executions. This model describes a data entity's data structure and its

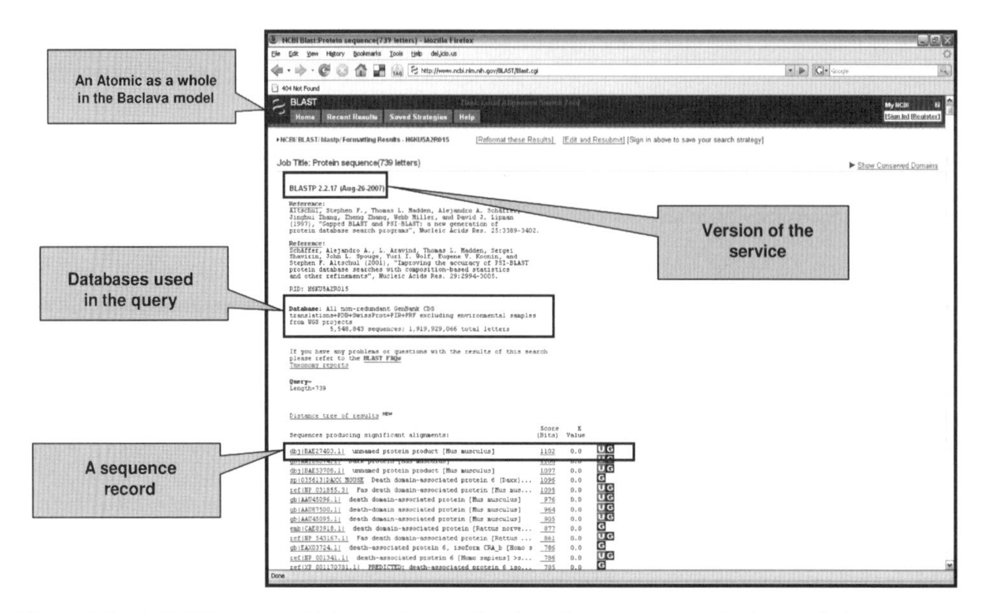

Figure 9.2 A BLAST report, which contains a collection of sequence records, is regarded as an atomic data product in the Baclava model. However, it is a collection at the domain level

MIME type. This information is from a data entity's corresponding port defined in the SCUFL model, e.g. the input of a particular processor.

A data entity can be either atomic or a collection. The Baclava describes a data entity at the *transportation* level to help the enactor to handle the implicit iterations, rather than at the *domain* level. This means the internal structure of a DataThing, either atomic or collection, is not exposed in the Baclava model, in order to adopt any services presented by the bioinformatics community (Oinn *et al.*, 2006).

A BLAST report is a typical example to illustrate this domain-independent Baclava model. A BLAST report, as shown in Figure 9.2, is produced by the BLAST service (Altschul *et al.*, 1990), which searches for similar sequences in a sequence database. At the domain level, this BLAST report is a 'collection' as it contains a collection of pairwise sequence alignments and associated links from the target sequence database. These pairwise sequence alignment records are, however, not exposed in the Baclava model. Rather, such a BLAST report is treated as an 'atomic' item by Baclava, when it is transferred between services during workflow execution.

These open-typing data models enable Taverna to be open to the heterogeneous services and data of the bioinformatics landscape. This, however, brings difficulty for bioinformaticians to interpret experiment data products at the domain level. The domain-independent models also impact on the kinds of identity that Taverna can allocate for a DataThing.

9.5 Building workflows in Taverna

This section describes the steps taken by bioinformaticians to describe and run workflows in the Taverna workbench.

9.5.1 Designing a SCUFL workflow

Bioinformaticians design a workflow by interacting with Taverna's user interface, see Figure 9.3. In the *advanced model explorer* (*AME*) window, they can either load an existing workflow from a local or remote workflow repository, or design a new workflow.

(a) They can search for services available by default in Taverna through the service panel or Feta. Feta is a semantic service discovery component, which describes each service held in Feta using ontological concepts. These descriptions include the tasks performed by a service, the types of input it takes, the types of output it produces etc. (Lord *et al.*, 2005).

(b) Researchers using the workbench may also use services that are not listed in the service panel or Feta. To do this they can add new services gathered anywhere on the Web to the service panel using the corresponding service import mechanisms, e.g. using the 'add new wsdl scavanger' option.

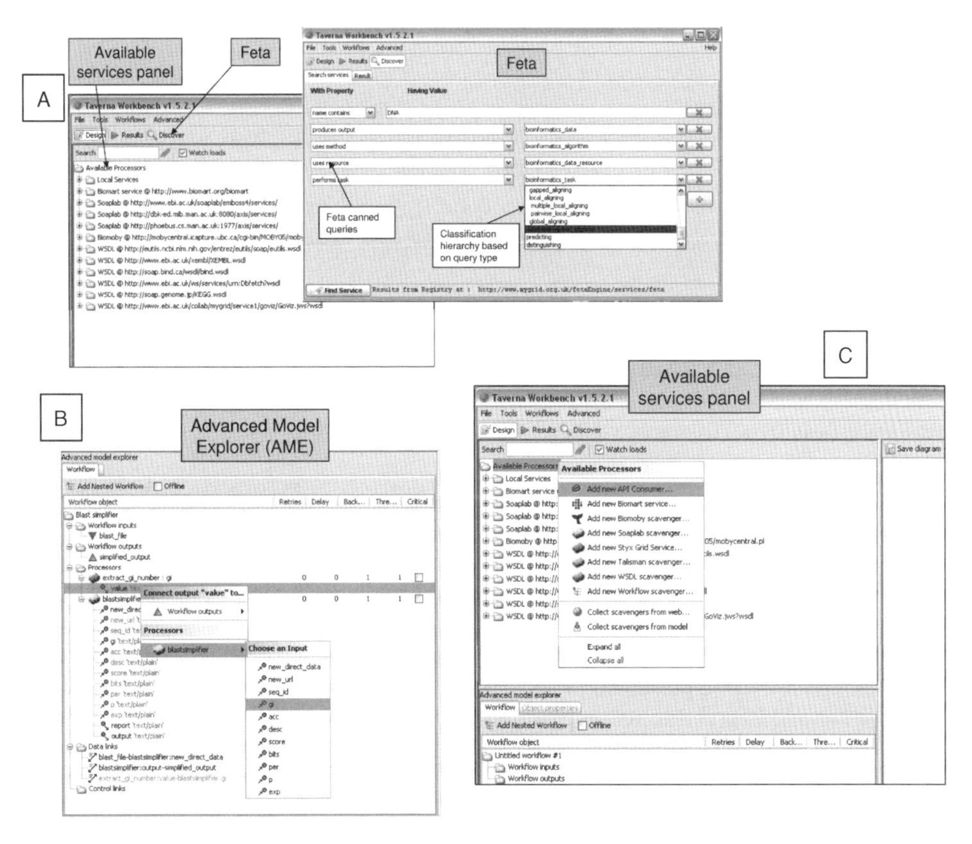

Figure 9.3 Taverna's user interface for supporting workflow design. Shown in this figure are three aspects of the interface: Feta (a), available services panel (b), and Advanced Model Explorer (c). Feta provides a means of search functionality for services. The available services panel allows users to import new services into the workbench. The AME provides the means of connecting services together to construct the desired workflow

(c) Once the appropriate services are found, they are imported into the AME window and any two services are connected together by the data passed between them. This corresponds to SCUFL's nature as a data-oriented workflow language.

(d) Due to the open-typing approach taken in Taverna, there is no verification that the right types of datum are passed between services during either workflow design or workflow execution. Workflow developers can import shim services into their workflows in order to cope with mismatches between the types of datum passed between two services. For example, the 'simplifier service' is a shim service that converts a BLAST report into a list of sequences.

(e) Workflow inputs and outputs need to be defined in the AME window and connected with the services. A diagrammatic representation of a workflow gives visual guidance for the workflow design.

(f) Once the workflow has been constructed, researchers may save the workflow using 'save' button, which subsequently prompts the user for a location in which to store the workflow. This results in a SCUFL workflow file, which describes the endpoints of the services (processors) used in the workflow and how they are connected with each other.

A workflow can be executed by selecting to 'run workflow' from the menu system. Users can then supply the workflow inputs in order to invoke the execution. This will start Taverna's workflow enactment engine (Oinn *et al.*, 2006), which manages the invocations of the domain services *for the bioinformaticians* and saves them from coping with the different styles of services, such as a local Java API, a SoapLab service or a WSDL Web service (Oinn *et al.*, 2006).

9.6 Workflow case study

This section describes a use case which is representative of the way Taverna is used to analyse biological data.

African trypanosomiasis, also known as sleeping sickness, is caused by the Trypanosoma genus of parasites. Infection in cattle is caused by *T. congolense* and *T. vivax*, and is a major restriction on cattle production in sub-Saharan Africa (Hill *et al.*, 2005). With no vaccine available, and with heavy expenditure on trypanocidal and vector control, trypanosomiasis is estimated to cost over four billion US dollars each year in direct costs and lost production (Hanotte and Ronin, 2003). Infection from these species of parasite often induces chronic disease (though it may be presented in an acute form), causing severe anaemia, weight loss, foetal abortion, cachexia and associated intermittent fever, oedema and general loss of condition (Hanotte and Ronin, 2003; Hill *et al.*, 2005). The effects of this disease commonly lead to the death of the animal.

Some breeds of African cattle such as N'Dama, however, are able to tolerate infections and remain productive (Hanotte and Ronin, 2003; Hill *et al.*, 2005). This trait is presumed to have arisen through natural selection acting on cattle exposed to trypanosome infection. Certain strains of mice have also been noted to exhibit similar traits in resistance when under infection. These mice strains differ in their resistance to *T. congolense* infection: C57BL/6(B6) mice survive significantly longer than AJ(A) or Balb/c(C) (Iraqi *et al.*, 2000; Koudandé, van

Arendonk and Koud, 2005). These mice act as a model for the modes of resistance in cattle, and therefore can assist researchers in identifying the candidate genes that may be responsible for controlling resistance in both the mice and cattle.

In order to understand the mechanisms that govern trypanosomiasis resistance in both these animals, we must be able to successfully bridge the gap between genotype and phenotype. The process of linking genotype and phenotype plays a crucial role in understanding the biological processes that contribute to an organism's response when under a disease state (Mitchell and McCray, 2003; Hedeler *et al.*, 2006).

Researchers have discovered single-gene lesions for a large number of simple Mendelian traits. It has, however, proved much more difficult to discover genes underlying genetically complex traits. The mechanisms controlling complex phenotype's can be extremely varied and complex. With a single phenotype, multiple biological processes and pathways may be involved, each of which may contain numerous genes. It is therefore essential to identify the true candidate genes in these pathways that contribute to the phenotype's expression. In order to do this researchers can use classical genetic mapping techniques, from the wet lab, to identify chromosomal regions that contain genes involved in the phenotypes expression. These regions, known as *quantitative trait loci* (*QTLs*), are now being used to aid in the discovery of candidate genes involved in both simple and complex phenotypic variation. Tens to hundreds of genes, however, may lie within even well defined QTLs. It is therefore vital that the identification, prioritization and functional testing of *candidate quantitative trait genes* (*QTGs*) is carried out systematically without bias introduced from prior assumptions about candidate genes or the biological processes that may be influencing the observed phenotype (de Buhr *et al.*, 2006).

To aid in this functional testing the analysis of the genes' expression may be studied using microarray technology. With the advent of microarrays, researchers are able to directly examine the expression of all genes from an organism, including those within QTL regions. Each microarray chip is capable of capturing gene expression levels on a genome-wide scale. This allows researchers to take a snapshot of the expression of all the genes from strategic points within an experiment. Each microarray investigation can yield thousands of genes that are shown to be differentially expressed in the data. This makes the selection of candidate genes extremely difficult, partly due to the large numbers involved, but the domain knowledge of the investigator can also prejudice selection of genes to be studied (a kinase expert will be biased towards studying kinases). By combining both QTL and microarray data, however, these two methods of candidate gene identification can together make it possible to identify a limited number of strong candidate genes that are both differentially expressed within a gene expression study and are also located within the QTL region. This allows researchers to identify true candidate genes that may be involved in the expression of the phenotype (de Buhr *et al.*, 2006).

The correlation of genotype and phenotype can be further enhanced by investigating directly at the level of biological pathways. The expression of genes within their biological pathways contributes to the expression of an observed phenotype. By investigating links between genotype and phenotype at the level of biological pathways, it is possible to obtain a global view of the processes that may contribute to the expression of the phenotype (Schadt, 2006). Additionally, the explicit identification of responding pathways naturally leads to experimental verification in the laboratory. We therefore opted to analyse QTL and gene expression data not directly at the level of genes but at the level of pathways (see Figure 9.4). Using this pathway-driven approach provides a driving force for functional discovery rather than gene discovery.

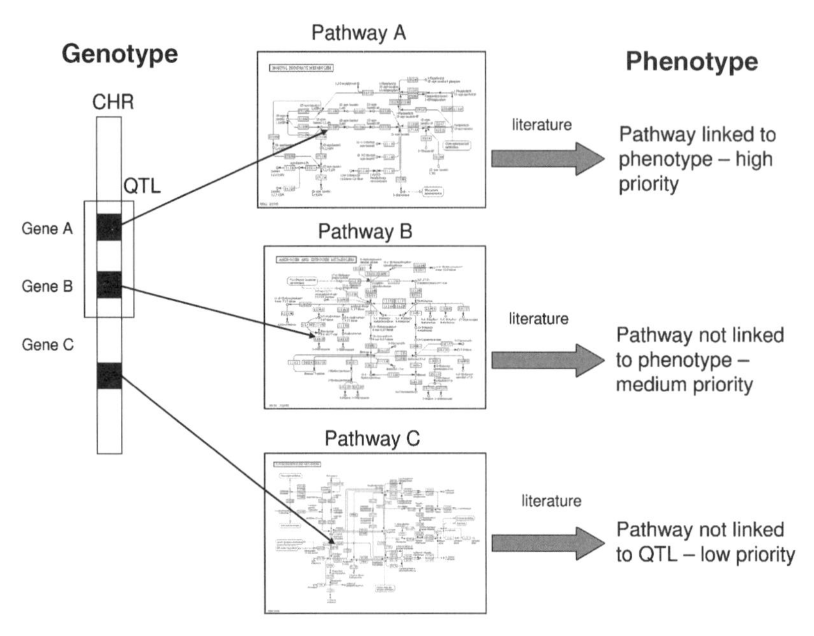

Figure 9.4 An illustration showing the prioritization of phenotype candidates, from the pathway-driven approach. (Abbreviations in the figure: CHR: chromosome, QTL: quantitative trait locus)

9.6.1 The bioinformatics task

For the process of identifying pathways and therefore candidate genes involved in trypanosomiasis resistance, we can begin by identifying all genes which reside in the chosen QTL regions. In order to determine that genes reside in the QTL region of choice, the physical boundaries of the QTL need to be determined. This can be conducted in a number of ways; however, we chose to map the genetic markers from the wet lab to the current release of the Ensembl Mouse Database. Those markers that cannot be mapped in Ensembl are given physical boundary values based on evidence in third-party databases, e.g. the MGI database or the NCBI gene database.

Each gene identified within the QTL region is annotated with its associated biological pathways, obtained from the KEGG pathway database (Kanehisa and Goto, 2000). This requires the researcher to send queries through the BioMart Web interface, using a chromosomal region as the chosen QTL region. For each QTL region, a separate query must be executed. Once all the genes are identified in the QTL, each gene is annotated with Entrez and UniProt identifiers. Currently KEGG has no means of linking directly to Ensembl gene identifiers. The use of Entrez and UniProt identifiers allows for cross-referencing to KEGG gene identifiers. These KEGG gene identifiers are then used to find all the biological pathways held in the KEGG database for each of the genes. The KEGG pathway identifiers can then be used to obtain textual descriptions of each of the pathways identified. These provide information that allows the researcher to search the current literature for any previously reported involvement of the pathway in the expression of the observed phenotype.

Pathway annotation is also carried out for the genes that are found to be differentially expressed in the microarray study of choice. These two sets of pathway data enable researchers to obtain a subset of common pathways that contain genes within the QTL region and genes that

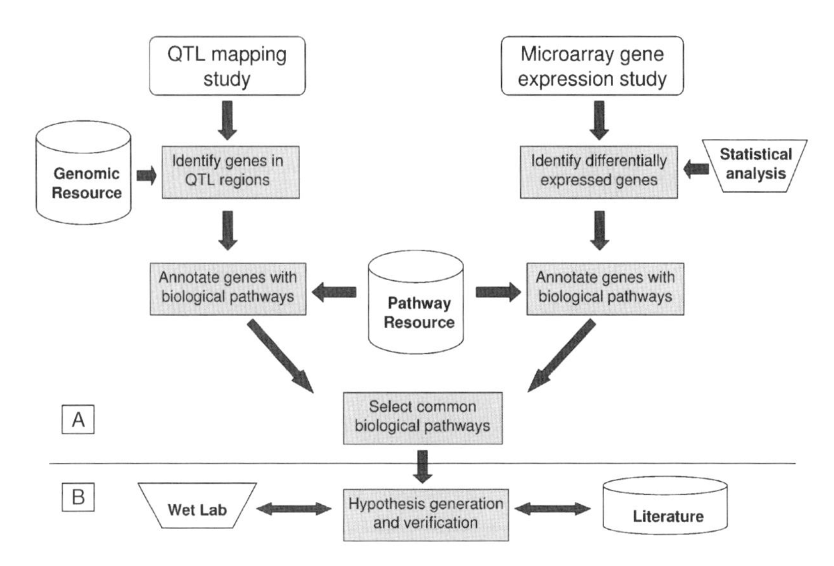

Figure 9.5 An illustration for the pathway-driven approach to genotype-phenotype correlations

are differentially expressed in the microarray study. By identifying those pathways common to both microarray and QTL data, we are able to obtain a much richer model of the processes that may be influencing the expression of the phenotype. This process is summarized in Figure 9.5.

In Figure 9.4 all pathways are differentially expressed in the microarray data. Those pathways that contain genes from the QTL region are assigned a higher priority (pathways A and B) than those with no link to the QTL region (pathway C). Higher-priority pathways are then ranked according to their involvement in the phenotype expression, based on literature evidence.

In Figure 9.5 the process of annotating candidate genes from microarray and QTL investigations with their biological pathways is shown. The pathways gathered from the two studies are compared and those common to both are extracted as the candidate pathways for further analysis. These pathways represent a set of hypotheses, in that the candidates are the hypothetical processes that may contribute to the expression of the observed phenotype. Subsequent verification is required for each pathway by wet lab investigation and literature searches. This approach is separated into two sections, distinguished by the dividing line between the selection of common pathways and the generation of hypotheses. The section labelled A represents the workflow side of the investigation; whilst the section labelled B represents verification of the hypotheses through wet lab experimentation and literature mining.

9.6.2 Current approaches and issues

There are numerous issues relating to the manual analysis of microarray and QTL data for correlating genotype and phenotype, including those previously identified by Stevens *et al.* (2004). These include difficulties of scale, user error during repetitive and tedious tasks and lack of recording of the analytic process. The scale of data generated by such high-throughput experiments has led some investigators to follow a hypothesis-driven approach (Kell, 2002). This occurs where the selection of candidate genes is based on some prior knowledge or assumption, and is often found in cases where a significant number of candidate genes have been shortlisted for further analysis. This process acts to reduce the number of genes that requires investigation. Although these techniques for candidate gene identification can detect true QTG, they run the risk of overlooking genes that have less obvious associations with the phenotype (Kell and Oliver, 2004). This problem is compounded by the complexity of multigenic traits, which can also lead to problems when attempting to identify the varied processes involved in the phenotype. For example, numerous processes can be involved in the control of parasitic infection, including the ability of the host to kill the parasite, mounting an appropriate immune response or control of host or parasite-induced damage. By making selections based on prior assumptions of what processes may be involved, the pathways, and therefore genes, that may actually be involved in the phenotype can either be overlooked or missed entirely.

A further complication is that the use of *ad hoc* methods for candidate gene identification are inherently difficult to replicate. These are compounded by poor documentation of the methods used to generate and capture the data from such investigations in published literature (Nature Editorial, 2006). An example is the widespread use of 'link integration' (Stein, 2003) in bioinformatics. This process of hyperlinking through any number of data resources exacerbates the problem of capturing the methods used for obtaining *in silico* results, since it is often difficult to identify the essential data in the chain of hyperlinked resources. The use of numerous software resources and bespoke scripts for carrying out large-scale analyses contributes to this problem of implicit methodologies.

From conducting a review of the current literature surrounding the analysis of microarray and QTL data (Fisher *et al.*, 2007), we were able to identify the specific issues facing the manual analysis of data, including the selection of candidate genes and pathways. These are listed below.

(a) Premature filtering of data sets to reduce the amount of information requiring investigation.

(b) Predominantly hypothesis-driven investigations instead of a complementary hypothesis and data-driven analysis.

(c) User bias introduced on data sets and results, leading to single-sided, expert-driven hypotheses.

(d) Implicit data analysis methodologies.

(e) Constant flux of data resources and software interfaces hindering the reproduction of searches and re-analysis of data.

(f) Error proliferation from any of the listed issues.

These issues relate directly to the nature of the bioinformatics landscape – with its heterogeneities in the data itself, analysis method, and software resources (including webpages and databases).

9.6.3 Constructing workflows

By replicating the original investigation methods in the form of workflows, we are able to pass data directly from one service to the next without the need for any interaction from researchers. This enables us to analyse the data in a much more efficient, reliable, un-biased, and explicit manner. The methodology provided through the workflows gathers the known pathways that

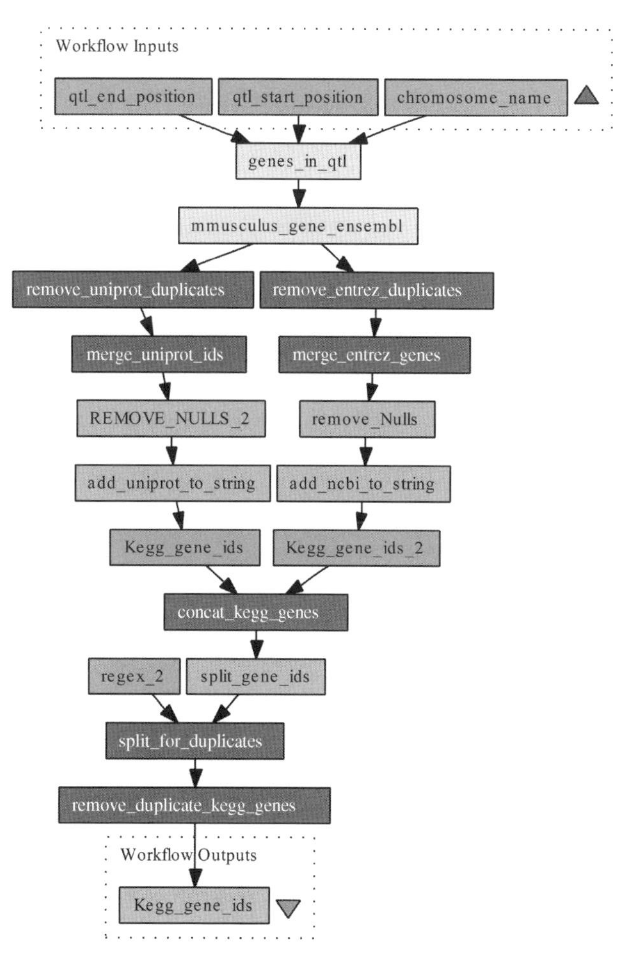

Figure 9.6 Annotation workflow to gather genes in a QTL region, and provide information on the pathways involved with a phenotype

intersect a QTL and those derived from a set of differentially expressed genes, returning them to the user.

The implementation of the pathway-driven approach consisted of three Taverna workflows. The first workflow constructed, `qtl_pathway`, was implemented to identify genes within a QTL region, and subsequently map them to pathways held in the KEGG pathway database. Lists of genes within a QTL are obtained from Ensembl, together with UniProt (Bairoch *et al.*, 2005) and Entrez gene (Maglott *et al.*, 2007) identifiers, enabling them to be cross-referenced to KEGG gene and pathway identifiers. A fragment of this workflow can be seen in Figure 9.6, which shows the mapping from the QTL region (label A) to KEGG gene identifiers (label C). This workflow represents an automated version of the manual methods required to perform such a task. This replication of the manual method is supplemented with an automation of the process of collating all information into single output files.

The process of collating the results into single output files is of importance for analysing results returned from the workflows. The format required by a number of bioinformatics services is in the format of an array, or list, of input values. The resulting output may be a list of outputs, with each output containing a list of gene or pathway identifiers. This mass of inter-connected data makes it difficult to interpret the results gathered from the workflows, where results can be amplified into the hundreds from just a few input values. This problem means that we must progress from the task of just data gathering to data gathering and management.

In order to resolve this, we applied a means of merging these multiple output files into single output files, through the use of shim services (Hull *et al.*, 2006). To assist the bioinformatician in the analysis of these workflow outputs, we constructed a simple relational flat-file system, where cross-references to identifiers held in additional files were stored with identifiers from the service that had been invoked. An example of this is the storage of KEGG genes and their KEGG pathway identifiers. The gene identifiers and pathway identifiers were obtained from separate services. To correlate these identifiers, however, the researcher would require a means of knowing which pathway identifiers were the product of a search with a gene identifier. We resolved this issue with a number of shim services that stored the KEGG gene id with all of its associated KEGG pathway identifiers in a tab-delimited format. This enabled the bioinformaticians to query via a gene or pathway identifier and obtain its corresponding gene or pathway identifier, e.g. a query for the Daxx gene, mmu:13163, would return 'mmu:13163 path:mmu04010'.

The workflow depicted in Figure 9.6 (as a subset of the complete workflow) requires a chromosome, and QTL start and stop positions in base pairs. The genes in this QTL region are then returned from Ensembl via a BioMart plug-in. These genes are subsequently annotated with UniProt and Entrez identifiers, start and end positions, Ensembl Transcript IDs, and Affymetrix probeset identifiers for the chips `Mouse430_2` and `Mouse430a_2`. The UniProt and Entrez IDs are submitted to the KEGG gene database, retrieving a list of KEGG gene IDs.

Genes that show differential expression in the microarray study are chosen for further analysis. This is done independently of the QTL in order to avoid biases. Permissive criteria should be used for selecting responding genes at this stage of data analysis. This is carried out in order to reduce the incidence of false negative results (which could result in missing one of the true QTGs). Any true negative results would later be discarded on correlation of the pathways with the observed phenotype.

The second workflow, `Probeset_to_pathway`, provided annotation of microarray probe set identifiers. Ensembl gene identifiers associated with Affymetrix probe sets were obtained from Ensembl. These genes are entered into a similar annotation workflow as that used for `qtl_pathway`. The Entrez and UniProt database identifiers were used to map the Affymetrix probe set identifiers to KEGG pathways.

To obtain the pathways that both intersect the QTL region and are present in the gene expression data, we can use a third workflow, named `common_pathways`, to obtain a list of KEGG pathway descriptions. Each of the pathways returned from the `common_pathways` workflow were investigated in turn.

9.6.4 Candidate genes involved in trypanosomiasis resistance

The systematic application of the workflows to the lists of genes identified in the chosen QTL regions identified a number of biological pathways that contained genes in both the QTL and the set of differentially regulated genes. The complete list of 344 genes, initially identified from the chosen Tir1 QTL region, was narrowed down significantly to just 32 candidate QT genes. Those pathways in which a high proportion of component genes showed differential expression following trypanosome challenge were prioritized for further analysis. One such pathway identified was the MapK signalling pathway. There are four genes from the MapK pathway within the Tir1 QTL: Daxx, TNF, Mapk13 and Map14. Of these, Daxx showed the strongest signal of differential expression at early time points and TNF has already been shown to be a poor candidate QTG (Naessens, 2006). Daxx was therefore chosen as the primary candidate QTG for further investigation.

Daxx is widely reported to be an enhancer for apoptosis (Yang and Khosravi-Far, 1997; Zuñiga and Motran, 2000). It is also reported that susceptible mice infected with trypanosomiasis show an increase in apoptosis (cell death) (Shi *et al.*, 2005). During the acute stage of trypanosome infection, a large number of leucocytes undergo apoptosis, as the immune response is modified to control the infection (Yan *et al.*, 2004). This pathway provides an example of the pathway labelled A in Figure 9.4, with the candidate gene being directly related to the QTL region and known, through literature, to be involved in the phenotype. The remaining candidate genes identified from these workflows are currently undergoing further investigation to establish their role in the trypanosomiasis resistance phenotype.

9.6.5 Workflows and the systematic approach

Since we have chosen to automatically pass data from one service into the next using workflows, we are now able to process a far greater volume of information than can be achieved using manual analysis. This in turn provides the opportunity to systematically analyse any results we obtain without the need to prematurely filter the data (for human convenience). An example of this triaging process was found where studies carried out by researchers on the Wellcome Trust Pathogen–Host Project (see Acknowledgments) had failed to identify Daxx as a candidate gene for trypanosomiasis resistance. This occurred when manually analysing the microarray and QTL data; researchers hypothesized that the mouse–cow syntenous QTL region may contain the same QT genes. It was later found through a systematic analysis that Daxx lay outside this region, and so the mouse QTL data were prematurely filtered based on researcher bias (although this does not preclude the discovery of other QT genes within this syntenous region). The use of a hypothesis-driven approach is essential for the construction of a scientifically sound investigation however, the use of a data-driven approach should also be considered. This would allow the experimental data to evolve in parallel to a given hypothesis and form its own hypotheses regardless of any previous assumptions (Kell and Oliver, 2004), as shown by this case.

Worthy of note is that the expression of genes and their subsequent pathways can be investigated with little to no prior knowledge, other than that of the selection of all candidate genes from the entire QTL region. This reduces the bias that may be encountered from traditional hypothesis/expert-driven approaches. By implementing the manually conducted experiments in the form of workflows, we have shown that an automated systematic approach reduces, if not eliminates, these biases whilst providing an explicit methodology for recording the processes involved. These explicit analysis pipelines increase the reproducibility of our methods and also provide a framework for which future experiments can adapt or update the underlying methods.

In using the Taverna workflow workbench we are able to state the services used and the parameters chosen at execution time. Specifying the processes in which these services interact with one another in the Scufl workflow language enables researchers to re-use and re-run experiments. An additional feature of the Taverna system is the capture of experimental provenance. The workflow parameters and individual results are captured in this execution log (Stevens, Zhao and Goble, 2007).

The generality of these workflows allows them to be re-used for the integration of mapping and microarray data in other cases than that of response to trypanosomiasis. Furthermore, the QTL gene annotation workflow may be utilized in projects that use the mouse model organism and do not have any gene expression data to back up findings. Likewise, the microarray gene annotation workflow may be used in studies with no quantitative evidence for the observed phenotype.

It should be noted that an unavoidable ascertainment bias is introduced into the methodology, in the form of Utilizing remote resources for candidate selection. The lack of pathway annotations limits the ability to narrow down the true candidate genes from the total genes identified in the QTL region, with the reliance on extant knowledge. With a rapid increase in the number of genes annotated with their pathways, however, the number of candidate QTG identified in subsequent analyses is sure to increase. The workflows described here provide the means to readily repeat the analysis.

The KEGG pathway database was chosen as the primary source of pathway information due to its being publicly available and containing a large set of biological pathway annotations. This results in a bias, relying on extant knowledge from a single data repository; however, this investigation was established as a proof of concept for the proposed methodology and, with further work, may be modified to query any number of pathway databases, provided they offer (Web) service functionality.

The workflows developed in this project are freely available for re-use – either by ourselves or others in future analyses. And so, by revisiting issues (a)–(f) outlined earlier in this section, we can show that by utilizing workflows within this investigation we have achieved the following.

(a) Successfully reduced the premature filtering of data sets, where we are now able to process all data systematically through the workflows.

(b) The systematic analysis of the gene expression and QTL data has supported a data-driven analysis approach.

(c) The use of a data-driven approach has enabled a number of novel hypotheses to be inferred from the workflow results, including the role of apoptosis and Daxx in trypanosomiasis resistance.

(d) The workflows have explicitly captured the data analysis methodologies used in this investigation.

(e) Capturing these data analysis methods enables the direct re-use of the workflows in subsequent investigations.

(f) The total number of errors within this investigation has been reduced as a whole from all of the issues addressed above.

9.7 Discussion

myGrid's Taverna, as well as workflow management systems in general, have supported a step change in how bioinformatics analyses are conducted. The industrialization of bioinformatics analysis has begun, just as the industrialization of much of biology's experimentation has already been accomplished. Industrial data production must be matched by industrial data analysis. In addition, as with any technological shift, how best to use technological innovations has to be understood. This industrialization is not simply mass production; it also changes the nature of the work; that is the methodology of data analysis.

Much bioinformatics is rooted in human-driven analysis through the manual transfer of data between services, delivered by webpages. In the days when single sequence analysis with one service was all that was necessary, this approach was adequate, but lacked support for easily attaining a level of rigour in recording what happened during an analysis. With today's increase in complexity – larger range of data types, larger quantities and more services needed in an analysis, such an approach does not scale. In these situations other consequences of human fallibility are also seen: biases in the pursuit of one data item in preference to another; premature triage to cope with too many data and others described in Section 9.6.

Taverna workflows can overcome many of these issues. Tedious repetition of data transfer can be automated. Large numbers of services can be coordinated. Workflows can be utterly systematic, thus avoiding issues of bias in selection of data items. There is no temptation to shed data to ease processing and thus risking false negative disposal. These consequences are typical of automation, be it workflow or bespoke scripts. Workflows, however, have other advantages. They are high level, allowing a bioinformatician to concentrate better on the bioinformatics and avoiding the 'plumbing'. They afford re-use much more easily than scripts, by allowing service substitution, extension of the workflow, fragmentation, data and parameter change etc. They are explicit about the nature of the analysis and, through common services such as provenance, afford better recording of an *in silico* experiment's events, data lineage and origins, etc.

Taverna also gives a means by which a bioinformatician can address the issues arising from the autonomy and consequent heterogeneity of data and tools that exist in a volatile knowledge environment. Taverna does not solve these issues, but provides a mechanism by which the issues can be addressed. One option is to require that the services that provide data and analysis tools conform to a particular type system that would hide the issues of heterogeneity. This would, in effect, mean providing a wide range of bioinformatics service in-house or necessitating a high 'activation energy' before any workflows could be developed. Taverna took the alternative route of being entirely open to any form of service. This immediately provides access to a large number of tools and data services (3500 at the time of writing). This openness avoids having to provide services or make them conform to some internal type model. This choice does, however, mean that heterogeneity is mostly left to the workflow composer to manage. The cost is still there, but it is lessened and itself distributed. Consequently, the cost of start-up is lower.

Taverna addresses, with ease, the factor of distribution. It makes the data and tools appear as if they are within the Taverna workbench. The bioinformaticians still have to use their skill to address basic bioinformatics issues of syntax transformation, extraction and mapping of identifiers between data resources. In a perfect world, this would not be the case, but the Taverna approach does exploit the target audience's skills.

In the case study, significant problems were encountered in all workflows when attempting to cross-reference between database identifiers. This matter, together with the naming conventions assigned to biological objects (Drăghici, Sellamuthu and Khatri, 2006), has proven to be a considerable barrier in bioinformatics involving distributed resources. In an attempt to resolve this, a single and explicit methodology has been provided by which this cross referencing was done. This methodology is captured within the workflows themselves.

In the initial use of Taverna (Stevens *et al.*, 2004), workflows were developed that were direct instantiations of human protocols. When developing genetic maps for Williams–Beuren syndrome (Stevens *et al.*, 2004), three workflows were developed that replicated the actions of the work a human bioinformatician performed in gathering data for human analysis. This task took two days and biological insight was gained over this time-span. The Taverna workflows gathered all these data (some 3.5 MB in numerous files (Stevens *et al.*, 2004)) within a 20 minute period. This left a bioinformatician with a near un-manageable set of data to interpret.

These data were crudely co-ordinated and simply created one problem whilst solving another. In light of this experience, the high-level methodology for the goals of workflows has changed. Data still have to be gathered, but this is not the end of the workflow. In the African tryoanosomiasis workflows reported here the strategy is now of gathering and managing data.

By using workflows the nature of the question asked of bioinformaticians and their collaborating biologists has changed. When there is simply a large number of gathered data they are in effect asked to 'make sense of these data'. The workflow approach, through mining the data for patterns and relationships, changes the nature of the question to 'does this pattern or these relationships make sense to you?'. Rather than requiring biologists to become data processors, their background and training are exploited by asking them to design analysis protocols (experiments) and then to assess the data mining outcomes. The patterns and relationships that are the outcomes of an analysis are really just a hypothesis that will prompt further analysis, either in the wet lab or through *in silico* experimentation.

As well as exploiting a biologist's background and training, the computer is exploited for the tasks that it is best at performing. Computers are good at being systematic and unbiased – two problems identified as being problems in analyses of microarray and QTL data.

This change in approach to analyses has led to a methodology for analysing microarray and QTL data. An approach has been developed, implemented as workflow, that exploits the systematics of a computational approach and couples this with an unbiased approach to evaluating a QTG's contribution to observed effects. The industrial, systematic approach enables premature triage and consequent false negatives to be avoided. A computer is only limited by its memory size, not its fragility, as is human memory. Finally, this industrial approach alters the granularity with which we look at the data – investigating pathways not gene products. This has gained biological insights into tolerance to African tryoanosomiasis infection (Fisher *et al.*, 2007). This is just one methodology for one data situation – similar effects in other data situations are expected.

In Taverna an analytical approach has been developed that manages to live with the heterogeneity, autonomy and volatility of the bioinformatics landscape. If bioinformatics were to cure its obvious ills of identity and other forms of heterogeneity, complex analyses of data would be far easier. Such a cure will not happen soon. Taverna provides a vehicle by which a bioinformatician can traverse the rough bioinformatics landscape. In doing so, the myGrid project has changed the nature of bioinformatics analysis.

Acknowledgements

The authors thank the whole myGrid consortium, software developers and its associated researchers. We would also like to thank the researchers of the Wellcome Trust Host–Pathogen Project – grant number GR066764MA. Paul Fisher is supported by the UK e-science EPSRC grant for myGrid GR/R67743.

References

Altschul, S. F., Gish, W., Miller, W., Myers, E. W. and Lipman, D. J. (1990), 'Basic local alignment search tool', *Journal of Molecular Biology* **215** (3), 403–410.

Apweiler, R., Bairoch, A., Wu, C. H., Barker, W. C., Boeckmann, B., Ferro, S., Gasteiger, E., Huang, H., Lopez, R., Magrane, M., Martin, M. J., Natale, D. A., O'Donovan, C., Redaschi, N. and Yeh, L.-S. L. (2004), 'UniProt: the universal protein knowledgebase', *Nucleic Acids Research* **32**, 115–119.

Bairoch, A., Apweiler, R., Wu, C. H., Barker, W. C., Boeckmann, B., Ferro, S., Gasteiger, E., Huang, H., Lopez, R., Magrane, M., Martin, M. J., Natale, D. A., O'Donovan, C., Redaschi, N. and Yeh, L. L. (2005), 'The universal protein resource (UniProt)', *Nucleic Acids Research* **33**, D154–D159.

Benson, D. A., Karsch-Mizrachi, I., Lipman, D. J., Ostell, J. and Wheeler, D. L. (2005), 'GenBank', *Nucleic Acids Research* **33**, 34–38.

Davidson, S., Overton, C. and Buneman, P. (1995), 'Challenges in integrating biological data sources', *Journal of Computational Biology* **2** (4), 557–572.

de Buhr, M., Mähler, M., Geffers, R., Westendorf, W. H. A., Lauber, J., Buer, J., Schlegelberger, B., Hedrich, H. and Bleich, A. (2006), 'Cd14, Gbp1, and Pla2g2a: three major candidate genes for experimental IBD identified by combining QTL and microarray analyses', *Physiological Genomics* **25** (3), 426–434.

Drăghici, S., Sellamuthu, S. and Khatri, P. (2006), 'Babel's tower revisited: a universal resource for cross-referencing across annotation databases', *Bioinformatics* **22** (23), 2934–2939.

Fisher, P., Hedeler, C., Wolstencroft, K., Hulme, H., Noyes, H., Kemp, S., Stevens, R. and Brass, A. (2007), 'A systematic strategy for large-scale analysis of genotype–phenotype correlations: identification of candidate genes involved in African trypanosomiasis', *Nucleic Acids Research* **35** (16), 5625–5633.

Frawley, W., Piatetsky-Shapiro, G. and Matheus, C. (1992), 'Knowledge discovery in databases: an overview', *AI Magazine* **13** (3), 57–70.

Galperin, M. Y. (2006), 'The molecular biology database collection: 2006 update', *Nucleic Acids Research* **34**, 3–5.

Hand, D., Mannila, H. and Smyth, P. (2001), *Principles of Data Mining*, MIT Press.

Hanotte, O. and Ronin, Y. (2003), 'Mapping of quantitative trait loci controlling trypanotolerance in a cross of tolerant West African N'Dama and susceptible East African Boran cattle', *Proceedings of the National Academy of Sciences* **100** (13), 7443–7448.

Hedeler, C., Paton, N., Behnke, J., Bradley, E., Hamshere, M. and Else, K. (2006), 'A classification of tasks for the systematic study of immune response using functional genomics data', *Parasitology* **132**, 157–167.

Hill, E., O'Gorman, G., Agaba, M., Gibson, J., Hanotte, O., Kemp, S., Naessens, J., Coussens, P. and MacHugh, D. (2005), 'Understanding bovine trypanosomiasis and trypanotolerance: the promise of functional genomics', *Veterinary Immunology and Immunopathology* **105** (3–4), 247–258.

Huang, L., Walker, D. W., Rana, O. F. and Huang, Y. (2006), Dynamic workflow management using performance data, *in* 'Proceedings of the 6th International Symposium on Cluster Computing and the Grid (CCGrid'06)', pp. 154–157.

Hull, D., Zolin, E., Bovykin, A., Horrocks, I., Sattler, U. and Stevens, R. (2006), Deciding semantic matching of stateless services, *in* 'Proceedings of the 21st National Conference on Artificial Intelligence and the 18th Innovative Applications of Artificial Intelligence Conference', Boston, MA, pp. 1319–1324.

Iraqi, F., Clapcott, S. J., Kumari, P., Haley, C. S., Kemp, S. J. and Teale, A. J. (2000), 'Fine mapping of trypanosomiasis resistance loci in murine advanced intercross lines', *Mammalian Genome* **11** (8), 645–648.

Kanehisa, M. and Goto, S. (2000), 'KEGG: Kyoto Encyclopedia of Genes and Genomes', *Nucleic Acids Research* **28** (1), 27–30.

Karp, P. (1995), 'A strategy for database interoperation', *Journal of Computational Biology* **2** (4), 573–586.

Kell, D. (2002), 'Genotype–phenotype mapping: genes as computer programs', *Trends Genetics* **18** (11), 555–559.

Kell, D. and Oliver, S. (2004), 'Here is the evidence, now what is the hypothesis?, the complementary roles of inductive and hypothesis-driven science in the post-genomic era', *Bioessays* **26**, 99–105.

Koudandé, O. D., van Arendonk, J. A. and Koud, F. (2005), 'Marker-assisted introgression of trypanotolerance QTL in mice', *Mammalian Genome* **16** (2), 112–119.

Lord, P., Alper, P., Wroe, C. and Goble, C. (2005), Feta: a light-weight architecture for user oriented semantic service discovery, *in* 'Proceedings of the 2nd European Semantic Web Conference', pp. 17–31.

Maglott, D., Ostell, J., Pruitt, K. D. and Tatusova, T. (2007), 'Entrez gene: gene-centered information at NCBI', *Nucleic Acids Research* **35**, D26–D31.

Mitchell, J. and McCray, A. (2003), 'From phenotype to genotype: issues in navigating the available information resources', *Methods of Information in Medicine* **42** (5), 557–563.

Naessens, J. (2006), 'Bovine trypanotolerance: a natural ability to prevent severe anaemia and haemophagocytic syndrome?', *International Journal for Parasitology* **36** (5), 521–528.

Nature Editorial (2006), 'Illuminating the black box', *Nature* **442** (7098), 1.

Oinn, T., Greenwood, M., Addis, M., Alpdemir, M. N., Ferris, J., Glover, K., Goble, C., Goderis, A., Hull, D., Marvin, D., Li, P., Lord, P., Pocock, M. R., Senger, M., Stevens, R., Wipat, A. and Wroe, C. (2006), 'Taverna: lessons in creating a workflow environment for the life sciences', *Concurrency and Computation: Practice and Experience* **18** (10), 1067–1100.

Schadt, E. (2006), 'Novel integrative genomics strategies to identify genes for complex traits', *Animal Genetics* **37** (1), 18–23.

Shi, M., Wei, G., Pan, W. and Tabel, H. (2005), 'Impaired Kupffer cells in highly susceptible mice infected with *Trypanosoma congolense*', *Infection and Immunity* **73** (12), 8393–8396.

Stein, L. (2002), 'Creating a bioinformatics nation', *Nature* **417** (6885), 119–120.

Stein, L. (2003), 'Integrating biological databases', *Nature Reviews Genetics* **4** (5), 337–345.

Stevens, R., Tipney, H. J., Wroe, C., Oinn, T., Senger, M., Lord, P., Goble, C., Brass, A. and Tassabehji, M. (2004), 'Exploring Williams–Beuren syndrome using my Grid', *Bioinformatics* **20**, i303–i310.

Stevens, R., Zhao, J. and Goble, C. (2007), 'Using provenance to manage knowledge of in silico experiments', *Briefings in Bioinformatics* **8** (3), 183–194.

Turi, D. (2006), Taverna workflows: syntax and semantics, Internal technical report, University of Manchester.

Yan, Y., Wang, M., Lemon, W. and You, M. (2004), 'Single nucleotide polymorphism (SNP) analysis of mouse quantitative trait loci for identification of candidate genes', *Journal of Medical Genetics* **41** (9), e111.

Yang, X. and Khosravi-Far, R. (1997), 'Daxx, a novel Fas-binding protein that activates JNK and apoptosis', *Cell* **89** (7), 1067–1076.

Zhao, J., Wroe, C., Goble, C., Stevens, R., Quan, D. and Greenwood, M. (2004), Using semantic Web technologies for representing e-science provenance, *in* 'Proceedings of the 3rd International Semantic Web Conference', Vol. 3298, Hiroshima, pp. 92–106.

Zuñiga, E. and Motran, C. (2000), '*Trypanosoma cruzi*-induced immunosuppression: B cells undergo spontaneous apoptosis and lipopolysaccharide (LPS) arrests their proliferation during acute infection', *Clinical and Experimental Immunology* **119** (3), 507–515.

10

Specification of distributed data mining workflows with DataMiningGrid

Dennis Wegener and **Michael May**

ABSTRACT

This chapter gives an evaluation of the benefits of grid-based technology from a data miner's perspective. It is focused on the DataMiningGrid, a standard-based and extensible environment for grid-enabling data mining applications. Three generic and very common data mining tasks were analysed: enhancing scalability by data partitioning; comparing classifier performance and parameter optimization. Grid-based data mining and the DataMiningGrid in particular emerge as a generic tool for enhancing the scalability of a large number of data mining applications. The basis for this broad applicability is the DataMiningGrid's *extensibility mechanism*. To support the scenarios described above, we have extended the original DataMiningGrid system by a set of new components.

10.1 Introduction

An important benefit of embedding data mining into a grid environment is *scalability*. Data mining is computationally intensive: when searching for patterns, most algorithms perform a costly search or an optimization routine that scales between $O(n)$ and $O(n^3)$ with the input data; where n, the number of data points, can be in the range from thousands to millions.

Improving scalability can be based on the fact that a number of algorithms are able to compute results on a subset of data in such a way that the results on the subset can be merged or aggregated to give an overall result. A well-known instance is the k-nearest-neighbor algorithm (Fix and Hodges, 1989), which can be partitioned into as many independent subproblems as there are instances to classify. Another class of algorithms repeatedly applies an algorithm to variations of the same data set, e.g. the bootstrap (Hastie, Tibshirani and Friedman, 2001). Again, each variation can be treated as an independent subproblem.

Data Mining Techniques in Grid Computing Environments Edited by Werner Dubitzky

In both cases, it is relatively easy to decompose the overall computation into a number of independent or nearly independent sub-computations, each of which can be executed on a different computing node. Thus, distributed computing potentially offers great savings in processing time and excellent scale-up with the numbers of machines involved. However, in the past it has been technically difficult, costly and time-consuming to set up a distributed computing environment, e.g. a PC cluster. A promise of grid technology is to facilitate this process by providing access to resources across institutions, e.g. a cluster running across the administrative domains of several universities.

A second benefit of grid technology is the ability to handle problems that are *inherently distributed*, e.g. when data sets are distributed across organizational boundaries and cannot be merged easily because of organizational barriers (e.g. firewalls, data policies etc.) or because of the number of data involved. Relying on the same principles of decomposability as discussed in the previous paragraph, the grid offers various opportunities to *ship* both algorithms as well as data across the grid, so that the inherently distributed nature of the problem can be addressed. Hence, grid computing (Foster, Kesselman and Tuecke, 2001) appears to be a promising candidate technology to apply for such scenarios. It is able to connect single workstations or entire clusters and provides facilities for data transport and sharing across organizational boundaries.

A number of environments for grid-enabling data mining tools have been proposed in recent years (Guedes, Meira and Ferreira, 2006; Alsairafi *et al.*, 2003; Ali and Taylor, 2005; Brezany, Janciak and Tjoa, 2005; Cannataro, Talia and Trunfio, 2002; García Adeva and Calvo, 2006; Cheung *et al.*, 2006; Talia, Trunfio and Verta, 2005; Stankovski *et al.*, 2008). To a varying degree, they address the scenarios described above. They aim at supporting the data miner by providing facilities to distribute the mining task while taking away the burden of becoming deeply involved with grid technology. An interesting trend that can be observed is the convergence on a set of standard technologies, including Globus Toolkit, *Web Service Resource Framework* (*WSRF*) on the grid side and Weka (Witten and Frank, 2005) on the data mining side.

The aim of this chapter is to give an evaluation of the benefits of grid-based technology from a data miner's perspective. It is focused on the DataMiningGrid (Stankovski *et al.*, 2008), a standard-based and extensible environment for grid-enabling data mining applications. Three generic and very common data mining tasks that can be performed with the help of the DataMiningGrid system were analysed:

(a) Enhancing scalability by data partitioning. This scenario is applicable for all algorithms that are able to compute results on a subset of the data.

(b) Comparing classifier performance on a given data set, a basic and often very time-consuming task.

(c) Parameter optimization, where the parameters of an algorithm are systematically varied to see which setting works best. This is another common and time-consuming task.

Each case study represents a broad class of data mining applications. Most non-trivial data mining applications will be based on at least one of the scenarios described, and often on a combination of them. Therefore, grid-based data mining in general and the DataMiningGrid in particular emerge not as a specialist technology, but as a general tool for enhancing the scalability of a large number of data mining applications.

The basis for this broad applicability is the DataMiningGrid's *extensibility mechanism*, which is the possibility of adding third-party data mining components to the grid environment without programmatically modifying the data mining component or grid environment (Wegener and May, 2007). To support the scenarios described above, we have extended the original DataMiningGrid system by a set of new components, e.g. in order to support splitting of input data or result processing in the grid environment.

10.2 DataMiningGrid environment

The DataMiningGrid project (Stankovski *et al.*, 2008) was an EU-funded project[1] with the objective of developing a generic system facilitating the development and deployment of grid-enabled data mining applications. The main outcome of the project was a collection of software components constituting the DataMiningGrid system. It is based on a *service-oriented architecture* (*SOA*) designed to meet the requirements of modern distributed data mining scenarios. Based on open-source technology like Globus Toolkit and compliant to common standards as WSRF and *Open Grid Services Architecture* (*OGSA*), the DataMiningGrid system forms an up-to-date platform for distributed data mining.

10.2.1 General architecture

The DataMiningGrid system is based on a layered architecture comprising the following layers: hard- and software resources, grid middleware, DataMiningGrid High-Level Services and client. The main user interface of the system is the Triana Workflow Editor and Manager (Taylor *et al.*, 2007), which was extended by specific components in the course of the project to facilitate access to the grid environment.

In the context of the DataMiningGrid system, an application is defined as a stand-alone executable file (e.g. C, Python, Java) that can be started from the command line. Grid-enabling an application requires a machine-readable description of the application. This is achieved through the *application description schema* (*ADS*). Details on the architecture of the system, the ADS and the grid-enabling process are described elsewhere (Stankovski *et al.*, 2008).

10.2.2 Grid environment

The set-up of a test bed can be seen as an integration of different computational components into the grid of a virtual organization. A typical DataMiningGrid grid environment contains one *head* site that runs unique middleware and high-level services (e.g. Resource Broker service, central MDS4 (Monitoring and Discovery) service, Information Integrator service), several *central* sites (which run Grid Resource Allocation and Management services and manage the aggregated resources) and many *worker* machines, usually orchestrated by a local scheduler, e.g. Condor (Litzkow and Livney, 1990).

10.2.3 Scalability

Scalability of a distributed data mining system is a key issue, and was one of the key requirements for the DataMiningGrid system (Stankovski *et al.*, 2008). Scalability characteristics of the DataMiningGrid have already been evaluated elsewhere on basis of different scenarios,

[1]The project was supported by the European Commission under grant number IST-2004-004475.

e.g. a text-classification scenario and a scenario running the Weka algorithms M5P, IBK and LWL (Stankovski *et al.*, 2008; Wegener and May, 2007). The experiments demonstrated that grid-enabled applications in the DataMiningGrid system can reach very good scalability and that the flexibility of the system does not result in a significant performance overhead (Stankovski *et al.*, 2008; Wegener and May, 2007). Instead of focusing again on experimental comparisons of scalability of the grid-enabled applications used within the scenarios, this chapter demonstrates how standard data mining scenarios can be applied to the DataMiningGrid system to enable users to benefit from the system's features.

10.2.4 Workflow environment

The workflow editor used to design and execute complex workflows in the DataMiningGrid system is based on Triana (Taylor *et al.*, 2007). It allows composing workflows in a drag-and-drop manner. Triana was extended by special components which allow access to and interaction with the DataMiningGrid grid environment. Figure 10.1 gives an overview of the user interface.

The workflow editor is the main user interface the data miner can utilize for constructing data mining tasks. The aim of this graphical environment is to provide a powerful, yet user-friendly, tool that hides the complexity of setting up a grid workflow. Thus, no programming or scripting effort is needed to set up distributed data mining tasks. In the following sections the

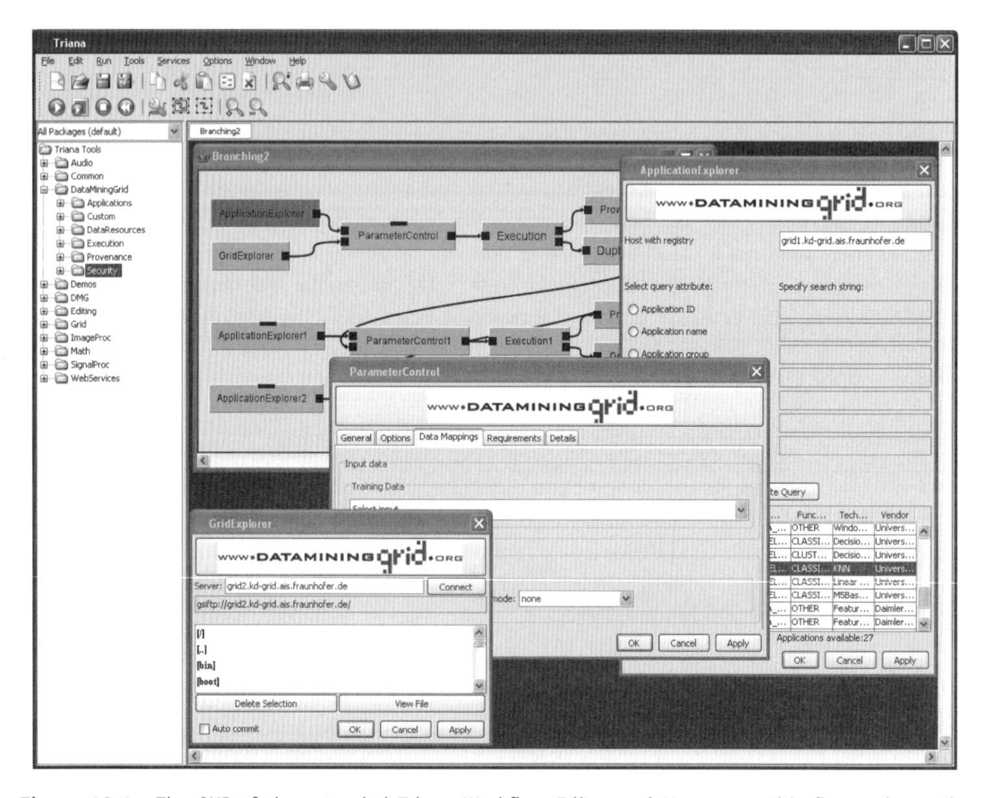

Figure 10.1 The GUI of the extended Triana Workflow Editor and Manager – this figure shows the treelike structure of units containing the DataMiningGrid extensions, a graphical workflow and the GUI of the DataMiningGrid units ApplicationExplorer, ParameterControl and GridExplorer

components of the DataMiningGrid Workflow Editor supporting various data mining scenarios are defined.

A component inside a Triana workflow is called a *unit*. Each unit, which can be seen as a wrapper component, refers to special operations. The Triana units are grouped in a treelike structure. In the user interface, units are split into several subgroups referring to their functionality, e.g. applications, data resources, execution, provenance and security.

The *generic job template* is a component of the DataMiningGrid system supporting the user in the construction of workflows and in the specification of grid jobs. The template consists of four different groups of units for application exploration, data selection, application control and execution.

10.3 Operations for workflow construction

A workflow in a data mining application is a series of operations consisting of data access and preparation, modeling, visualization, and deployment. Such a workflow can typically be seen as a series of data transformations, starting with the input data and having a data mining model as final output. In a distributed environment, additional operations are related to splitting a data mining task into sub-problems that can be computed in a distributed manner, and selecting an appropriate distribution strategy. This section describes the facilities the DataMiningGrid offers for defining workflows.

Workflows in Triana are data driven. Control structures are not part of the workflow language but are handled by special components – e.g., looping and branching are handled by specific Triana units and are not directly supported by the workflow language (Shields, 2007; Taylor *et al.*, 2007).

In the DataMiningGrid system the client-side components are implemented as extensions to the Triana Workflow Editor and Manager. The workflow can be constructed by using (a) the standard units provided by Triana and (b) the DataMiningGrid extensions. By using and combining these units, workflows performing many different operations can be defined.

10.3.1 Chaining

A typical data mining task spans a series of steps from data preparation to analysis and visualization that have to be executed in sequence. Chaining is the concatenation of the execution of different algorithms in a serial manner. The result data of an algorithm execution are used by the next execution in the series.

Concatenation of analysis tasks is achieved by using multiple generic job templates inside the Triana workflow. Each generic job template represents a single analysis step with a (not necessarily different) application. The output of the previous task can be used as input for the next task. Different tasks can run on different machines. It is up to the user to decide whether it makes sense to run a second task on the same or on a different machine, which could mean transferring the results of the first task over the network.

10.3.2 Looping

Looping means the repeated execution of an algorithm. The algorithm can run on the same (or different) data or with the same (or different) parameter settings until a condition is fulfilled.

The DataMiningGrid system provides different ways of performing loops. Triana contains a *Loop unit* that controls repeated execution of a subworkflow (Shields, 2007; Taylor *et al.*,

2007). Additionally, it provides a loop functionality when grouping units. The DataMiningGrid components do not directly provide loops but parameter sweeps (see Section 10.3.6). Depending on the kind of loop to be performed, one or more of these choices are possible.

10.3.3 Branching

Branching means letting a program flow follow several parallel executions paths from a specified point or following a special execution path depending on a condition.

In a workflow there are different possibilities for branching without a condition. A user would, e.g. divide the workflow into two (or more) branches after the execution of an application is finished (e.g. pre-processing) and start the parallel execution of other applications. After each application execution the Execution unit returns a URI as reference to the results' location in the grid. As inputs and outputs of grid jobs are file based, they can be accessed at any later step of the workflow. To set up a branch, the GridURI output node of the Execution unit has to be cloned. Analogously, branching can be performed at various steps of the workflow (provided the unit after the branches start has a clonable output node). In addition, Triana contains a Duplicator unit that enables workflow branching (Taylor *et al.*, 2007). This unit can be used if there is no clonable output node at a unit. The Duplicator unit's function is to duplicate any object received at the input node.

Triana also provides a unit supporting conditional branching, which is designed for executing a designated part of the workflow dependent on a condition (Shields, 2007; Taylor *et al.*, 2007).

10.3.4 Shipping algorithms

Shipping of algorithms means sending the algorithm to the machine where the data it operates on are located. This is one of the major options in setting up a distributed data mining application. It is required when, as is often the case, no pre-configured pool of machines is available that already has the data mining functionality installed. The option to ship algorithms to data allows for flexibility in the selection of machines and reduces the overhead in setting up the data mining environment. This is especially important when the data naturally exist in a distributed manner and it is not possible to merge them. This may be the case when data sets are too large to be transferred without significant overhead or when, e.g., security policies prevent the data being moved.

The execution system of the DataMiningGrid system allows us to ship algorithms. The executable file that belongs to the application is transferred to the execution machine at each application execution. If the algorithm is to be shipped to the data to avoid copying files among different sites, the machine where the data is located has to be selected as the execution machine. If a job is submitted to a *Grid Resource Allocation Manager* (*GRAM*)[2] that is connected to a Condor cluster, it is not possible to specify exactly on which machine of the cluster a job should run. By selecting the appropriate execution mode, the job can be submitted and processed either on the GRAM itself or on the cluster where the clusterware is responsible for the resource management.

10.3.5 Shipping data

Shipping of data means sending the data to the machine where the algorithm that processes them is running. This operation is important in cases where either the full data set is partitioned

[2]A GRAM enables the user to access the grid in order to run, terminate and monitor jobs remotely.

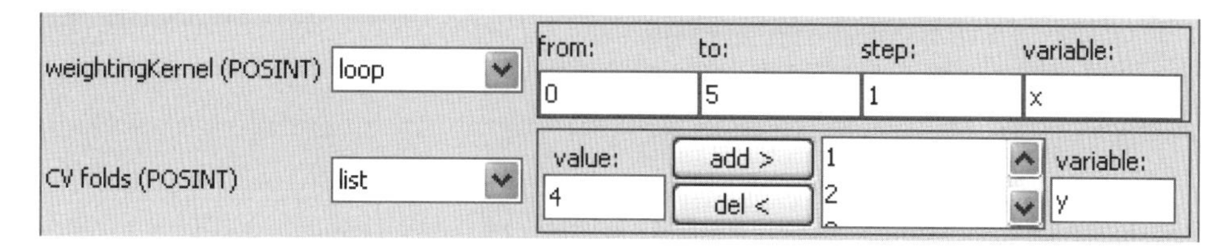

Figure 10.2 Parameter sweep – parameter loops and lists can be specified in the GUI

to facilitate distributed computation (e.g. application of k-NN) or where the same source data set is moved to different machines and repeatedly analysed, for instance for ensemble learning.

Each time an application is executed in the DataMiningGrid environment the input data (consisting of one or more files) for the application are copied to a local work directory on the execution machine (when using the DataMiningGrid unit built-in functionality). If only the data and not the algorithm are to be shipped (e.g. for copyright reasons) the machine can be specified where the job should run (see Section 10.3.4).

10.3.6 Parameter variation

Parameter variation means executing the same algorithm with different input data and different parameter settings. To set up a parameter variation as a distributed computation requires shipping the data and the algorithms.

The DataMiningGrid system provides the possibility of using parameter sweeps, which means that a loop (for numeric values) or a list (for any data type) can be specified for each option of the application as well as a loop for the input files or directories (see Figure 10.2, showing parts of the GUI). With this approach it is possible to submit hundreds of jobs at the same time.

When performing a sweep the system handles the naming of the result files automatically to ensure that results are not overwritten when they are collected in the result directory.

10.3.7 Parallelization

Parallelization means distributing subtasks of an algorithm and executing them in parallel. The system supports the parallel or concurrent execution of applications at the same time, e.g. by performing a parameter sweep (Section 10.3.6) and the parallel execution of workflow branches (Section 10.3.3).

In general, parallelization of a single algorithm is difficult on the grid because of the communication overheads involved. The QosCosGrid[3] project is currently investigating whether grids can be used for parallel computing purposes. Thus, the DataMiningGrid environment does not support the parallelization of the execution of a single algorithm. If the application itself takes care of the parallelization process, then an integration is possible.

10.4 Extensibility

Data mining tasks are highly diverse. In many cases the functionality already provided by a system will not be sufficient for the task at hand. This makes extensibility an important feature.

[3] QosCosGrid Web site: www.QosCosGrid.com

In the context of the DataMiningGrid system, there are two types of extensibility: on the client side (local extensibility), and on the grid environment, e.g. the inclusion of new grid-enabled applications.

Local extensibility A local extension of the DataMiningGrid system is a client side extension. The Triana workflow editor, which is the main client of the system, can be easily extended by new workflow components. Such components – implemented as Triana units – could, for instance, be viewer or inspection components, which may or may not directly interact with the grid environment.

Extensibility of the grid environment The requirement for extensibility of the grid environment requests the following:

- *Extensibility without platform modification.* A data miner who wants to make use of a grid-enabled data mining platform typically does not have any knowledge about the details of the underlying system. Therefore, he or she does not want to change – and even might not be capable of changing – any components or program code of the data mining platform. Additionally, the data miner may not want to be dependent on a grid platform developer.
- *Extensibility without algorithm modification.* The data miner can reuse his or her favorite algorithms and combine them with third-party methods to address complex data mining tasks.

New applications can be grid-enabled with the DataMiningGrid system without platform and algorithm modification. In the following, a simple scenario of extending the system by a decision tree application is described.

The grid-enabling process requires an executable file that contains the application and an application description file (an instance of the ADS). In this example, the Weka-J4.8 decision tree algorithm from Weka v3.4.3 was grid-enabled. The Weka implementation of the algorithm does not use flag-value pairs for specifying command-line calls, so a wrapper component that translates the command-line call into the required format is needed for the integration. The sources together with the wrapper were packaged into an executable jar file.

The application description file has to be created according to the ADS and contains information about the application that is, e.g., used within a grid-wide application registry. It contains the application name, group and domain and the technique used, as well as data about the executable file and the class to call, information about the application's options, inputs and outputs, and system requirements. The data input for the J4.8 application is a file on which the training is performed; the output is the result file Weka creates. The ADS can be provided either manually or by using the DataMiningGrid Application Enabler, which is a wizard style webpage that guides through the process of grid-enabling. Once the executable file and the application description file are made available in the grid environment, the application can be used with the client side components.

Using this approach, we have grid-enabled a number of new applications. In particular, these are the algorithms for the scenarios used in the case studies (see Section 10.5) as well as a number of helper-applications responsible for input file and result processing. For example, in the k-NN scenario (see Section 10.5.2) there is a need for two additional applications that process the input and output of the k-NN application in order to split the input file into several small files and to combine the results after the k-NN algorithm executions are completed.

Similar applications for the result processing were grid-enabled for the classifier comparison scenario (see Section 10.5.3), in which an overall result had to be computed by collecting the individual classifications in a single file, and the parameter optimization scenario (see Section 10.5.4), where the application was responsible for combining the results of the different algorithm executions in order to compute the result of the parameter optimization.

10.5 Case studies

10.5.1 Evaluation criteria and experimental methodology

In the previous section the major features at the disposal of the data miner for specifying complex data mining tasks were described. This section presents the analysis of how three generic and very common data mining tasks can be performed with the help of the DataMiningGrid system.

(a) In the first case study it was investigated how scalability can be increased by partitioning the data. The data are automatically partitioned and distributed to different nodes. This scenario is applicable for all algorithms that are able to compute results on a subset of the data.

(b) In the second case study a number of different data mining algorithms were applied to the same data set, and their predictive performance was compared. Both the data and the algorithms were automatically distributed to different nodes. Comparing classifier performance is a basic and often very time-consuming task in data mining.

(c) In the third case study the parameters of an algorithm were systematically varied to see which setting works best. Both the algorithms and the data were automatically distributed. This is another common and time-consuming task.

Each case study was illustrated with one or more specific algorithms taken from Weka (Witten and Frank, 2005). It should be stressed, however, that due to the extensibility mechanism of the DataMiningGrid each case study represents a broad class of data mining applications, and instead of the Weka algorithms almost any third-party or user-supplied algorithm can be used. Most non-trivial data mining applications will be based on at least one of the scenarios described, and often on a combination of them. Therefore, grid-based data mining and the DataMiningGrid in particular emerge not as a specialist technology, but as a general tool for enhancing the scalability of a large number of data mining applications.

Table 10.1 shows which workflow operations were needed to run the scenarios in the grid environment.

10.5.2 Partitioning data

This section describes two tasks based on partitioning the data: k-NN and bagging. The scenario for this case study is the following. A large number of data are to be analysed. The large data set is not processed on a single machine, but is split into several smaller data sets. These smaller

Table 10.1 Scenarios

Mechanism	Partitioning data	Classifier comparison	Parameter optimization
Chaining	yes	yes	yes
Looping	no	no	no
Branching	no	yes (InputURI and OutputFiles)	no
Shipping algorithms	(yes)	(yes)	(yes)
Shipping data	yes	(yes)	(yes)
Parameter variation	yes (InputFiles)	no	yes (Options)
Parallelization	yes (Jobs)	yes (Applications)	yes (Jobs)

data sets are distributed and analysed in parallel in the grid environment. The results of the analysis are combined afterwards.

k-NN (Fix and Hodges, 1989) is a simple yet powerful classification and regression algorithm. It performs a local approximation, taking the k nearest neighbours of an instance into account for classification or regression. In the experiment a grid-enabled Weka algorithm is used for the k-NN task, together with two applications for input file splitting and result aggregation (see Section 10.4).

Three sub-tasks are executed. The first one takes the large input file and splits it into a user-defined number of partitions. The second one is the execution of the k-NN algorithm. A file sweep is used, which results in a separate execution of the k-NN algorithm for each of the partitions in parallel. The distribution of files to computing nodes is done transparently to the user. The final step is to combine the results of the several distributed runs.

With a similar set-up of components also ensemble learning, e.g. bagging (Hastie, Tibshirani and Friedman, 2001), can be performed. An ensemble or committee of classifiers is a set of classifiers individually trained on some data set. The individual decisions are combined to classify new examples. Combining of individual results is performed typically by weighted or unweighted voting.

For bagging, a bootstrap sampling on the input data is performed, i.e. a sampling with replacement from the input data set of size n, creating n new sample data sets. Since these new sample data sets contain some duplicated instances (because of sampling with replacement), each one differs from the original data set. A learning task on each new data set using some classifier, e.g. a decision tree, and a majority vote as the final classification is performed. This approach has proven to be very successful for unstable classifiers such as decision trees.

To implement this strategy, a new application was needed that performs the sampling with replacement, creating a set of new files in a directory. This application was integrated analogously to the split application for k-NN. A data mining algorithm, e.g. a decision tree, then works on the files in the directory, sending each job to a different machine. Finally, a second component is needed that reads the individual results and performs the majority vote.

These two examples demonstrate that a computationally intensive data mining task can often be decomposed into simpler sub-tasks by partitioning the data and analysing them in parallel on different nodes. In this case, the basic algorithm does not need to be modified. Since there is no need for communication between the sub-tasks (except for gathering the final results), the computational overhead for distributed computation is low and excellent scalability can be achieved.

10.5.3 Classifier comparison scenario

A typical task in data mining is to asses the performance of several classifiers on the same data set. It is it often not clear in advance which one will perform best, and the data miner has to carry out a number of experiments. This task can be very time-consuming if many algorithms have to be compared and the data set is large. Performing the evaluation on a grid speeds up the evaluation, since evaluation can be done in parallel.

This learning experiment can be seen as a combination of multiple algorithms on the same data. The result of the different algorithm runs is combined into a global file, so that results can be compared. The following Weka algorithms (learners) were used: model trees, k-NN and locally weighted learning.

In order to compute an overall result of the learning experiment, the results of the different algorithm executions had to be combined with the help of another grid-enabled application (see Section 10.4).

In this experiment three application executions of the learning algorithms were performed, each consisting of the selection of the learner, the control of the application parameters (all operate on the same input data) and the execution. After the learning tasks were completed the last component combined the results of the learners to an overall result.

10.5.4 Parameter optimization

A further typical task in data mining is parameter optimization. Some algorithms have a rich set of options to be set, and selection of an option can have a high impact on the performance of the algorithm. In many cases the best setting has to be found experimentally by systematically exploring various settings.

A parameter optimization experiment consists of different runs of the same algorithm on the same input data but with different parameter settings. For the parameters that are to be optimized a range of values is used. Each change of a parameter value means a separate run of the algorithm. The best combination of parameters is found by comparing and evaluating the results of the different runs.

The experiment consisted of the execution of a Weka algorithm for the parameter optimization task. The parameter optimization was performed for four parameters. The results were aggregated by another grid-enabled application (see Section 10.4).

In the experiment a single execution of the application on the input data was performed. Within this run a parameter sweep, specified though the ParameterControl unit, was used for the parameter to be optimized. The ParameterControl component (see Figure 10.2) saves the user a lot of time in specifying the tasks, since the individual jobs are generated automatically. Depending on the number of possible values for the four parameters to be optimized, a very large number of jobs were generated. After the parameter optimization task was completed, the results were transferred into a single directory and the result files were collected and evaluated for processing the overall result of the experiment.

10.6 Discussion and related work

Today several environments for grid-enabled data mining exist. Such systems are, e.g., Anteater (Guedes, Meira and Ferreira, 2006), Discovery Net (Alsairafi *et al.*, 2003), FAEHIM (Ali and Taylor, 2005), GridMiner (Brezany, Janciak and Tjoa, 2005), Knowledge Grid

(Cannataro, Talia and Trunfio, 2002), Pimiento (García Adeva and Calvo, 2006), SODDM (Cheung *et al.*, 2006), Weka4WS (Talia, Trunfio and Verta, 2005) etc. All of these systems are – to a varying degree – capable of distributed data mining in the grid and have already been comprehensively reviewed and compared, based on the architectural perspective (Stankovski *et al.*, 2008). Also, a comparison with the focus on the extensibility requirement was given (Wegener and May, 2007).

The study has demonstrated that the capability of the DataMiningGrid system, extended by new grid enabled applications, is sufficient to carry out all data mining scenarios described. There was no need for the integration of new workflow operations. Because the scenarios can be adapted to run in the DataMiningGrid grid environment, the data miner benefits from all the advantages the grid technology provides, among them a user-friendly set-up of experiments, a decrease in runtime or other benefits like a massive storage volume, and an integrated security concept. For more details on performance evaluation, the reader is referred to the literature (Stankovski *et al.*, 2008).

It should be noted that the conversion of an existing algorithm into a distributed one was in all cases straightforward. Additional components were needed just for splitting a large data set into pieces, gathering results or performing a simple vote. The potential of using grid technology for data mining lies in its simple distribution scheme for providing performance gains. The data miner will not need deep grid knowledge or to know – as required for parallel computing – how to set up his or her own algorithm in a way that makes it suitable for grid execution. The DataMiningGrid addresses his or her needs by providing an extension mechanism that makes it possible to grid-enable his or her favorite algorithm without writing any code, by simply providing metadata that can even be specified via a webpage.

10.7 Open issues

The success of grid-enabled data mining technology depends on whether it is taken up in real-world data mining projects. For this to happen, there is a number of open issues. First, case studies, success stories and systematic evaluations of one or more systems from a data mining perspective are rare so far. These case studies are needed to convince data miners to consider the use of grid technology for their real-world data mining applications. Second, a comparison of grid-based platforms with more traditional platforms as well as new developments for distributed computing are needed. Third, on the architectural side, the combination of grid technology with mobile devices is a potentially very promising area, especially in scenarios where the data mining is done inside these mobile devices themselves.

10.8 Conclusions

In this chapter the DataMiningGrid system was evaluated from a problem-oriented data mining perspective. The capability of the DataMiningGrid workflow environment was discussed and workflow operations important for common data mining tasks, e.g. looping, branching and more special operations such as shipping of data or algorithms, were defined. Then, in a number of case studies, the capability of the DataMiningGrid system was demonstrated.

A desirable future extension of the DataMiningGrid system is full compatibility with the CRISP-DM standard (CRoss-Industry Standard Process for Data Mining) (Shearer, 2000), a process model for data mining projects. The CRISP-DM process model is organized in different phases: business understanding, data understanding, data preparation, modelling, evaluation and deployment.

In general, in the DataMiningGrid system each single data mining task performed refers to the execution of a grid-enabled application, which results in separated job executions in the grid. The workflow focuses on the grid-based aspects. However, it is possible to map the grid-enabled application in the DataMiningGrid system to the CRISP-DM tasks. The mapping is based on the metadata description of the applications. The DataMiningGrid system in principle covers all CRISP-DM phases, although it was not designed to follow the CRISP-DM structure and does not support the different tasks by templates or wizards yet. Providing this CRISP-DM perspective would lead to a view of the data mining process that is more natural for the data miner.

References

Ali, A. S. and Taylor, I. J. (2005), Web services composition for distributed data mining, *in* 'Proceedings of the 2005 International Conference on Parallel Processing Workshops (ICPPW'05)', IEEE Computer Society, Washington, DC, pp. 11–18.

Alsairafi, S., Emmanouil, F.-S., Ghanem, M., Giannadakis, N., Guo, Y., Kalaitzopoulos, D., Osmond, M., Rowe, A., Syed, J. and Wendel, P. (2003), 'The design of Discovery Net: towards open grid services for knowledge discovery', *International Journal of High Performance Computing Applications* **17** (3), 297–315.

Brezany, P., Janciak, I. and Tjoa, A. M. (2005), GridMiner: a fundamental infrastructure for building intelligent grid systems, *in* 'Proceedings of the 2005 IEEE/WIC/ACM International Conference on Web Intelligence (WI'05)', IEEE Computer Society, Washington, DC, pp. 150–156.

Cannataro, M., Talia, D. and Trunfio, P. (2002), 'Distributed data mining on the grid', *Future Generation Computer Systems* **18** (8), 1101–1112.

Cheung, W. K., Zhang, X.-F., Wong, H.-F., Liu, J., Luo, Z.-W. and Tong, F. C. (2006), 'Service-oriented distributed data mining', *IEEE Internet Computing* **10** (4), 44–54.

Fix, E. and Hodges J., Jr. (1989), 'Discriminatory analysis, nonparametric discrimination: consistency properties', *International Statistical Review* **57** (3), 238–247.

Foster, I., Kesselman, C. and Tuecke, S. (2001), 'The anatomy of the grid: enabling scalable virtual organizations', *International Journal of High Performance Computing Applications* **15** (3), 200–222.

García Adeva, J. and Calvo, R. A. (2006), 'Mining text with Pimiento', *IEEE Internet Computing* **10** (4), 27–35.

Guedes, D., Meira W., Jr. and Ferreira, R. (2006), 'Anteater: a service-oriented architecture for high-performance data mining', *IEEE Internet Computing* **10** (4), 36–43.

Hastie, T., Tibshirani, R. and Friedman, J. (2001), *The Elements of Machine Learning*, Springer, New York.

Litzkow, M. and Livney, M. (1990), Experience with the Condor distributed batch system, *in* 'IEEE Workshop on Experimental Distributed Systems', pp. 97–100.

Shearer, C. (2000), 'The CRISP-DM model: the new blueprint for data mining', *Journal of Data Warehousing* **5** (4), 13–22.

Shields, M. (2007), Control- versus data-driven workflows, *in* I. Taylor, E. Deelman, D. Gannon and M. Shields, eds, 'Workflows for e-Science', Springer, New York, pp. 167–173.

Stankovski, V., Swain, M., Kravtsov, V., Niessen, T., Wegener, D., Kindermann, J. and Dubitzky, W. (2008), 'Grid-enabling data mining applications with DataMiningGrid: an architectural perspective', *Future Generation Computer Systems* **24**, 259–279.

Talia, D., Trunfio, P. and Verta, O. (2005), Weka4WS: a WSRF-enabled Weka toolkit for distributed data mining on grids, *in* 'Proceedings of the 9th European Conference on Principles and Practice of Knowledge Discovery in Databases (PKDD 2005)', Vol. 3721 of *LNAI*, Springer, Porto, pp. 309–320.

Taylor, I., Shields, M., Wang, I. and Harrison, A. (2007), The Triana workflow environment: architecture and applications, *in* I. Taylor, E. Deelman, D. Gannon and M. Shields, eds, 'Workflows for e-Science', Springer, New York, pp. 320–339.

Wegener, D. and May, M. (2007), Extensibility of grid-enabled data mining platforms: a case study, *in* 'Proceedings of the 5th International Workshop on Data Mining Standards, Services and Platforms (KDD 2007)', San Jose, USA, pp. 13–22.

Witten, I. H. and Frank, E. (2005), *Data Mining: Practical Machine Learning Tools and Techniques*, Morgan Kaufmann Series in Data Management Systems, 2nd edn, Morgan Kaufmann.

11 Anteater: Service-oriented data mining

Renato A. Ferreira, Dorgival O. Guedes and **Wagner Meira Jr.**

ABSTRACT

Data mining focuses on extracting useful information from large volumes of data, and thus has been the centre of great attention in recent years. Building scalable, extensible and easy-to-use data mining systems, however, has proved to be a hard task. This chapter discusses Anteater, a service-oriented architecture for data mining, which relies on Web services to achieve extensibility, offers simple abstractions for users and supports computationally intensive processing on large amounts of data. Anteater relies on Anthill, a run-time system for irregular, data intensive, iterative distributed applications, to achieve high performance. Data mining algorithms are irregular because the computation is irregularly distributed over the input data, which greatly complicates the parallelization. It is data intensive, for it deals with potentially enormous data sets. And it is iterative as many of the algorithms make multiple passes over the input in a sort of a refining process to come up with the output. The combination of a Web service architecture and a parallel programming environment provides a rich environment for exploring different levels of distributed processing with good scalability. Anteater is operational and being used by the Brazilian government to analyse government expenditure, public health and public safety policies. Feedback from Anteater users has been very positive regarding the system usability, flexibility and impact. Concerning the parallel performance, experiments have shown the data mining algorithms within Anteater to scale linearly to large number of compute nodes.

11.1 Introduction

Recent trends in computer technology have yielded both a continuous drop of the cost of data storage and a continuous increase in the sophistication of equipments and algorithms that collect and store data. Therefore, collecting and storing multiple-gigabytes of data is very common nowadays. Analysing these huge data sets in their raw form is rapidly becoming impractical, so data mining techniques have increased in popularity recently as a means of collecting meaningful summarized information from them.

Data Mining Techniques in Grid Computing Environments Edited by Werner Dubitzky

The Brazilian federal government employs seven large systems that control most of its executive activities, including budget, expenditures and personnel. The data from these systems are analysed by thousands of civil servants for the sake of verification, auditing and policy enhancement. These civil servants, who usually do not have expertise in data mining techniques, are spread across hundreds of organizations and have their own interests and obligations in terms of extracting information from data. In terms of infrastructure, their demands require a significant amount of computational power, given the diversity and number of potential users of data mining techniques. Finally, it is important to highlight that it is very hard to predict the demand imposed by data mining techniques, first because the computational cost of these techniques is data dependent, and second because the process itself is iterative, and the number of iterations depends on the data, the analyst's expertise and the task being performed.

Notice that even the fastest sequential algorithms may not be sufficient to summarize the volumes of data being currently stored in reasonable time, so there is a need for distributed algorithms and scalable solutions. Besides this, providing end users with interfaces they may use effectively without expert knowledge of the algorithms is still a challenge.

At the same time, *service-oriented architectures* (*SOAs*) are becoming one of the main paradigms for distributed computing. In this context, services provide access to programs, databases and business processes whose semantics are defined by a language interpretable by machines in a way that abstracts implementation details. This makes services accessible over the network in a platform-independent way (Perrey and Lycett, 2003).

This chapter presents an experience in the application of the SOA paradigm to the development of a real system, the Anteater, a data mining service platform for decision support. The use of an SOA enables the activation of resources on demand: resources are made available by participants as independent services that may be accessed in a standardized way. This solution is shown to achieve extensibility and scalability in a level not available in other data mining systems.

An analysis of the data mining process indicates a series of aspects that must be addressed by such an architecture. These aspects include the large volume of data, the computational cost of the algorithms and the diversity of the user population. Most of the existing systems do not address all these concerns adequately. Simple client/server solutions, with data kept in the data mining server or an associated database server, have scalability limitations when multiple large databases and a large number of users are considered. Further, these solutions rely on clients that demand significant computational resources, which may not be widely available, as is the case for the Brazilian government. On the other hand, the use of an SOA in Anteater offers facilities for the addition of large databases and the expansion of the computing capacity in a simple and transparent way, by just advertising new services according to the chosen interface.

Much work has been done in developing parallel data mining algorithms over the years. The main limitation in several of these algorithms is that they have been shown not to scale well to a very large number of processors. Several data mining algorithms were, however, designed, implemented and validated using the Anthill runtime framework (Ferreira *et al.*, 2005), and these have shown to exhibit high parallel efficiency (Veloso *et al.*, 2004). In the context of Anteater, where the work has to be done by clusters of workstations overnight, the availability of such algorithms is key to the effective use of the system.

Other solutions have been proposed for distributed data mining services. (Krishnaswamy, Zaslavsky and Loke, 2001) describe a general model for the area and provide a good revision of the field. Other efforts have used Web services to build a data mining environment, e.g. the GridMiner (Brezany *et al.*, 2003) and the Knowledge Grid (Cannataro and Talia, 2003). In general, these solutions focused on coarse-grain parallelism and extensibility. Besides addressing, through Anthill, the scalability and extensibility problems with a fine grain of parallelism, Anteater also addresses the problem of providing users with a simple interface that abstracts the technical details of the algorithms. The architecture allows users to choose from different visual metaphors to analyse the results of a mining task, reducing the need for training end users in data mining specifics. Moreover, all interaction is performed through the Web, that is, a user needs just a browser to have access to Anteater resources.

The features provided by Anteater made it a good solution for the Brazilian scenario. The system's proposed architecture was accepted by the user community as matching the current computational infrastructure available in the large government processing centres. Also, the model of operation, including its user interface and selection of algorithms, was accepted by the users as meaningful for a variety of investigation scenarios these government officials were pursuing. It is currently being used by the government to process more than one million records of public expenditure data and over 200 000 records from the National Health Services on haemodialysis procedures, and by the state government to mine almost two million payroll records and a public safety database with approximately 500 000 records of 911 calls. In all cases, users are experts in their application areas, with just minimal training on data mining principles. Measurements of the impact of Anteater for its current user base is hard to quantify, but the feedback the system has received has been very positive, specially when considering that the user base is expanding as more government officials are looking to adopt Anteater as their data analysis platform.

11.2 The architecture

The Anteater architecture is based on the standard life-cycle of *knowledge discovery in databases* (*KDD*) (McElroy, 2000). In short, a user of a data mining system must be able to understand the available data sets, select the data upon which he or she wants to work, select the kind of algorithm to be applied to the data and visualize the results in an intuitive way, possibly returning to any of the previous steps for new queries. This means the process is iterative by nature, with the user defining new tasks as a result of the analysis of previous ones. Since each task may be data and computationally intensive, the system must be able to withstand long periods of heavy load.

The Anteater was designed as a set of distributed components that offer their services through well defined interfaces, so they can be used as needed to meet the application needs. The identification of the components was based on major points of the KDD problem that should be addressed by the architecture: provision of access to a data set, execution of a mining task and visualization of results. Each of these constitutes a problem in its own right and requires a set of functionalities to be exposed to the user as services.

The architecture of Anteater is shown in Figure 11.1. All components and interactions are highlighted there. The user interface is implemented over a webpage, and processed within the application server.

The *data server* (*DS*) is responsible for the access interface to data sets and all associated metadata. The metadata, in XML, describe the type of each attribute, whether they represent continuous or categorized entities, and whether the data were the result of some previous mining task, among other things. The data are stored in raw form to reduce processing costs. The DS offers services for retrieving descriptions of the data sets and their metadata, transferring data

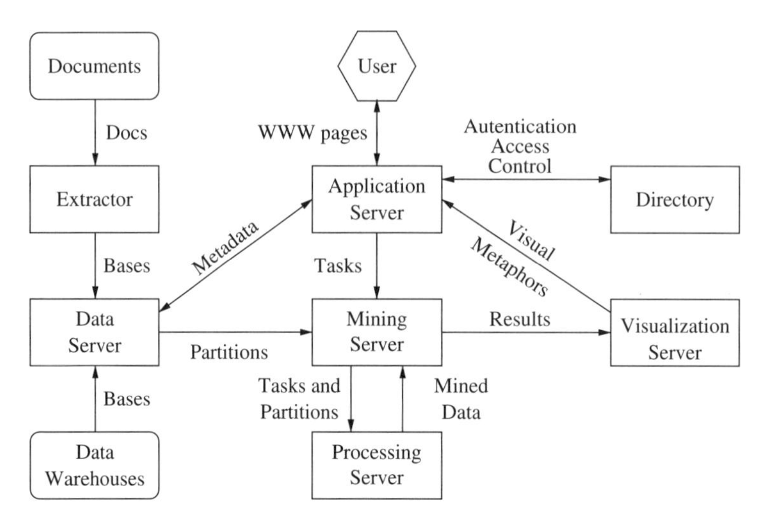

Figure 11.1 The Anteater architecture

sets from one DS to another and executing SQL-based queries on the data. Data may be fed into the DS from existing data warehouses when they have already been preprocessed, cleaned and organized, or they may be collected from other sources – Web documents, for example. In the case of Web documents, Anteater offers a set of extraction tools that can collect data from unstructured objects (free text from webpages). In either case (structured transactional data or unstructured Web documents), the input of data into a DS is a back-office operation that must be directly controlled by an operator. It is important to highlight here that the DS and the *processing server* are actually collocated within the distributed environment. Actual data transfers are avoided as much as possible at runtime, mostly sending just metadata from one machine to the other. Moving large data sets for processing is very time consuming, incurring a severe penalty on performance.

The *mining server* (*MS*) is responsible for composing the data mining task for processing. To accomplish the task, the MS relies on metadata stored in the DS that describes the kind of datum as well as specific parameters that are expected by each mining algorithm. It also maintains information about the output format for the several algorithms and the input format for the several visualizations available in Anteater. The MS also keeps track of the execution status of the tasks and is responsible for submitting the tasks for execution on the processing servers.

The *processing server* (*PS*), as mentioned earlier, is implemented using Anthill (a more detailed description is provided in Section 11.3). It runs on a distributed environment and is collocated with the DS. The tasks built on the MS are submitted for execution on the PS, which executes the implementations of data mining algorithms. It retrieves the requested data, processes it and produces new data sets as output. Local data for each PS are retrieved from the local instance of the DS. Local results are also stored locally as much as possible. Tasks are executed asynchronously; when they are started, the user receives an identifier to be used in any further communication regarding that task.

The *visualization server* (*VS*) creates the visual metaphors presented to the user. It receives a data set as input (usually the result of a previous mining task) and operates on the data to produce a visual representation. The matching of metaphors to data is partially controlled by the metadata, which will tell how columns of a data set should be matched to elements of the metaphor, for example. Section 11.5 discusses the visualization process.

Users do not interact directly with all servers. This is the responsibility of the *application server* (*AS*), which is the controlling element in Anteater. It controls user interaction and composes the HTML documents that will be returned as the result of any user action. When a user starts to build a data mining task the AS looks for the available data servers, queries them about their data sets and presents this information to the user along with the metadata, if requested. The user selects the data set and may further define a subset of it by highlighting columns or by providing expressions to select records according to some criteria. After this the AS identifies the available mining servers and their algorithms. When the algorithm is chosen and its arguments defined by the user, the task is ready to be processed. It is added to the user's list of tasks and can be scheduled for execution, edited for any updates or selected for visualization when complete. When this happens, the AS informs the visualization server about the metaphor and the data set. The VS builds the visual representation of the data under this metaphor and hands it back to the AS for presentation.

Throughout the process, the AS is responsible for the access control (determining which data sets a user may manipulate, which mining/visualization servers he or she may use, and which metaphors may be used for each kind of data set). This is all controlled by service and metadata definitions, which are kept in a distributed directory along with user credentials. When a user is properly authenticated, the AS applies preset rules that determine which data sets are visible to this user based on his or her credentials.

Components communicate with each other using the SOAP protocol. In this way, new servers may be added easily, as long as they implement the same service interface. By the same reasoning, a user may choose any server to use, as long as the application server allows it. Extending the services is simple: a new data set may be made available by just instantiating a DS for it; compute power may be added by either adding more nodes to existing clusters or by adding a new cluster as an independent mining server; new algorithms and metaphors may be added by just advertising them.

An experiment was performed to illustrate the scalability and extensibility aspects of Anteater. The Anteater system was configured with one server (Pentium 4, 3.0 GHz, 1 GB DDR 400 RAM, Fast Ethernet, running Linux kernel 2.6.7) and multiple clients (same configuration) were used as needed to reach certain loads. This load was artificially generated, based on the characterization of real user sessions monitored in Anteater installations. This configuration with a single server where all services were instantiated had a maximum throughput of approximately 150 000 requests per second with 500 simulated users, when roughly 60 per cent of the requests were served in less than 3 seconds. Another instance of the server was then added in such a way that MS and VS executed on two different machines with same configuration. Performance approximately doubled: throughput saturated a little above 300 000 requests per second with 1100 simulated users, also with 60 per cent of requests answered under 3 seconds.

11.3 Runtime framework

Building data mining applications that may efficiently exploit parallelism while maintaining good performance is a challenge. In this scenario, given their size, data sets are usually distributed across several machines in the system to improve access bandwidth. Usually, for such applications, the resulting data are many times smaller than the input. Moving data to where the processing is about to take place is often inefficient. The alternative is to carry out the computation where the data resides. The success of this approach depends on the application being divided into portions that may be instantiated on different nodes of the system for execution. Each of these portions performs part of the transformation on the data starting from the input data set and proceeding until the resulting data set is produced.

The filter-stream programming model was originally proposed for Active Disks (Acharya, Uysal and Satlz, 1998) exactly to allow applications to be organized in this way. Later, the concept was extended as a programming model suitable for a grid environment, and the Datacutter runtime system was developed to support this model (Beynon *et al.,* 2000*a*, 2000*b*; Spencer *et al., 2002*; Beynon *et al.,* 2002).

In Datacutter, filters are the representation of each stage of the computation, where data are transformed, and streams are abstractions for communication which allow fixed-size untyped data buffers to be transferred from one filter to the next. Creating an application that runs on Datacutter is a process referred to as filter decomposition. In this process, the application is modelled as a data flow computation and then broken into a network of filters, creating task parallelism as in a pipeline. At execution time, multiple (transparent) copies of each of the filters that compose the application are instantiated on several machines of the system and the streams are connected from sources to destinations. The transparent copy mechanism allows every stage of the pipeline to be replicated over many nodes of a parallel machine and the data that go through this stage can be partitioned by them, creating data parallelism.

For several data-intensive applications, a 'natural' filter decomposition yields cyclic graphs (Veloso *et al.*, 2004), where the execution consisted of multiple iterations over the filters. Such applications start with data representing an initial set of possible solutions, and as these pass down the filters new candidate solutions are created. Those, in turn, need to be passed through the network to be processed themselves. Also, this behaviour leads to asynchronous executions, in the sense that several solutions (possibly from different iterations) can be tested simultaneously at runtime.

In an attempt to exploit maximum parallelism in the applications, an extension to the Datacutter was proposed (Ferreira *et al.*, 2005) that allows the use of all three possibilities discussed above: task parallelism, data parallelism and asynchrony. By dividing the computation into multiple pipeline stages, each one replicated multiple times, a very fine-grained parallelism can be achieved and, since all this is happening asynchronously, the execution can be mostly bottleneck free. In order to reduce latency, the grain of the parallelism should be defined by the application designer at runtime.

Three important issues arise from this proposed model that could not be addressed with the Datacutter:

(a) Given the situation where the data are partitioned across multiple transparent copies, sometimes it is necessary for a certain data block to reach one specific copy of a stage of the pipeline.

(b) In order to simplify the communication, these distributed stages often might benefit from some shared state.

(c) Because of the cyclic nature of the application decomposition, it is impossible for each isolated transparent copy to detect whether the overall computation has finished.

The Anthill runtime framework addresses these issues extending Datacutter's filter stream model as described in the following sections.

11.3.1 Labelled stream

In Anthill, each filter (a stage of the application's pipeline) is actually executed on several different nodes on the distributed environment. Each copy is different from the others in the sense that each one handles a distinct and independent portion of the space defined by its internal state. In iterative data intensive applications, it is often necessary that data created during an iteration be sent to a specific copy of a filter to be processed, because of data dependences, for example. The labelled stream abstraction is designed to provide a convenient way for the application to customize the routing of message buffers to specific copies of the receiving filter.

Which copy should handle any particular data buffer may depend on the message itself. In the labelled stream, instead of just sending the message m down the stream, the application now sends a tuple $\langle l, m \rangle$, where l is the label associated with the message. Each labelled stream also has a hash function, which is called, with the tuple as parameter, for each message that must traverse the stream. The output of this hash function determines the particular filter copy to which this message should be delivered.

This mechanism gives the application total control over the routing of its messages. Because the hash function is called at runtime, the actual routing decision is taken individually for each message and can change dynamically as the execution progresses. This feature conveniently allows dynamic reconfiguration, which is particularly useful to balance the load on dynamic and irregular applications. The hash function may also be slightly relaxed, in the sense that its output does not have to be one single copy. Instead, it can output a set of filter copies. In this case, a message can be replicated and sent to multiple instances, or even broadcast to all of them. This is particularly useful for applications in which one single input data element influences several output data elements.

11.3.2 Global persistent storage

One of the problems of replicating pipeline stages over many nodes on a distributed environment is that often these stages have an internal state that must be maintained and that changes as more and more data pass down the pipeline. Replicating the entire state on all instances of the stage is problematic, for often the entire state is larger than the local memory on a single node. Also, it would lead to an extra step for merging the partially completed replicas of the state into one coherent output for that stage. In Anthill, therefore, even when a filter is replicated across many nodes, there is only a single copy of each state variable available in the global memory. In many cases, however, some state information may need to be accessed or even updated by different filter copies. This is particularly true for situations where the applications are dynamically reconfiguring themselves to balance the workload, or during failures, when the system may be automatically recovering from the crash of one copy. Anthill must provide a mechanism that allows the set of transparent copies of any filter to share global state.

One basic aspect of the global state is that information that is part of it must be available in different filter copies as computation progresses, maybe by migrating from one copy to another as needed. Considering the fault tolerance scenario, another interesting feature must be added to this global state: it must be stable, in the sense that it must have transactional properties. Once a change is committed to global state, it has to be guaranteed even in the presence of faults.

Anthill provides a global space that is similar to Linda tuple spaces (Gelernter *et al.*, 1985; Carriero, Gelernter and Leichter, 1986) to maintain the state. Such structure seems very convenient: whenever a filter copy updates a shared data element, it is updated on the tuple space. A copy of it is maintained on the same filter for performance reasons, while other copies may be forwarded to other hosts for safekeeping. The system then allows some degree of fault tolerance, in the sense that once a copy of a data element is safely stored on a different host the update is assumed to be committed to stable storage.

11.3.3 Termination detection

In Anthill, applications are modelled as general directed graphs. As long as such graphs remain acyclic, termination detection for any application is straightforward: whenever data in a stream ends, the filter reading from it is notified in the same way as a process is notified about the end of a file it is reading. When the filter finishes processing any outstanding tasks it may propagate the same end of stream information to its outgoing streams and terminate. When application graphs have cycles, however, the problem is not that simple. In this case the filters in a cycle cannot decide by themselves whether a stream has ended or not. When this happens, Anthill must detect the termination and break the loop by sending an end-of-stream message.

For the sake of the termination protocol, a filter copy at one end of a stream is connected to all copies of the filter at the other end of this stream. The algorithm works in rounds, when some filter copy suspects computation is over. Each copy keeps a round counter `R` that is used by the protocol. A special process, called the process leader, is responsible for collecting information and reaching the final decision about whether or not to terminate the application.

Three types of message are exchanged in the protocol.

- Copies that suspect that termination has been reached send `SUSPECT(R)` to their neighbours (in both directions) stating they suspect termination in round `R`.
- When a copy has received `SUSPECT(R)` messages from all its neighbours, it notifies the process leader using a `TERMINATE(R)` message, also identifying the round number.
- If the leader collects `TERMINATE` messages from *all* filter copies for the same round it broadcasts an `END` message back to them.

Besides the round counter, `R`, each filter copy keeps a list of the neighbours that suspect the same termination round `R` has been reached. The core of the protocol is illustrated by the extended *finite state machine* (*FSM*) in Figure 11.2. Each filter copy may be in one of two states: *running* or *suspecting termination*.

While a filter is computing or has data in its input streams still to be read it remains in the *running* state and does not propagate messages for the termination protocol. If a filter copy has been idle for some time it moves to the *suspecting termination* state and notifies all its neighbours, sending them a `SUSPECT` message with its current round number. It keeps the list of suspected neighbours it collected while in *running* state, since they were considered to be in the same termination round as itself.

In either state, if a copy receives a `SUSPECT(R')` message from another node for its current round `(R'$=$R)` it adds the sender to its list of suspected nodes; if it is for a newer round `(R'$>$R)` it clears this list before adding the sender (the only one known to be in that round so far) and updates `R`. If it receives an application message from another filter, it removes

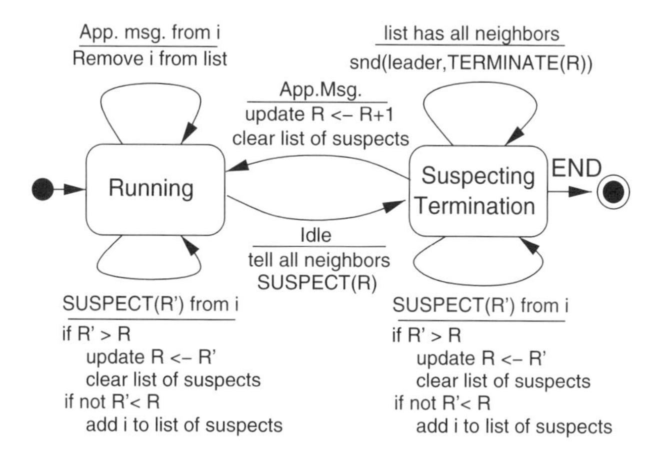

Figure 11.2 Termination algorithm (FSM)

the sender from its list of neighbours suspecting termination, since it is obviously computing. If the receiver was in the *suspecting* state, it goes back to the *computing* state, updates its round counter and clears the list of suspecting nodes.

Whenever a copy in the *suspecting termination* state has collected SUSPECT messages from all its neighbors for a given round, there is a widespread suspicion that termination has been reached (although this may be true for just the vicinity of this copy, and not for all filters in the cycle). At that point the copy sends a TERMINATE(R) message to the process leader. The copy remains in the *suspecting* state, since termination must still be confirmed by the leader.

The process leader, in turn, must keep track of the most recent termination round it has heard of (R). Whenever it receives a TERMINATE(R′) message from a filter copy, it must compare R and R′: if R′ is lower, the message is simply discarded, since it relates to a round that is already known to have passed; if R′ is larger than R, a new round has started and R is not relevant anymore, so the list of terminated copies must be cleared and just the sender of that message must be added to it; finally, if they are equal, another filter copy has joined the group of processes that suspect termination was reached, so it must be added to the list. When this list is complete with all processes the leader can declare that the application has ended. At that point it broadcasts the END message and all processes take steps towards termination. In the filters, this causes the re-initialization of the termination protocol and the closing of the stream selected by the user with an end-of-stream notification.

11.3.4 Application of the model

To illustrate the application of the Anthill programming model, this section describes the parallel implementation of the ID3 decision tree algorithm (Mitchell, 1997). In a decision tree, the leaf nodes are the individual data elements. The internal nodes contain an attribute and each descending pointer encodes a possible value for the attribute mapped on the node that will distinguish the descendants. The depth of such a tree is the maximum number of questions about attribute values that need to be asked in order to find one single element on the data. The basic idea of the ID3 algorithm is to use a top-down and greedy search on the data to find

```
1   p.atr = None
2   p.instance = None
3   p.dataset = T
4   while(∃p ∈ Partitions)
5      Partitions− = p
6      q = {t ∈ p.dataset ∧ p.atr = p.instance}
7      ∀a ∈ Attributes
8         ∀v ∈ Values(a)
9            prob_c = |{t∈q∧t.a=t.v∧t.c=c}| / |{t∈q∧t.a=t.v}|
10           info_{a,v} = Σ_{c∈Classes} − prob_c * log(prob_c)
11     ∀a ∈ Attributes
12        prob_v = |{t∈q∧t.a=t.v}| / |q|
13        gain_a = Σ_{v∈Values(a)} info_{a,v} * prob_v
14     disc = a|gain_a isMaximum
15     ∀v ∈ Values(disc)
16        p.atr = disc
17        p.instance = v
18        p.dataset = q
19        Partitions+ = p
```

Figure 11.3 ID3 algorithm

the most discriminating attribute on each level of the tree. For the sake of the filter definition, three main tasks can be distinguished to insert a node in the decision tree.

(a) For each value of each attribute, count the number of instances that have this value.

(b) Compute the information gain of each attribute.

(c) Find the attribute with the highest information gain.

The starting point of the ID3 algorithm is a set of m tuples, containing instances of n attributes and one out of c possible classes. Each attribute a may assume va values. The tree generation process is based on discriminants. Each discriminant is a test on an specific attribute that is used to divide a set of tuples into two or more subsets (depending on the number of different values that occur for the discriminant). Initially, there is no discriminant and the partition is the whole set of tuples. Then, as new discriminants are found, the algorithm recursively partitions the set of tuples into subsets, until all tuples in one partition belong to the same class.

The pseudo-code of the algorithm is presented in Figure 11.3. Lines 1–3 are the initialization of the *Partitions*, the first one being the entire database. The loop in Line 4 will execute until there are no new partitions. For each of them (Line 5), it selects the tuples that will compose the partition q in Line 6. Then it computes the information (using an entropy metric) for each attribute and instance value in lines 7–10. Lines 11–13 compute the information gain for each attribute and, in the final step, find the attribute that yields maximum information gain and inserts the corresponding partitions into *Partitions*. In terms of parallelization, there are two multi-target reductions in lines 10 and 13, which identify boundaries among filters. The processing is divided into three filters. The first filter, named *Counter*, performs the operations associated with lines 4–9, and is responsible for counting the number of instances that each value of each attribute has. The second filter, *Attribute*, performs the operations associated with lines 10–12, which corresponds to computing the information gain on each of the attributes tested on the previous filter. The third, *Decision*, performs the remaining operations, which corresponds to communicating the decision of the appropriate attribute to back to the first filter, where the process will continue, selecting new discriminating attributes for each of the produced classes.

In the algorithm above all three dimensions of parallelism described earlier are exploited. In the data parallelism dimension, several instances of the *Counter* filter can run, each processing a subset of the entire database. The granularity of the partition can be very fine. On the task parallelism dimension, there is the pipeline of filters. The third dimension, asynchrony, is accomplished as different filter copies are processing multiple nodes of the decision tree simultaneously and there are no idle filter copies.

11.4 Parallel algorithms for data mining

This section describes and evaluates the implementation of three data mining algorithms using Anthill's programming model, focusing on their efficiency and scalability. The experiments were run on cluster with 16 PCs connected using switched Fast Ethernet. Each node was a 3 GHz Pentium IV with 1 GB main memory, running Linux 2.6.

11.4.1 Decision trees

1A decision tree identifies a sequence of attributes that must be inspected in order for any data element to be separately identified. The algorithm and its parallel implementation were thoroughly described in Section 11.3.4.

The experiments for ID3 were executed with the *Decision* filter alone on a separate node. The other nodes ran both *Counter* and *Attribute* filters.

Synthetic data sets, as described by Zaki, Ho and Agrawal (1999), were used to evaluate the performance of the implementation. In particular, two classification functions with different complexities were used, F2 and F7, as identified in this work. Function F2 is simpler and produces smaller decision trees when compared with function F7. Table 11.1 shows the characteristics of the data sets generated for the two functions. The notation Fx-Ay-DzK is used to denote a data set with function x, containing y attributes and $z * 1000$ instances.

Figure 11.4 shows the speed-ups for data sets F2 and F7. The executions of the F2 data set show better speed-ups than those using F7, and both show super-linear behaviour. F2 has better speed-ups because it is simpler and demands less memory, which seems to affect the speed-up significantly. In order to understand the super-linear speed-ups, a detailed analysis of the processor cache usage is performed as the number of processors varied. The number of cache misses is measured using the *performance application programming interface* (*PAPI*) (Mucci, 2001) for each configuration. Figure 11.5 shows that there is a substantial drop in the number

Table 11.1 Data set characteristics

Data set	DB size (MB)	No. of levels	Max leaves/level
F2-A32-D1000K	172	8	5612
F7-A32-D750K	129	8	8195

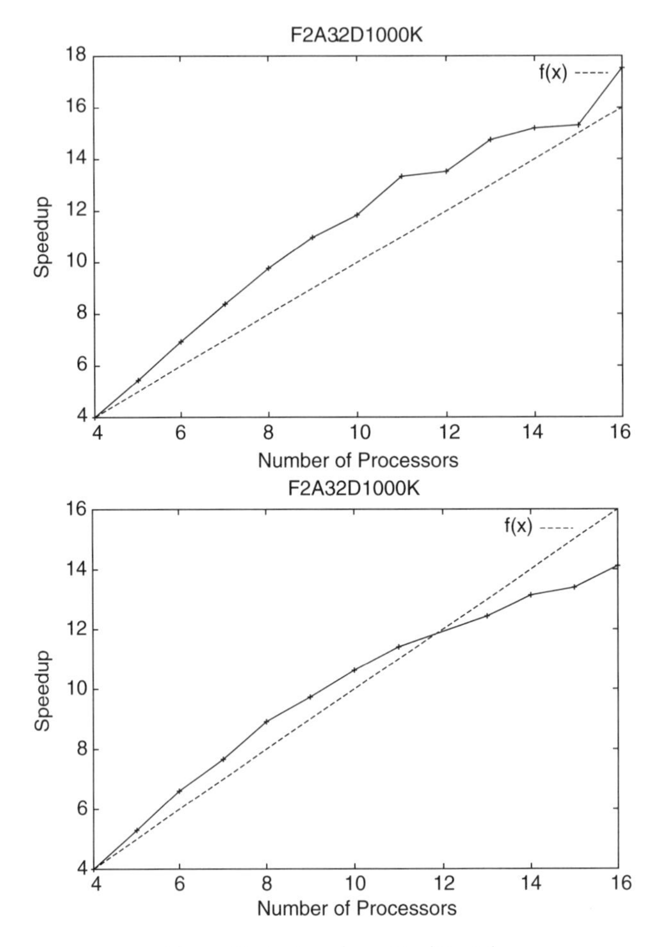

Figure 11.4 Speed-up ID3 (F2, F7)

of cache misses as processors are added, which is expected, and explains the super-linear behaviour. For instance, increasing the number of processors for F2, from four to 14 (a factor of 3.5), resulted in a reduction of the total number of cache misses by a factor of 11.

The next experiments focus the analysis on three criteria that help to demonstrate how the various aspects of Anthill collaborate to produce the observed speed-ups.

Task analysis

Each task in the algorithm is associated with determining the discriminant for a decision tree node. Asynchrony arises by overlapping the processing of several tree nodes, which may belong to the same tree level or not. Notice that all tasks from the same tree level are independent and the parallelism in this case is trivial, and has been exploited in other contexts. The goal is to verify whether tasks from more than one level are executed simultaneously, thus exploiting all the potential parallelism present in the algorithm. To evaluate the level of asynchrony the number of active tasks from each tree level needs to be plotted over time. Figure 11.6 shows

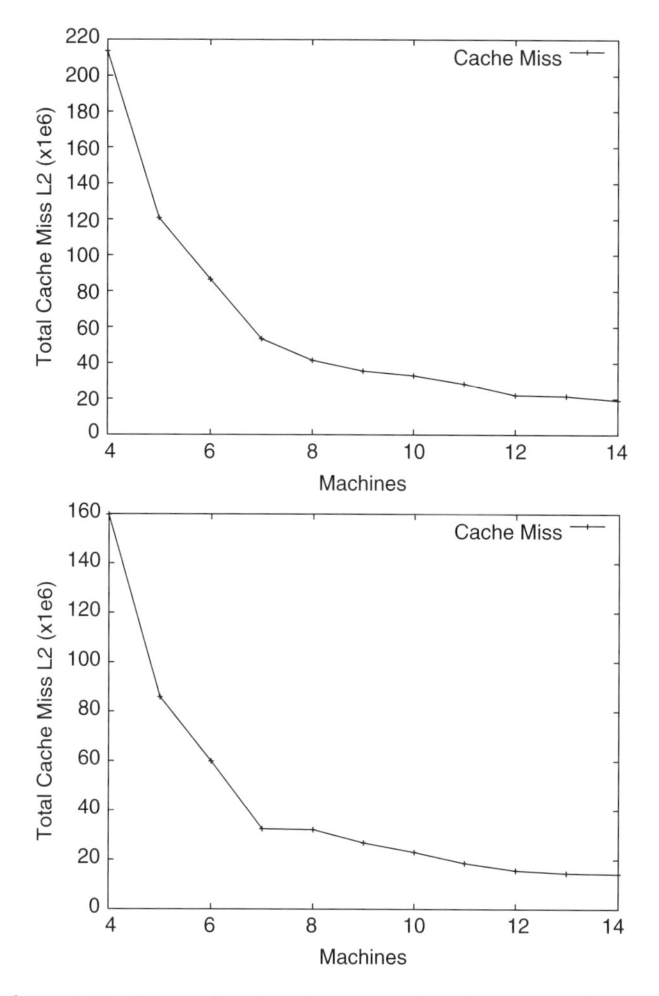

Figure 11.5 ID3: total cache misses × number of processors (F2, F7)

the task behaviour during the execution running on 16 processors using F7 as input. Tasks from more than one tree level are seen as overlapping during the whole experiment (e.g. during the last two seconds), explaining the efficiency of the algorithm.

Filter analysis

Table 11.2 shows the break-down of the task execution time per filter. Executions from nine to 12 nodes are considered, using F7 as the input. The majority of the processing time (over 95%) of the task occurs in the *Counter* filter.

Confirming the higher demand imposed by the *Counter* filter, the message counters for the same configurations are presented in Table 11.3. It shows how the number of data to be processed decreases as the computation goes from the *Counter* to the *Decision* filter. Finally, it is interesting to notice that there is a significant amount of parallelism to explore, since the elapsed time for executing the tasks is usually larger than the elapsed time in each processor

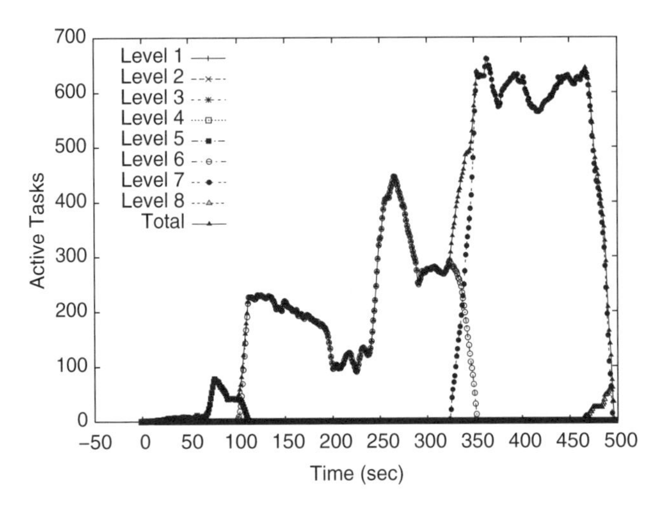

Figure 11.6 Active tasks × execution time

(Figure 11.7), which demonstrates that a larger number of processors would allow even better results.

It is interesting to note that, despite the high variability in the demand imposed by the different filters, the application performs very well. Other task parallelization schemes, such as simple pipelining, may suffer significant performance degradation as a result of such imbalance, but this Anthill parallelization was not affected at all.

Filter instance analysis

The last collection of experiments for this applications evaluates the performance of the filter instances, in order to determine the imbalance that is generated by the data skewness, by the labelled stream or by both. One basic metric for such evaluation is the variability of the execution times of each filter instance, since they vary significantly among tasks, and comparing them does not quantify the load imbalance among instances. To accomplish this, the standard deviation of the execution times for the filter instances is calculated for each task and for each filter.

The resulting values are summarized in Table 11.4, which shows a slight increase of the variability with the number of instances, as expected, since the load assigned to each instance is reduced, and skewness may have a stronger impact. Also, the relative standard deviations are quite high, although there is a compensating effect taking place, that is, the instance that

Table 11.2 Time in each filter (F7-A32-D750K)

Number of processors	Counter	Attribute	Decision
9	96.17	3.36	0.46
10	95.27	4.20	0.52
11	96.13	3.30	0.56
12	96.10	3.36	0.52

Table 11.3 Messages sent (F7-A32-D750K)

Number of processors	Counter	Attribute	Decision
9	36785592	408649	15185
10	40872880	408649	15185
11	44960168	408649	15185
12	49047456	408649	15185

works more for a given task works less later, showing that the labelled stream presents good performance as both a load distribution and a balancing mechanism.

11.4.2 Clustering

Cluster analysis algorithms partition objects in groups that are similar regarding a user-defined similarity criterion (MacQueen, 1967). This section discusses the parallelization of a popular clustering algorithm, k-means.

The algorithm is based on the concept of centroids, which represent the objects that compose each cluster. Each iteration of the algorithm assigns each object to the closest centroid, updating its value properly. The algorithm ends when no object changes cluster or a maximum number of iterations is reached.

Since there is a single task to identify clusters, there is no task graph. The algorithm is implemented using two reduction filters: assigner and centroid calculator. The assigner holds the objects that must be clustered based on the centroids and determines which centroid is the closest to each object. The list of objects assigned to each cluster is then sent to the centroid calculator, which recalculates the centroids.

The evaluation of the clustering algorithm is based on two synthetic data sets, containing 400 000 and 800 000 points to be clustered. Each point has 50 dimensions. As before, the experiments evaluate both scale-up and speed-up. The scale-up is linear and close to unity, that is, the application scales perfectly in the intervals considered. The execution times and

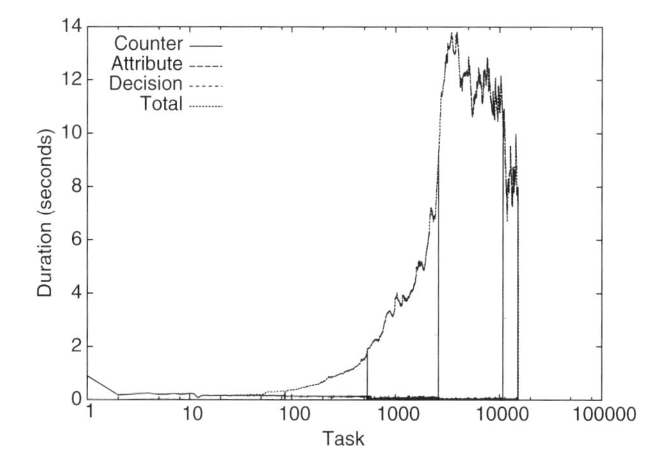

Figure 11.7 Elapsed time per filter and overall

Table 11.4 Standard deviation of execution times of filter instances

No. of processors	Counter filters	Attribute filters
9	40.82%	28.45%
10	45.10%	28.59%
11	46.59%	29.65%
12	50.39%	29.79%

respective speed-ups are plotted in Figure 11.8, which shows the time per iteration of the algorithm for the two data sets and the speed-up when varying the number of processors. The speed-up is almost linear for the 400 000-point data set and is super-linear for the 800 000-point data set. This super-linear behaviour comes from the reduction of memory requirements as the number of processors increases.

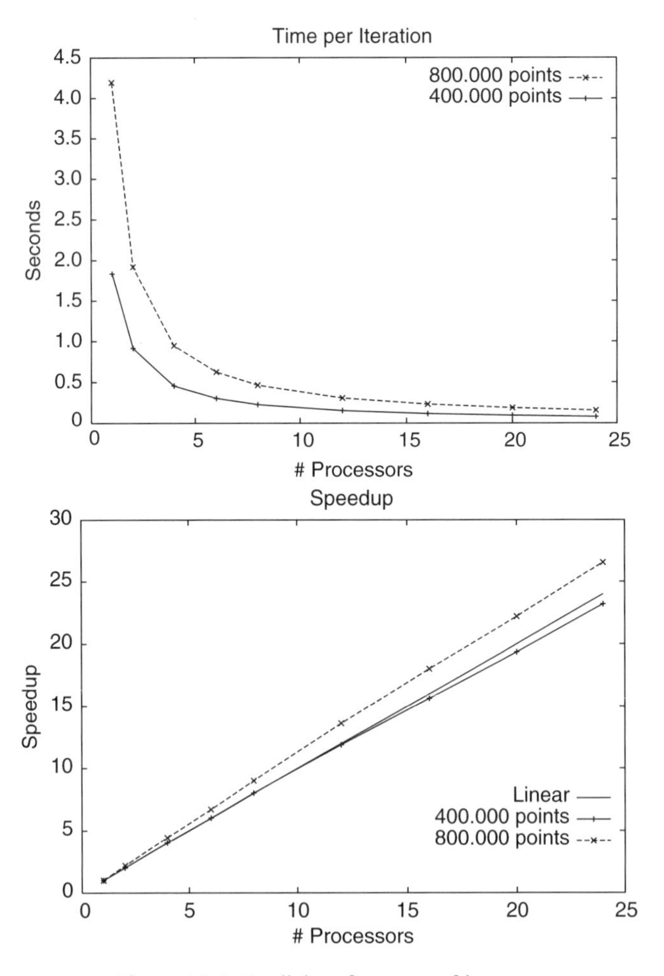

Figure 11.8 Parallel performance of k-means

11.5 Visual metaphors

The goal with Anteater was to create a data mining system where users do not have to know details about the algorithms and their related concepts. To achieve this, the interface should be streamlined, dealing with most of the technical details transparently, and results should be shown in a way the user could easily understand with little training.

The technical details related to individual algorithms and metaphors are handled by the application server, as previously mentioned, based on a set of rules defined for each data set, algorithm and visual metaphor. The user is only called to participate in the process by defining values for the arguments that are dependent on his or her goals.

The presentation of results through the appropriate metaphors is another important element of Anteater. Each class of algorithms must be assigned adequate metaphors. The visual interface for association rules is used in this section as an example. An association rule mining task finds rules that satisfy some criteria defined by the user, often the minimum support and confidence. The number of rules produced by the algorithm that meet these constraints may be very large (up to hundreds of thousands, even). The user must then use filters and other metrics of interest to find the most interesting rules in the set.

As observed by Fayyad, Grinstein and Wierse (2002), a visualization system for mining association rules should (i) help users to find the most interesting rules in a large set and (ii) help them understand these rules. The first requirement deals with presenting rules in a way that makes the interesting ones stand out, and it has already been addressed in the literature. The second is related to the complexity of rules themselves, and has not been given much attention so far (Shneiderman, 2002).

The main visual metaphor in the Anteater for association rules is shown in Figure 11.9. The display is clearly composed of four different elements, which are discussed here. To address the problem of identification of interesting rules, the metaphor shows rules as points on a matrix (centre of the display). The perpendicular axis represent two metrics of interest (for example, support and confidence, or lift and leverage). The users may dynamically select which metrics to use for each axis, varying them as they see fit in each case, by selecting them in the two panes on the bottom left area of the display. They may also apply filters to the result of the data mining task to highlight particular subsets of rules, for example, indicating that only rules that have a given item in the consequent should be displayed (top left corner).

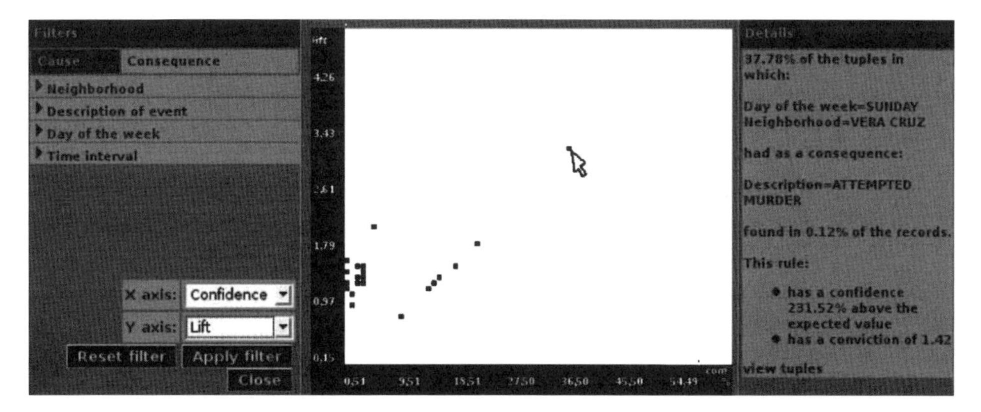

Figure 11.9 The visualization of association rules

The use of the interface can be illustrated through the analysis of an interesting outcome of a case study. As mentioned, Anteater is being applied to 911 calls of a major Brazilian city, Belo Horizonte. One of the mining tasks performed by users tried to understand the reasons behind attempted murders. In particular, they looked for relationships among neighbourhoods, type of crime, day of the week and time of the day. These attributes are shown in the left-hand part of the display, where it is also possible to filter the output by attribute instances. In the display shown, these filters were used to select association rules containing homicides.

The rule shown on the right states that among the records in the database '22.29% of the crimes that happened on Sundays from 10:30 a.m. to 4:30 p.m. were attempted murders'. Another rule not shown says that '37.78% of the crimes that happened on Sundays in the Vera Cruz area (a slum) are related to attempted murders'. Both rules have a quite low popularity (0.54 and 0.12 per cent, respectively) being hidden in a first analysis. However, they become clearly visible when the interface is configured to show lifts (theirs are 95.57 and 231.52 per cent above the expectation, respectively). If the rule is not enough, users can ask for a sample of the records that led to the rule, or the complete set of related records. After checking the records associated with these rules, users found that most of these crimes were in the neighbourhood of pubs, and, more interesting, happened on days when there were major soccer games. Although this finding may seem intuitive, it is remarkable how fast and precisely users got to this evidence. Experience shows that users can handle this information after just a basic training session.

A previous version of Anteater used an antecedent/consequent matrix, where lines represented antecedents and columns represented consequents in the rules, and markers in the matrix varied their colour and size based on the values of the metrics of interest for each rule. Such a metaphor required users to absorb a large amount of information about the algorithm before they could start to interact with the system, and faced strong user resistance.

User behaviour in Anteater installations has been monitored and the current metaphor has been well accepted even by novice users. It provides enough information that the user can understand the workings of the algorithm as the sessions progress. In a coming version of the system it will be possible for the user to annotate the screen as the session progresses, highlighting rules so they can be easily found again in the matrix after the display is rearranged using a different pair of metrics, for example. This should improve even more the way users handle rules during an interactive session.

11.6 Case studies

It is interesting to note the diversity of databases that were mined within Anteater, as well as users' backgrounds. In particular, case studies show, among others, police officers, social scientists, physicians and public accountants as users. This section presents some findings from the usage of Anteater in public health and government expenses.

The public health study, as mentioned, focused on haemodialysis. Health insurance in Brazil is public and free for all. If we consider just haemodialysis, it costs about 750 million dollars per year, for approximately 50 thousand patients. The goal of the study is to qualify and quantify the efficiency of the treatment as well as the medical histories that lead to haemodialysis. One intriguing statistics is that about 40 per cent of the haemodialysis treatment forms have no associated diagnostics, which does not make much sense, given the cost, complexity and impact on the life of the patient. Thus, given the haemodialysis records associated with undefined diagnostics, users used Anteater to mine the similarities among them. In this case, two significant patterns emerged: there were some providers and some cities where such lack of information was common. In the case of specific providers, they were audited. In the case of the cities, it was verified that they had local providers but did not have enough medical experts to evaluate the patients, resulting in general practitioners being in charge of the task, who had not been aware of the importance of providing the proper information.

Regarding expenditures, government users look for fraudulent patterns in expense records. The focus is usually on purchases where product costs were above the average market price. In this case, they looked for the reasons behind expensive prices. As a result, they were able to nail down, with very good precision, two common behaviours. The first pattern is associated with biased relationships between government buyers and providers, that is, buyers always buy from the same providers. The second case detected provider cartels, where high prices were associated with the same group of sellers, regardless of the location of the buyer. In both cases, the findings were immediately sent to the auditors.

On the other hand, there are some cases where the system was not effectively used. There are some reasons for this. The first reason is the quality and (or) the level of detail of the data. For instance, it was not possible to evaluate the state government expenditure data because they used a very abstract taxonomy to classify products. Two expense records, although assigned to the same class, might be associated with very different products. The second reason was the lack of demand for data mining techniques; that is, the use of information technology in some organizations was quite sparse and there were neither personnel nor a well defined demand for data mining solutions. This situation has been improving though; as successful case studies are reported, some groups of users often reconsider their decisions.

11.7 Future developments

Two aspects can be highlighted as far as future developments are concerned. On one hand, there are the aspects related to the novel data sets being created and the tools for processing and analysing them. On the other hand, grid infrastructure is becoming a reality and exploiting such frameworks for running data analysis algorithms is both convenient, given the unprecedented computing powers it yields, as well as a new challenge for the developers.

In terms of data processing, it is said that data are not information, information is not knowledge and knowledge is not wisdom. Unprecedented numbers of data are being currently collected and stored due to advances in our technology. We live in the era of information. The effort of converting data into information, and then into knowledge, lies within the realm of data mining. With these fast growing data sets, both in size and in complexity, more sophisticated analyses are becoming an urgent necessity. Novel algorithms that are capable of handling these bigger and more sophisticated data sets are in high demand as well as novel algorithms capable of extracting more complex relationships from the data: for example, efforts towards analysing time series events, mining graphs or other complex relationships of data, and handling more complex data, such as multimedia files.

In terms of the emerging computational infrastructure provided by the grid initiatives, several aspects need to be addressed. Achieving high performance in such systems has been demonstrated by means of expert system programming skills as well as careful implementation of the algorithms. Means of facilitating the development of applications in this environment is crucial, as it would enable the application experts to exploit the benefits of the platform while concentrating on the applications. Other aspects are related to access to remote data, involving issues ranging from security to efficient query and retrieval of data.

11.8 Conclusions and future work

In this chapter we have presented the Anteater data mining service architecture, implemented as a service-oriented architecture. The use of Web services makes it easily extensible, since new resources (data sets, servers and algorithms) may be added by just advertising them to the application servers.

The run-time support for the system is Anthill, a framework that was developed to support the efficient implementation of a significant class of applications on heterogeneous distributed environments. The Anthill programming model has also been described, which was the approach used to implement the data mining algorithms in the context of Anteater. This programming model, however, is general, and can be applied to a large class of applications.

The experimental results show that three applications designed using the strategy and executed with Anthill runtime system scale almost linearly to a large number of nodes. Although the results were limited to the number of compute nodes we had available for experimentation, there is no particular reason for the applications to change their behaviour for larger configurations.

Another important aspect of the project is the development of visual metaphors that help users from other areas of expertise to understand the output of data mining algorithms.

The architecture has been implemented and is operational, being used by different branches of the Brazilian government to mine data from such areas as public servant payroll, public expenditure records, public safety (911 records), public health (national haemodialysis programs) and others.

Besides expanding the installed base, efforts currently under way include the development of a distributed authentication and access control mechanism, a parallelizing compiler for Anthill, of schedulers for tasks among and within mining servers and of a semantically rich solution for high-performance I/O and storage.

A demo version of the service is available on the project's Web site.[1] Anteater is free software; the sources and a live CD are also available through the Web site.

References

Acharya, A., Uysal, M. and Satlz, J. (1998), Active Disks: programming model, algorithms and evaluation, *in* 'ASPLOS VIII: Proceedings of the International Conference on Architectural Support for programming Languages and Operating Systems', ACM Press, pp. 81–91.

Beynon, M., Chang, C., Çatalyürek, U., Kurç, T., Sussman, A., Andrade, H., Ferreira, R. and Saltz, J. (2002), 'Processing large-scale multidimensional data in parallel and distributed environments', *Parallel Computing* **28**(5), 827–859.

Beynon, M., Ferreira, R., Kurc, T. M., Sussman, A. and Saltz, J. H. (2000*a*), Datacutter: middleware for filtering very large scientific datasets on archival storage systems, *in* 'IEEE Symposium on Mass Storage Systems', pp. 119–134.

[1] http://tamandua.speed.dcc.ufmg.br

Beynon, M., Kurc, T., Sussman, A. and Saltz, J. (2000*b*), Design of a framework for data-intensive wide-area applications, *in* 'HCW: Proceedings of the Heterogeneous Computing Workshop', IEEE Computer Society Press, pp. 116–130.

Brezany, P., Hofer, J., Tjoa, A. M. and Woehrer, A. (2003), GridMiner: an infrastructure for data mining on computational grids, *in* 'Conference on Advanced Computing, Grid Applications and eResearch', APAC.

Cannataro, M. and Talia, D. (2003), 'The Knowledge Grid', *Communications of the ACM* **46**(1), 89–93.

Carriero, N., Gelernter, D. and Leichter, J. (1986), Distributed data structures in Linda, *in* 'Conference Record of the 13th Annual ACM Symposium on Principles of Programming Languages', pp. 236–242.

Fayyad, U., Grinstein, G. G. and Wierse, A., eds (2002), *Information Visualization in Data Mining and Knowledge Discovery*, Morgan Kaufmann.

Ferreira, R., Meira, W. Jr., Guedes, D., Drummond, L., Coutinho, B., Teodoro, G., Tavares, T., Araujo, R. and Ferreira, G. (2005), Anthill: a scalable run-time environment for data mining applications, *in* 'SBAC-PAD 2005: Proceedings of the 17th Brazilian Symposium on Computer Architecture and High-Performance Computing', SBC.

Gelernter, D., Carriero, N., Chandran, C. and Chang, S. (1985), Parallel programming in Linda, *in* 'Proceedings of the 1985 International Conference on Parallel Processing (ICPP'85)', IEEE, University Park, PA, pp. 255–263.

Krishnaswamy, S., Zaslavsky, A. B. and Loke, S. W. (2001), 'Towards data mining services on the Internet with a multiple service provider model: an XML-based approach', *Journal of Electronic Commerce* **2**(3), 103–130.

MacQueen, J. B. (1967), Some methods for classification and analysis of multivariate observations, *in* '5th Berkeley Symposium on Mathematical Statistics and Probability', University of California Press, Berkeley, pp. 281–297.

McElroy, M. W. (2000), 'The new knowledge management', *Journal of the KMCI* **1**(1), 43–67.

Mitchell, T. M., ed. (1997), *Machine Learning*, McGraw-Hill.

Mucci, P. (2001), 'The performance API PAPI', White Paper of the University of Tennessee.

Perrey, R. and Lycett, M. (2003), Service-oriented architecture, *in* 'SAINT-W'03: Proceedings of the 2003 Symposium on Applications and the Internet Workshops', IEEE Computer Society, Washington, DC, p. 116.

Shneiderman, B. (2002), 'Inventing discovery tools: combining information visualization with data mining', *Information Visualization* **1**, 5–12.

Spencer, M., Ferreira, R., Beynon, M., Kurc, T., Catalyurek, U., Sussman, A. and Saltz, J. (2002), Executing multiple pipelined data analysis operations in the grid, *in* 'Proceedings of the 2002 ACM/IEEE conference on Supercomputing', IEEE Computer Society Press, pp. 1–18.

Veloso, A., Meira, W. Jr., Ferreira, R., Neto, D. G. and Parthasarathy, S. (2004), Asynchronous and anticipatory filter-stream based parallel algorithm for frequent itemset mining., *in* 'PKDD 2004: Proceedings of the 8th European Conference on Principles and Practice of Knowledge Discovery in Databases', Vol. 3202 of *Lecture Notes in Computer Science*, Springer, pp. 422–433.

Zaki, M. J., Ho, C.-T. and Agrawal, R. (1999), Parallel classification for data mining on shared-memory multiprocessors, *in* 'ICDE '99: Proceedings of the 15th International Conference on Data Engineering', IEEE Computer Society, Washington, DC, p. 198.

12

DMGA: A generic brokering-based Data Mining Grid Architecture

Alberto Sánchez, María S. Pérez, Pierre Gueant, José M. Peña and **Pilar Herrero**

ABSTRACT

The concept of a *data mining grid* has become one of the most challenging topics in both the grid and data mining areas. Indeed, grid environments seem to be an answer to the great demand of computational power and data facilities required by current data mining applications.

The *Data Mining Grid Architecture* (*DMGA*) is a generic brokering-based architecture for deploying data mining services in a grid. This approach presents two different composition models: (i) horizontal composition, which offers workflow capabilities, and (ii) vertical composition, for increasing the performance of inherently parallel data mining services. This scheme is specially significant to those services accessing a large volume of data, which can be distributed through diverse locations.

This chapter describes DMGA and addresses both kinds of composition. Additionally, two use cases are shown with the aim of demonstrating the advantages of combining service functionalities and processing distributed data in data-intensive applications.

12.1 Introduction

Nowadays the amount of information required by many business and scientific applications is unmanageable. Data mining has become a critical task for understanding these large volumes of data. This paradigm has evolved in several directions, trying to tackle limitations of original data mining systems. As a paradigmatic example, the distributed data mining field aims at mining distributed data by means of the use of distributed resources.

New data-intensive applications have arisen with the aim of facing ambitious challenges, such as climate modeling (Natural Environment Research Council, 2006) or earthquake prediction (Guo *et al.*, 2005). The complexity properties of data from these applications do not fit with centralized and monolithic data mining systems, mainly due to the following reasons.

Data Mining Techniques in Grid Computing Environments Edited by Werner Dubitzky

(a) The data mining process can be inefficient, due to the execution of complex data mining tasks.

(b) There are distributed data that cannot be integrated into a single database due to technical or privacy restrictions.

Distributed data mining systems offer a framework that can speed up the execution of data mining tasks and enable data distribution. However, this kind of system does not meet the demands of data-intensive applications. The most significant drawbacks are the following.

- Current scenarios involve the use of heterogeneous resources. This includes storage systems, computing systems, data access mechanisms and policies. Heterogeneity affects not only the infrastructure, but also the data themselves. Different kinds of data format and data from heterogeneous sources make efficient data management more difficult.
- In some scenarios, data mining systems must be able to tackle huge volumes of data, from terabytes to petabytes, e.g. weather forecast, climate change prediction and fluid dynamics.
- It is required to access multiple databases and data repositories, in general, because no single database is able to hold all data required by an application.
- In a generic scenario, multiple databases do not belong to the same institution and they are not situated at the same location, but geographically distributed.
- It is possible to use local copies of the whole data set or subsets to increase the performance of some steps of the data mining process.
- Business databases or data sets may be updated frequently, which implies replication and coherency problems.

Distributed data mining systems do not address all these requirements through a general and complete solution. Therefore, it is necessary to deploy data mining processes in adequate infrastructures, which could belong to several organizations. Grid computing can help to overcome all these limitations, since it provides an integration framework of heterogeneous and geographically distributed computing systems and data resources, possibly owned and managed by diverse organizations (Foster *et al.*, 2002).

Several initiatives have emerged that revolve around the concept of a *data mining grid*. A data mining grid could be defined as a grid computing environment whose main purpose lies in the analysis of data by means of technology from data mining and related fields. DMGA (Sánchez *et al.*, 2004; Pérez *et al.*, 2007) is one of these proposals. It consists of a flexible architecture for providing data mining services in grid environments. Two kind of service composition can be performed by using DMGA, horizontal and vertical.

12.2 DMGA overview

As we have mentioned in the previous section, a grid can be a suitable computing infrastructure in which data mining applications are deployed and executed. However, in order to make this 'marriage' between grid and data mining successful, some problems and requirements need to be addressed, namely the following.

(a) Some of the data mining applications require access to large volumes of data.

(b) Data may reside in geographically distributed locations.

(c) Data may belong to several organizations.

(d) Every organization may apply its own security policies.

DMGA tries to address these issues by means of the definition of both a set of grid services for data mining and usage patterns for their composition. This architecture is compliant with the current vision of grids, which is based on the *Open Grid Services Architecture* (*OGSA*). OGSA provides support for the creation, maintenance and lifecycle of services offered by different *virtual organizations* (*VOs*) (Foster *et al.*, 2002).

DMGA is a generic architecture, which defines services reflecting the main stage in a data mining process: pre-processing, data mining and post-processing. Figure 12.1 shows an overview of DMGA. DMGA extends OGSA by the definition of new data mining services. Such services use basic grid services, of which data-related services are the most relevant for DMGA.

The three services at the top, which are designed to support data mining applications, are a central element of the DMGA architecture and are usually linked to data mining techniques and algorithms. For example, the *AprioriG* service provides the Apriori algorithm functionality in a grid environment.

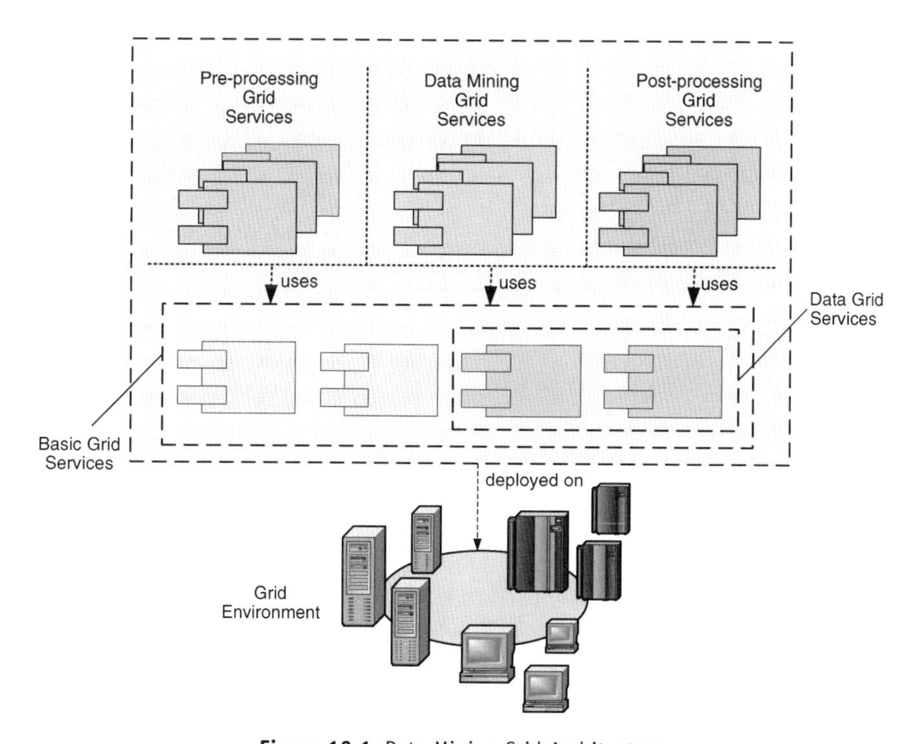

Figure 12.1 Data Mining Grid Architecture

DMGA not only provides new specialized data mining services, but also offers a framework in which the composition of several grid services helps to deploy an enhanced solution to a data mining problem. For instance, if the data mining scenario requires the transfer of files in the pre-processing stage, DMGA enables the use of a GridFTP-based service, which will be combined with other services with the aim of solving the complete data mining process.

Therefore, DMGA must provide the capability of composition of services, that is, the ability to create workflows, which allows several services to be scheduled in a flexible and efficient manner. The service composition can be performed in two ways.

(a) *Horizontal composition.* Different functional services are combined. The functions of these services are different. This composition depends on the decomposition of the overall data mining process into stages and is characterized by the DMGA. This kind of composition is dealt with in Section 12.3.

(b) *Vertical composition.* Several outputs of the same replicated services are combined. Such services have the same functionality, but access a different data portion. Data can be geographically distributed. Each service is tied to a specific data division. This composition is based on how the data mining algorithm can be executed in parallel. The main goal of the vertical composition is enhancing the performance of the data mining process. The vertical composition of services is addressed in Section 12.4.

An implementation of the DMGA architecture, named WekaG (Pérez *et al.*, 2005), has been developed. This implementation is based on Weka (Witten and Frank, 2002), a well known tool for developing machine learning algorithms. Weka contains tools for all data mining phases: data pre-processing, data mining tasks (e.g. classification, regression, clustering, association rules) and data post-processing.

Although WekaG constitutes a useful tool for data mining, it is possible to extend this functionality for several libraries and new algorithms. Therefore, a more complete implementation of the DMGA can be composed of the deployment of several data mining libraries by means of grid services. Figure 12.2 shows the generic deployment scenario of the DMGA. The set of DMGA data mining services uses Weka or other libraries for providing the core data mining functionality. As underlying infrastructure, a *Web Services Resource Framework* (*WSRF*) (Graham *et al.*, 2006) implementation is required. The Globus Toolkit (Foster and Kesselman, 1997) has been chosen for WekaG because it is the most comprehensive and known grid middleware.

12.3 Horizontal composition

A complete data mining task can involve the collaboration of different functional services. In general, pre-processing, data mining and post-processing functions must be performed for completing such a task. Typically, the output of one service corresponds to the input of the subsequent service, and so on (see Figure 12.3).

For composing two services, it is possible that an adapter may be required. This adapter is responsible for performing the required adjustments so that the output of one service can be used as input to the other.

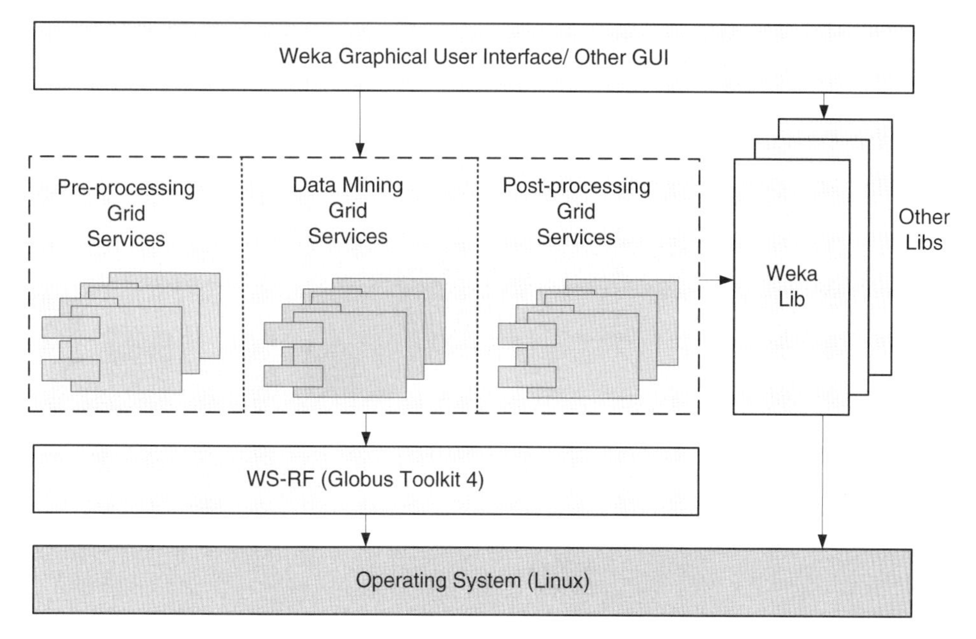

Figure 12.2 Generic DMGA deployment scenario

One common example of horizontal composition is the combination of a data mining algorithm service and a data transfer service, to move data to a specific location and after that to run a specific algorithm on it. This scenario is shown in Figure 12.4. Such a scenario is divided into a client-side and a server-side. The client-side includes the WekaG user interface, which is exactly the same as the Weka user interface but with access to additional functions. On the other hand, the server-side contains the grid services and the capabilities for composing them. Some adapters may be required.

As discussed in Section 12.5, some of the capabilities exposed to the grid by the server side can be forwarded to a centralized component, like a broker.

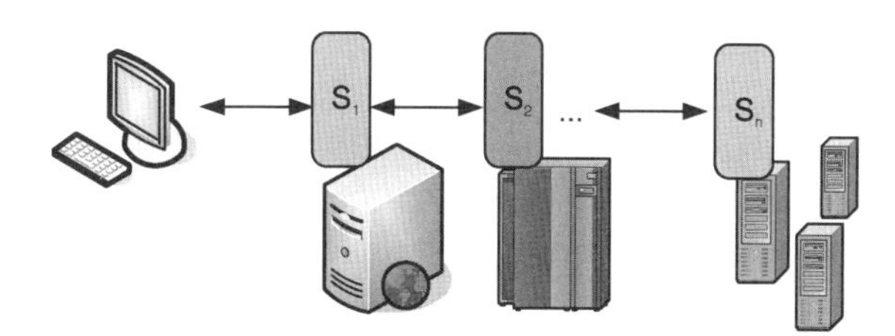

Figure 12.3 Horizontal composition

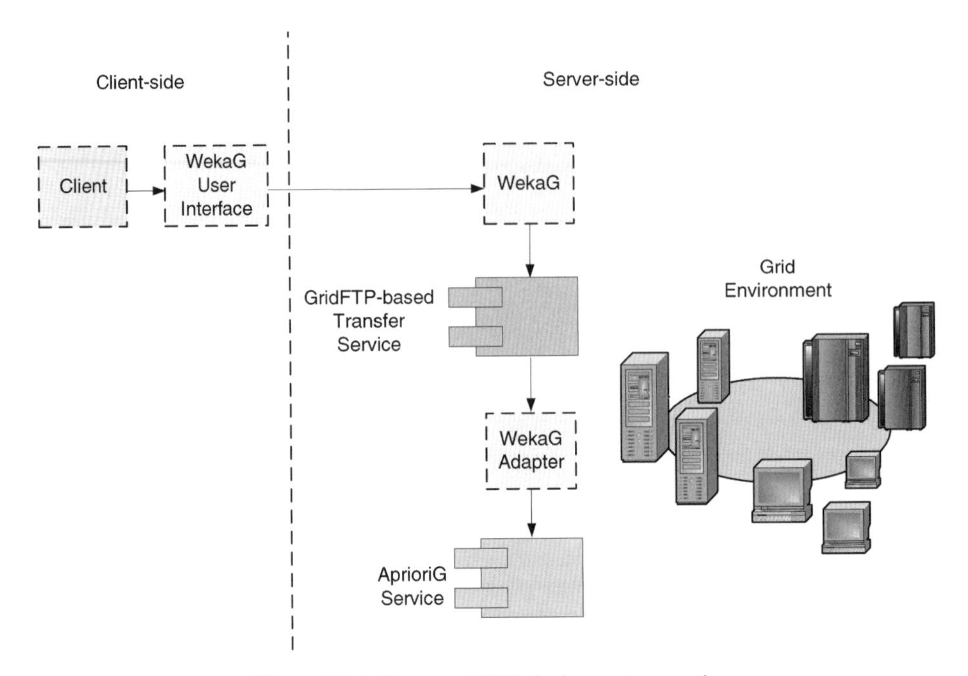

Figure 12.4 Common DMGA deployment scenario

12.4 Vertical composition

The vertical composition refers to the combined use of several services that provide the same functionality. As current data mining problems involve the analysis of large numbers of data, data can be geographically distributed, allowing each service that forms a vertical composition to access a different data portion. Therefore, as can be seen in Figure 12.5, each service instance accesses different views of the same data set depending on data division. Then, results from every resource are combined to compute the answer, which is then transferred to the client.

In this sense, the data division and distribution have influence on this kind of composition and therefore on its performance. Since the goal of the vertical composition is to enhance the performance of the data mining process, different alternatives of data division exist.

- *Fair data division.* To simplify computation tasks, the data may be partitioned into homogeneous segments. In this case, each service accesses a similar portion of data without previous classification.
- *Data division based on non-functional properties.* In this case, the non-functional properties of grid resources, such as computational power, are taken into account to partition and distribute data. Thus, it is possible to assign data blocks of changeable size to each grid element to optimize the joint data mining process. In a heterogeneous environment, if data blocks of the same size are transferred to the resources, optimal overall performance may not be obtained, since the slowest resource is usually the limiting factor. To avoid this, the

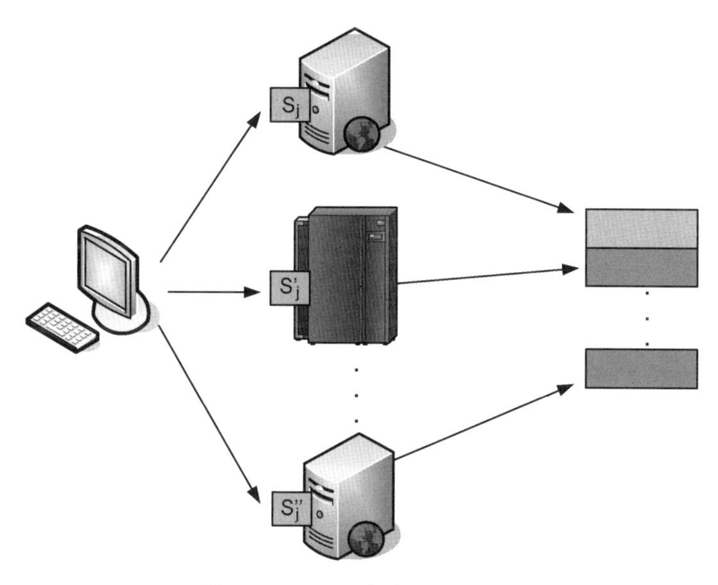

Figure 12.5 Vertical composition

data size assignment for each resource must be calculated taking into account the processing features of each resource. That is, a relation among the data size each resource can process during the same time must be calculated. In this way, no elements have to wait for the rest of element, providing an efficient process. Figure 12.6 shows an example of this kind of division.

By using vertical composition it is possible to improve the performance of a data mining process in a parallel way. However, since a grid can be used by several users, it is required to have a global knowledge of the whole grid behaviour to take advantage of the grid. Thus,

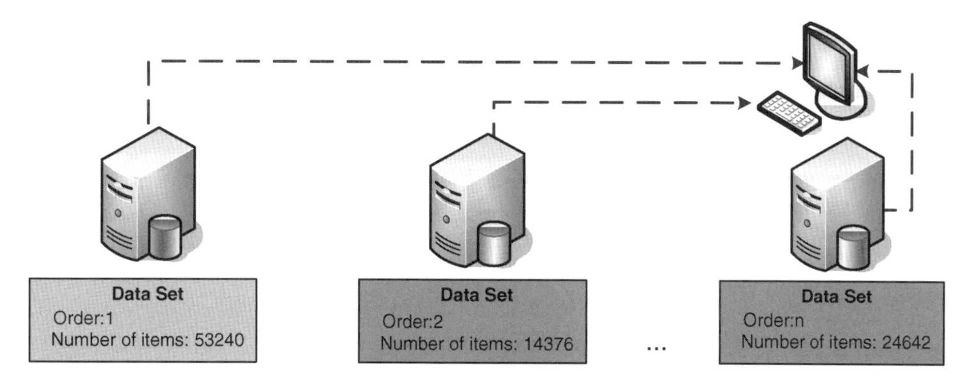

Figure 12.6 Variable data size to carry out a vertical composition-based data mining process

whether resources should be allocated in a static or client-based fashion is often difficult to determine in the absence of information on resources, their utilization and the requests of other users. Therefore, a centralized element that has a complete and global view of the resources and clients of the grid is required. The next section describes this approach.

12.5 The need for brokering

Grid computing does not follow a centralized approach. Grids are organized in VOs, each having a local administration and security policy. Each VO controls the access to its own resources. As a consequence, a grid is a highly dynamic environment. Nevertheless, a grid service must be able to access resources distributed across several VOs. Moreover, a grid service must be aware of the availability of these resources. To maintain a global state of the resources needed by a grid service, a centralized element called a broker is needed.

The broker is vital for any grid infrastructure because its operation and performance determine to what extent the user requirements are met and how efficiently the underlying resources are utilized. A broker consists of three basic components, linked to key functionalities present in most of the brokers (Venugopal, Buyya and Winton, 2004).

(a) *A resource discovery system.* This system collects information about the grid nodes. It must be aware of available services, such as a particular software library, or a statistical computation grid service. It must also be aware of hardware details, the CPU architecture, the amount of installed memory or the presence of a specific device. It must finally be aware of the current state of a node, providing parameters such as available disk space, CPU utilization, external bandwidth or free memory.

(b) *A matchmaking system.* Based on the information provided by the resource discovery system, the matchmaking system controls the selection of candidate nodes, matching all the client's requirements.

(c) *A decision-making system.* Following a pre-established logic or algorithm, the system chooses the best nodes from the previously selected candidates.

For example, a client asks the broker for five nodes with at least 500 MB disk space and the MPI library installed. The matchmaking system queries the discovered resource data and filters all the nodes with at least 500 MB available disk space and MPI installed. Finally, the decision-making system chooses the most suitable nodes from this list according to the defined policies. For instance, the lowest CPU-loaded nodes could be selected if the pre-established algorithm tries to optimize the workload. Depending on the application, it can be more suitable to establish this algorithm in the broker, or to let the client define it.

The brokering concept can be applied to the DMGA in two ways. One lets the client choose the decision-making algorithm. Here, the broker need not be aware of the specifics of the data mining algorithms. A generic brokering system could be used, but this solution has mainly a drawback. Clients must provide all the logic, making them harder to design.

A second way includes the decision-making logic into the broker itself. The next section presents an optimized broker following this idea.

12.6 Brokering-based data mining grid architecture

A DMGA-enabled broker must take into account both the vertical and horizontal composition. As a consequence, the DMGA-enabled broker has a matchmaking and decision-making system involving two dimensions: vertical and horizontal. The horizontal dimension selects the appropriate service in each data mining stage: pre-processing, data mining and post-processing. The vertical dimension depends on the specific service previously selected by the horizontal stage. For instance, in the pre-processing stage, a typical operation is data filtering. The horizontal dimension discovers and chooses the required data sets across several virtual organizations, according to their consistency or replication level. Then, the vertical dimension parallelizes the data pre-processing, separating data into subsets and distributing pre-processing depending on data proximity, available bandwidth or CPU availability criteria.

This design, shown in Figure 12.7, allows more client transparency concerning decision-making algorithms. It also offers optimizations such as query result caching. Finally, it allows the design of simple lightweight data mining clients, taking advantage of the grid, but hiding its complexity. This kind of brokering provides simplicity and transparency to the client, as most of the operations are performed by the broker instead of the client. Thus, the inherent characteristics of grid environments, such as heterogeneity, distributed resources, scalability and the involvement of multiple administrative domains, are solved by the broker in a way that is transparent to the client.

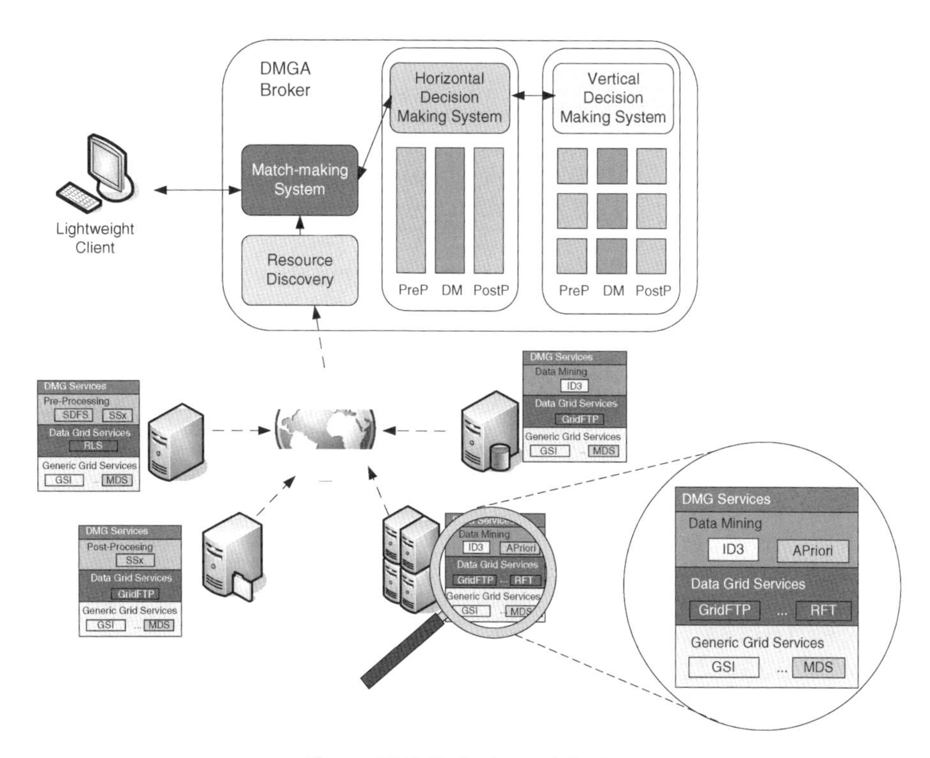

Figure 12.7 Brokering architecture

12.7 Use cases: Apriori, ID3 and J4.8 algorithms

DMGA offers a framework that facilitates the deployment of data mining grid services in a flexible and easy-to-use way. Depending on the application, the horizontal composition, the vertical composition or a combination of both can be used. This section shows two use cases, adopting the horizontal and vertical composition approach, respectively.

12.7.1 Horizontal composition use case: Apriori

In order to evaluate the horizontal composition, the Apriori technique is deployed on a grid environment. Once the client has contacted the broker and the broker has chosen the suitable services installed on some grid server, the functionality of the application is divided between the client and the server side. The client part includes the user interface and the server is responsible for running the Apriori algorithm.

The Apriori algorithm (Agrawal, Imielinski and Swami, 1993) is a traditional sequential algorithm whose main aim is to extract association rules from data (Han, Karypis and Kumar, 1997). In association mining, the most performance-critical step is candidate generation. In large databases, candidate item sets do not fit in memory and in some cases would require more storage size than the actual database. Therefore, a single server that can satisfy such extreme memory requirements may be hard to come by.

The Apriori algorithm can be executed in DMGA through the horizontal composition of two services: AprioriG and a GridFTP-based transfer service (Allcock *et al.*, 2001). The first service is intended to build the association rules and the second service serializes object transfer between the nodes involved. GridFTP has been chosen because it is integrated within Globus Toolkit, the de facto standard for grid computing, and supports parallel data transfer, which enhances the performance. In this implementation we have used the object serialization in order to store and retrieve the state of the required objects. The serialization allows Weka to support marshalling and unmarshalling, which enables accessing remote objects. Like the Apriori class, most of the Weka classes are serializable.

WekaG includes the AprioriG service, which implements the capabilities of the Apriori algorithm in a grid environment. The main function of the AprioriG service is named `buildAssociation`. This function produces association rules from a data set. The following piece of code describes the Web Services Description Language portType or interface of such grid service:

```
<wsdl:portType name="AprioriGPortType">
<wsdl:operation name="buildAssociations">
<wsdl:input message="impl:buildAssociationsRequest"
name="buildAssociationsRequest"/>
<wsdl:output message=Simpl:buildAssociationsResponseT
name="buildAssociationsResponse"/>
</wsdl:operation>
</wsdl:portType>
```

The client module is responsible for two related tasks.

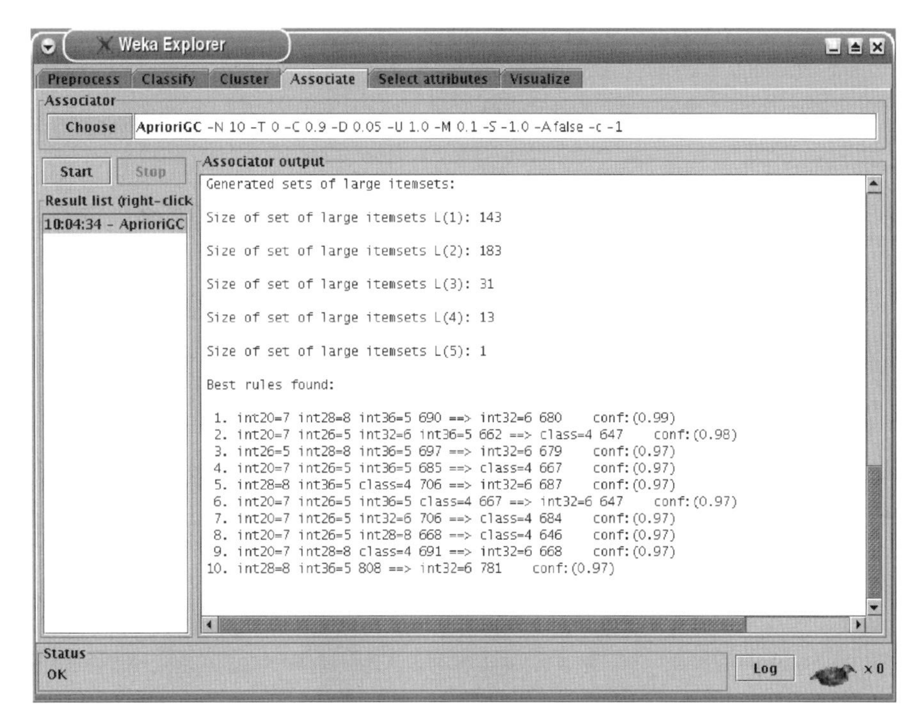

Figure 12.8 AprioriG demonstration

(a) Asking the GridFTP service for the transfer of data.

(b) Invoking the `buildAssociation` function of the AprioriG service.

Instead of overloading the client with this functionality, the broker is responsible for solving the composition of services and providing the server link to the client side.

Figure 12.8 shows an example of execution of the Apriori algorithm over a sample data set. The algorithm is executed by the AprioriG service, whose results are sent back to the graphical user interface.

We have evaluated the performance of the AprioriG service in comparison with the Apriori algorithm implemented in Weka. Before analysing the results of this evaluation, it is important to describe the scenario in which the tests have been performed.

As client, a computer with an Intel Pentium III, 735 MHz, processor and 640 MB RAM was used, and the following two servers have been evaluated.

- *Server 1.* Processor Intel Pentium IV, 3.20 GHz; RAM capacity 512 MB; network bandwidth 10 MB/s.
- *Server 2.* Processor Intel XEON, 2.40 GHz; RAM capacity 1 GB; network bandwidth 100 MB/s.

In order to run the tests, three representative data files were chosen from the UCI Repository of Machine Learning Databases and Domain Theories (Blake and Merz, 1998).

- *splice.arff.* File size: 512 KB. Although the size is medium, it requires higher processing time than the other two files.
- *m_feat_pixel.arff.* File size: 953 KB. It requires medium processing time.
- *contact_lenses.arff.* File size: 2.9 KB. It requires a short processing time.

The following parameters were measured:

- T_c. Time for running the Apriori algorithm, without using the grid environment.
- T_S. Time for completing the AprioriG service in the server.
- T_{cn}. Connection time, which corresponds to the time for obtaining a service reference and the creation of a service instance.
- T_{put}. Time for transferring serialized objects, from client to server.
- T_r. Time for creating and initializing a WSRF resource.
- T_{get}. Time for transferring serialized objects, from server to client.
- T_{gc}. Total time for running the AprioriG service. This includes the connection time, both transfer times and the creation of the WSRF resource, linked to the AprioriG service.

Table 12.1 shows the time for running the Apriori algorithm locally, without using DMGA.

Tables 12.2 and 12.3 show the times for running the grid-based Apriori service in both servers.

The comparison between Tables 12.1 and 12.2 shows that in those cases where the data mining process requires many process cycles the time spent for a standard workstation doubles the time spent for a more powerful server that provides this grid service.

Additionally, we can compare the results of Tables 12.2 and 12.3. In this case, the transfer times are shorter in the second case, since the network bandwidth of the second server is higher. On the other hand, Server 1 has a faster processor. Thus, the process time is shorter in this case.

As the final conclusion of this evaluation, it is noticeable that as the use case complexity increases the use of the grid technology provides more benefits. In simple cases involving low computational and data complexity, adopting a grid approach may not be beneficial, as the overheads of the grid become the dominant factor.

Table 12.1 Time for running the Apriori algorithm in the client

File name	T_c
splice	364.455 s
m_feat_pixel	148.038 s
contact_lenses	0.012 s

Table 12.2 Time for running the AprioriG service in Server 1

File name	T_{cg}	T_s	T_{cn}	T_{put}	T_r	T_{get}
splice	154.568 s	124.732 s	13.544 s	1.937 s	0.569 s	13.352 s
m_feat_pixel	137.597 s	59.613 s	15.687 s	3.198 s	0.926 s	31.919 s
contact_lenses	18.134 s	0.005 s	14.274 s	2.318 s	0.582 s	0.555 s

12.7.2 Vertical composition use cases: ID3 and J4.8

There are several data mining algorithms that could naturally benefit from data parallelism and thus from the vertical composition approach. ID3 is one of these algorithms.

ID3 is a simple tree induction algorithm (Quinlan, 1979). It is based on the information gain concept, which measures how well a given attribute separates training examples into targeted classes. Information gain is based on the entropy measures, from information theory.

$$I = -\sum_{i \in A} p(i) log(p(i)) \tag{12.1}$$

In Equation (12.1), A refers to the set of all partitions in the information space, and $p(i)$ is the probability, estimated as the proportion of instances in partition i. Information gain is computed as the difference in the entropy before and after selecting a given criteria for information space partitioning.

The output of ID3 is a model in the form of a classification tree. The leaf nodes of the decision tree contain the class name, whereas a non-leaf node is a decision node. The decision node represents an attribute test with each branch (to another decision tree) being a possible value of the attribute.

ID3 was extended in the C4.5 algorithm (Quinlan, 1993), providing mechanisms to deal with continuous and missing values. A significant improvement arises also from the possibility of pruning trees once they have been created. J4.8 (Witten and Frank, 2002) is an implementation of the C4.5 algorithm provided by the Weka Toolkit.

The main structure of both algorithms is, approximately, as follows.

(a) If all instances belong to the same class, then finish.

(b) For all possible features, compute the information gain achieved when using this feature (or condition based on this feature) in order to partition the instances.

(c) Select the attribute (or condition) to divide the instances into two or more groups.

(d) Go back to step (a).

Table 12.3 Time for running the AprioriG service in Server 2

File name	T_{cg}	T_s	T_{cn}	T_{put}	T_r	T_{get}
splice	158.986 s	142.013 s	9.513 s	1.254 s	0.653 s	10.914 s
m_feat_pixel	132.330 s	65.684 s	10.003 s	2.263 s	0.943 s	24.460 s
contact_lenses	11.751	0.005 s	9.857 s	1.662 s	0.601 s	0.338 s

Additional post-processing tasks could be performed. In the case of C4.5/J4.8, it would be induction pruning the induction tree.

Performing distributed rule induction

There are different alternatives to carry out distributed tree induction. A global classification model was built based on a vertical (feature-based) distribution of the data set (Prodromidis, Chan and Stolfo, 2000). Each partial model is combined using meta-learning techniques (Chan and Stolfo, 1996).

Horizontal partitioning (instance based) is proposed by Provost and Hennessy (1996). These independent rule induction algorithms use a subset of the instances, but learnt from the same feature space. This approach takes advantage of the invariant partitioning property, that guarantees that any rule that is satisfactory on the entire data set will be found by one of the local induction algorithms. Simple cooperation assures that only rules that are satisfactory on the entire data set will be found. This approach has been exported for independent agent learning collaboration (Nunes and Oliveira, 2003).

Kufrin (1997) proposes optimizations to produce final rules in parallel following different strategies, linked not only to how data are distributed, but also to how different tasks are carried out.

- *Tree pruning.* Following an instance-based strategy for pruning, in tree pruning all nodes participate in the pruning of a single rule by determining the contribution of their block of data to the coverage of the whole data set.
- *Selection of rules.* This implies combining the rules locally produced in a single rule set. This option could require considerable network bandwidth (to distribute local rules). In the case where rule sets are reduced, class-by-class rule selection is feasible otherwise, restricted information could be transmitted using the minimum description length principle. In this case estimated values can be used to guide the selection process.
- *Rule set evaluation.* A simple, rather effective approach for concurrent execution is instance-based parallelism using a static assignment of training cases to nodes.

Specific parallel induction algorithms have been proposed, with the idea not to reuse classical ID3-derived structures but with the goal of efficient distributed execution. SLIQ (Mehta, Agrawal and Rissanen, 1996) uses a breadth-first growth strategy, in contrast to the classical depth-first approach of ID3 and relatives. With this organization it needs only one pass over the data set per node level. Both approaches do not deal with regular local inducers and merge/combine results afterwards; instead, the structure of the algorithm is itself different. SLIQ solves the problem of repeatedly sorting the data at every node to arrive at the best splits for numeric attributes by introducing a separate, pre-sorted list for each attribute and a common, memory-resident lookup table, called the class list. SPRINT (Shafer, Agrawal and Mehta, 1996) extends the SLIQ structure to provide improved memory management in the class list.

Distributed model induction alternatives

This section presents a discussion about the possible alternatives to adapt induction algorithms to work on distributed environments. These algorithms construct a rule-based model from a data set with multiple instances (rows in data table) each of them includes the values of different features or attributes (columns in the data table). The alternatives are presented, considering an induction algorithm (similar to ID3, CART, C4.5 or J4.8) with iterative data set partition. This means that data are scanned to identify the best partition criterion (a Boolean condition based on a feature), then instances are split into two or more data sets and the process is repeated until a finishing criterion is met.

The parallel execution of this type of rule induction algorithm can be performed following two possible partitioning criteria: (i) by features (attributes), or (ii) by instances (items). Both alternatives are considered tightly coupled coordination schemes. Tightly coupled coordination implies that internal induction steps are actually performed in parallel, thus a significant amount of communication is required. Another possible distributed division can be made according to the interaction degree among independent vertical grid services. This possibility is based on a loosely coupled interaction. In this case, vertical grid services only exchange high-level information extracted from data. The amount of information exchanged would be minimal (although not in all cases), but the induced model could be less accurate and more complex.

Feature-based tightly coupled partitioning allows independent grid services to scan all the instances to compute partitioning criteria, for example computing information gain based on information entropy. Once information gain is computed, the best criterion is selected. To perform this task all the services should exchange the best local criterion and a coordinated election should identify the best overall criterion. Once best criterion is selected the original service should provide the list of instances (actually instance IDs) into which data are partitioned. When original data are replicated, the services can carry out this task once the selected criterion is known. The algorithm behaviour would be almost the same as in the centralized version, except that internal information is gathered in parallel.

The instance-based tightly coupled alternative requires the exchange of more detailed information. The information gain cannot be computed separately, and the alternatives are either to propagate $p(i)$ values and to compute information gain in a centralized way, or to use other partitioning criteria based on separable information (statistical significance, like the CART tree classifier (Breiman *et al.*, 1984)). In this case also, the actual induction algorithm is performed by a central node, based on parallel retrieved information.

Loosely coupled alternatives do not change intermediate information; instead, the final induced models are sent. Feature-based variants use the same instance space with independent features on each partition. Local services induce a complete rule set, which is able to classify all instances. As these rule sets are induced with partial information, the best instances are not combined to simplify the model and to purge redundant conditions. Complexity reduction can be carried out in a centralized way, selecting the best conditions using, for instance, greedy algorithms, information system theory (e.g. rough sets (Pawlak, 1991)) or truth maintenance systems.

The instance-based loosely coupled method would perform differently. In this case, all the features are shared among all the services, but these services do not use the same instances. In this case, the models extracted locally are not necessarily redundant, but cannot be inaccurate for unseen instances. To deal with this problem at a centralized level, all models are combined. As they are extracted from the different instances, a new data set can be created, selecting new instances (could be sampled from the original data) and using the output from each local classifier as a new feature. The new induction algorithm executes a consensus classifier from the decision provided by each service. The rationale behind this approach is taken from meta-learning theory (Chan and Stolfo, 1993, 1996), or from (adopting a machine learning perspective) Wolpert's stacked generalizator (Wolpert, 1992).

Vertical J4.8 algorithm

To illustrate a possible implementation of a rule induction algorithm in DMGA, the following implementation has been developed. DMGA has been equipped with a vertical composition for induction algorithms. This composition is a hybrid loosely coupled method based on Wolpert's stacked generalizator (Wolpert, 1992).

In this case, independent models are computed by local nodes. This task is performed by the `J4.8DistributedInducer` service, which is actually a wrapper for regular J4.8 in Weka. This service induces models from all the available data (instances and features) in this node. The data source could be all shared data from a given virtual organization related to a particular subject. The selection of appropriate data restricted to security and privilege policies.

The performance of the general algorithms is, actually, independent from both instance or feature partitioning. The instances belonging to particular node can be also stored in other nodes (but this is not required). The set of features could be the same, partially overlapped or completely different. The idea is to use all the data accessible from a computation node to construct a model; this model should be as accurate as possible, at least in the context of this particular node.

Once models are extracted, they are submitted to a coordination node. This node combines local models to produce a global result. In order to carry out this task, this node stores another data set, called *join-dataset*. This data set includes unseen instances (instances not used to train local models). These instances have all the features ever used in any of the nodes. It could be possible that some values of the instances are unknown.

Finally, a new data set is constructed by a service called `ModelCombiner`. This new data set is produced applying each of the models to the *join-dataset*. This new data set has the same instances as the *join-dataset*, but instead of the original features, new features are constructed based on the prediction from each model. Each instance has a predicted label, for each of the local models. This new data set is named *prediction-dataset*.

A new rule induction algorithms is executed on the *prediction-dataset* by means of the `J4.8GeneralizatorInducer` service. The model extracted from this later induction algorithm merges the predictions performed by the local models in a global model.

The selection of the best data partition and the best instance selection is a high-level decision which is part of the broker responsibilities. The decision should consider aspects such as communication costs, data set dimensionality (number of features times number of instances), the effect of noise and missing values, execution-time restrictions, accuracy and solution quality metrics.

12.8 Related work

DMGA has been implemented by means of WekaG, which is based on Weka. However, this architecture could be implemented with other different data mining libraries. Some other Weka-based toolkits, such as WekaG, have been developed with the aim of taking advantage of grid infrastructures. Some of the best known projects are GridWeka (Khoussainov, Zuo and Kushmerick, 2004), Weka4WS (Talia, Trunfio and Verta, 2005) and FAEHIM (Ali and Taylor, 2005).

GridWeka (Khoussainov, Zuo and Kushmerick, 2004) is a project of the University of Dublin to support Weka in grid infrastructures. The used infrastructure is an ad hoc grid lacking usual grid features, such as security and resource management, and it is not service oriented. This suggests that, in contrast to DMGA, GridWeka is not a true grid architecture or system.

Weka4WS (Talia, Trunfio and Verta, 2005) is a Weka-based toolkit for providing data mining on grid. Just like DMGA, it is service oriented and it uses the innovative WSRF to merge grid services with Web services. By means of WSRF it is possible to access remote data mining services and manage their execution. The main disadvantage of Weka4WS is that it is limited to the algorithms provided by Weka. Furthermore, Weka4WS does not address service composition, which allows the system to use services together in order to increase its usability and performance.

Federated Analysis Environment for Heterogeneous Intelligent Mining (FAEHIM) (Ali and Taylor, 2005) is another data mining toolkit for the grid. Nowadays it supports only some data mining activities derived from Weka, such as classification, clustering and plotting modules. These activities are deployed by using services that enable its integration within existing applications. Although it is similar to Weka4WS, it does facilitate workflows by combining services through horizontal composition. DMGA addresses not only the horizontal composition but also the vertical composition to increase the performance of data mining applications.

12.9 Conclusions

DMGA is a framework that provides a flexible integration of existing and newly developed data pre-processing, data mining applications and data post-processing into grid infrastructures.

DMGA enables the composition of services, both horizontal and vertical. The horizontal composition provides workflow capabilities, whereas the vertical composition allows for the optimization of data access. In order to demonstrate the benefits of both kinds of composition, two use cases have been shown.

The first one, a grid-based Apriori service, combines the functionality of a specific algorithm and a generic GridFTP-based data transfer service. The adaptation between the two functions is performed by the DMGA implementation.

The vertical composition enables the optimization of determined algorithms, such as the ID3 and J4.8 algorithms. Several alternatives have been presented for the vertical composition. These approaches are compatible with the two data division alternatives existing in DMGA, both fair and non-functional-property-based data division. In the case of the first kind of division, data are homogeneously distributed among all the nodes. Thus, if vertical distribution is selected a homogeneous number of features is divided among such nodes, and if horizontal distribution is chosen a homogeneous number of instances is distributed through the grid. The broker simply acts as a gateway between the application interface and the data servers, knowing the list of available servers. In the case of the second kind of data division, both vertical and horizontal distribution are not homogeneous, but based on the non-functional properties of the nodes. The broker knows the data distribution among the nodes, according to the their performance features.

DMGA has been implemented by means of WekaG. However, DMGA has been conceived to be used with any kind of data mining library.

References

Agrawal, R., Imielinski, T. and Swami, A. (1993), Mining association rules between sets of items in large databases, *in* 'Proceedings of the 1993 ACM SIGMOD International Conference on Management of Data', Vol. 22, pp. 207–216.

Ali, A. S. and Taylor, I. J. (2005), Web services composition for distributed data mining, *in* 'Proceedings of the 2005 International Conference on Parallel Processing Workshops (ICPPW'05)', pp. 11–18.

Allcock, W., Bester, J., Bresnahan, J., Chervenak, A., Liming, L. and Tuecke, S. (2001), 'GridFTP: Protocol Extensions to FTP for the Grid', http://www-fp. mcs.anl.gov/dsl/GridFTP-Protocol-RFC-Draft.pdf

Blake, C. and Merz, C. (1998), 'UCI Repository of Machine Learning Databases', http://archive.ics.uci.edu/ml/

Breiman, L., Friedman, J., Olshen, A. and Stone, C. J. (1984), *Classification and Regression Trees*, Wadsworth, Belmont, CA.

Chan, P. K. and Stolfo, S. J. (1993), Experiments in multistrategy learning by meta-learning, *in* 'Proceedings of the 2nd International Conference on Information and Knowledge Management', Washington, DC, pp. 314–323.

Chan, P. K. and Stolfo, S. J. (1996), Sharing learned models among remote database partitions by local meta-learning, *in* 'Proceedings of 2nd International Conference on Knowledge Discovery and Data Mining', pp. 2–7.

Foster, I. and Kesselman, C. (1997), 'Globus: a metacomputing infrastructure toolkit', *The International Journal of Supercomputer Applications and High Performance Computing* **11** (2), 115–128.

Foster, I., Kesselman, C., Nick, J. M. and Tuecke, S. (2002), 'The Physiology of the grid: an Open Grid Services Architecture for Distributed Systems integration', http://www.globus.org/research/papers/ogsa.pdf

Graham, S., Karmarkar, A., Mischkinsky, J., Robinson, I. and Sedukhin, I. (2006), 'Web Service Resource Framework', http://docs.oasis-open.org/wsrf/wsrf-ws_resource-1.2-spec-pr-01.pdf

Guo, Y., Liu, J. G., Ghanem, M., Mish, K., Curcin, V., Haselwimmer, C., Sotiriou, D., Muraleetharan, K. K. and Taylor, L. (2005), Bridging the macro and micro: a computing intensive earthquake study using Discovery Net, *in* 'Proceedings of the 2005 ACM/IEEE conference on Supercomputing (SC'05)', IEEE Computer Society, Washington, DC, p. 68.

Han, E.-H., Karypis, G. and Kumar, V. (1997), Scalable parallel data mining for association rules, *in* 'Proceedings of the 1997 ACM SIGMOD International Conference on Management of Data (SIGMOD'97)', ACM, New York, NY, pp. 277–288.

Khoussainov, R., Zuo, X. and Kushmerick, N. (2004), 'Grid-Enabled Weka: a Toolkit for Machine Learning on the Grid', http://www.ercim.org/publication/Ercim_News/enw59/khussainov.html

Kufrin, R. (1997), Generating C4.5 production rules in parallel, *in* 'Proceedings of the 14th National Conference on Artificial Intelligence', AAAI Press/MIT Press, Providence, RI, pp. 565–570.

Mehta, M., Agrawal, R. and Rissanen, J. (1996), SLIQ: A fast scalable classifier for data mining, in P. M. G. Apers, M. Bouzeghirub and G. Gardarin, eds, 'Advances in Database Technology – EDBT'96, Fifth International Conference on Extending Database Technology, Avighwn, France, March 25–29, 1996, Proceedings', Springer, pp. 18–32.

Natural Environment Research Council (2006), 'Grid ENabled Integrated Earth System Model', http://www.genie.ac.uk

Nunes, L. and Oliveira, E. (2003), 'Cooperative learning using advice exchange', *Lecture Notes in Artificial Intelligence – Hot Topics Sub-Series* **5** (2636), 33–48.

Pawlak, Z. (1991), *Rough Sets – Theoretical Aspects of Reasoning About Data*, Kluwer.

Pérez, M. S., Sánchez, A., Herrero, P., Robles, V. and Peña, J. M. (2005), Adapting the Weka data mining toolkit to a grid-based environment, *in* 'AWIC, Lecture Notes in Computer Science 3528', pp. 492–497.

Pérez, M. S., Sánchez, A., Robles, V., Herrero, P. and Peña, J. M. (2007), 'Design and implementation of a data mining grid-aware architecture', *Future Generation Computer Systems* **23** (1), 42–47.

Prodromidis, A., Chan, P. and Stolfo, S. (2000), Meta-learning in distributed data mining systems: issues and approaches, *in* H. Kargupta and P. Chan, eds, 'Advances of Distributed Data Mining', AAAI/MIT Press, pp. 79–117.

Provost, F. J. and Hennessy, D. N. (1996), Scaling up: distributed machine learning with cooperation, *in* 'AAAI/IAAI, Vol. 1', pp. 74–79.

Quinlan, J. (1979), Discovering rules by induction from large collections of examples, *in* D. Michie, ed., 'Expert Systems in the Micro-Electronic Age', Edinburgh University Press, pp. 168–201.

Quinlan, J. (1993), *C4.5: Programs for Machine Learning*, Morgan Kaufmann, Los Altos, CA.

Sánchez, A., Sánchez, J. M. P., Pérez, M. S., Robles, V. and Herrero, P. (2004), Improving distributed data mining techniques by means of a grid infrastructure, *in* 'OTM Workshops. LNCS 3292', pp. 111–122.

Shafer, J. C., Agrawal, R. and Mehta, M. (1996), SPRINT: a scalable parallel classifier for data mining, *in* 'Proceedings of the 22nd International Conference on Very Large Databases (VLDB'96)', pp. 544–555.

Talia, D., Trunfio, P. and Verta, O. (2005), Weka4WS: a WSRF-enabled Weka toolkit for distributed data mining on grids inproceedings of PICDD 2005, Springer, pp. 309–320.

Venugopal, S., Buyya, R. and Winton, L. (2004), A grid service broker for scheduling distributed data-oriented applications on global grids, *in* 'Proceedings of the second Workshop on Middleware for Grid Computing (MGC'04)', ACM, New York, NY, pp. 75–80.

Witten, I. H. and Frank, E. (2002), 'Data mining: practical machine learning tools and techniques with Java implementations', *SIGMOD Record* **31** (1), 76–77.

Wolpert, D. H. (1992), 'Stacked generalization', *Neural Networks* **5**, 241–259.

13

Grid-based data mining with the Environmental Scenario Search Engine (ESSE)

Mikhail Zhizhin, Alexey Poyda, Dmitry Mishin, Dmitry Medvedev, Eric Kihn and **Vassily Lyutsarev**

ABSTRACT

The increasing data volumes from today's collection systems and the need of the scientific community to include an integrated and authoritative representation of the natural environment in their analysis requires a new approach to data mining, management and access.

The natural environment includes elements from multiple domains such as space, terrestrial weather, oceans and terrain. Systems such as the *Global Change Master Directory* (*GCMD*) from NASA[1] or the *Master Environmental Library* (*MEL*) from the DMSO[2] and others provide the ability to search metadata by keywords, the result being a set of links to archived environmental data sets distributed across the network, but they are unable to search for specific patterns within the data themselves.

The environmental modelling community has begun to develop several archives of continuous environmental representations. These archives contain a complete view of the Earth system parameters on a regular grid for a considerable period of time. The numerical models used to reproduce environmental parameters take all available observational data as initial conditions, so the resulting petabyte-size data sets may be considered as an authoritative high-resolution representation of terrestrial weather during the last 50 years (Kalnay *et al.*, 1996; Uppala *et al.*, 2005).

This chapter describes the *Environmental Scenario Search Engine* (*ESSE*) for data grids, which provides uniform access to heterogeneous distributed environmental data archives and allows the use of human linguistic terms while querying the data. A set of related software tools leverages the ESSE capabilities to integrate and explore environmental data in a new and seamless way. The ESSE software is available for download[3] from the Internet.

[1] http://gcmd.nasa.gov
[2] https://mel.dmso.mil/
[3] https://esse.wdcb.ru/

Data Mining Techniques in Grid Computing Environments Edited by Werner Dubitzky

13.1 Environmental data source: NCEP/NCAR reanalysis data set

We shall demonstrate the functionality of the ESSE on data from the first global weather reanalysis project, which was created in the late 1990s by the *National Center for Environmental Protection* (*NCEP*) and the *National Center for Atmospheric Research* (*NCAR*) (Kalnay *et al.*, 1996). It continues to be extended with current data in near-real time, so that its products are available from 1948 to the present. The NCEP/NCAR 50-year reanalysis uses a frozen modern global data assimilation system, which is identical to the global weather forecast system implemented operationally at NCEP on January 1995, except for the lower horizontal resolution (about 210 km) as compared with the data operationally used for weather global weather forecast (currently less than 100 km). The horizontal resolution of a global atmosphere circulation model is related to the spacing between grid points for grid point models or the number of waves that can be resolved for spectral models.

Modern global circulation models have a number of degrees of freedom of the order of 10^7. For example, a latitude–longitude model with a typical resolution of 1° and 20 vertical levels would have $360 \times 180 \times 20 = 1.3 \times 10^6$ grid points. At each grid point the model has to carry the values of at least four prognostic variables (two horizontal wind components, temperature, moisture), and the surface pressure for each column, giving over five million variables that need to be given an initial value. For any given time window of 6 hours, there are typically 10–100 thousand observations of the atmosphere, two orders of magnitude less than the number of degrees of freedom of the model.

For this reason, a short-range forecast is often used as the first guess in the operational data assimilation systems analysis cycles. Instead of 'direct' spatial interpolation of raw observations to gridded fields for the model input, each 6 hour data assimilation cycle for a global model takes as a background the field from the model's past 6 hour forecast made for the given time moment. Then the background grid data are interpolated to the locations of the raw observations taken with the 6 hour assimilation cycle and the difference between the interpolated forecast values and raw observations is propagated back with some weighting to correct the background grid values. The resulting field, sometimes called a *nowcast*, can be stored in the model output data set or used as an input for the next model analysis cycle. When such an assimilation is carried out for the historical observations using all the raw data available from the long-term archives, the nowcast output data set is called a reanalysis. We can refer to several terabyte-size long-term reanalysis data sets, among them NCEP/NCAR and ECMWF ERA-40 (Uppala *et al.*, 2005) global reanalysis products, and the North American Regional Reanalysis data set (Mesinger *et al.*, 2006).

In the NCEP/NCAR reanalysis data set there are about 80 different variables, (including geopotential height, temperature, relative humidity, U and V wind components etc.) available from 1 January 1948 till present with output every 6 hours in several different coordinate systems roughly equivalent to a 2.5×2.5 degree geographical latitude–longitude grid with 17 vertical layers from 1000 to 10 Mbar. They are organized as different subgroups of the model output binary data files in the data archive. Most of the project's TB-sized outputs are stored in the GRIB[4] format (a format for meteorological data exchange) some output files are in the NetCDF[5] format (an array-based scientific data format and network protocol).

[4]http://dss.ucar.edu/docs/formats/grib/gribdoc/

[5]http://www.unidata.ucar.edu/software/netcdf/

For use in interactive data mining applications, we have converted the NCEP/NCAR reanalysis project data into tables in a relational database. The database contains four-byte numeric values on a reference 2.5×2.5 degree grid for all the parameters. When the reanalysis data are given on a Gaussian grid, we apply Akima bivariate interpolation (Akima, 1978) to produce the values for the database.

13.2 Fuzzy search engine

The base data model in our study is the vector-valued time-series

$$\mathbf{X} = \{\mathbf{x}(t_1), \ldots, \mathbf{x}(t_N)\},$$

$$\mathbf{x}(t_i) = (x_1(t_i), \ldots, x_M(t_i)),$$

where N is the number of time samples, and M is the number of observed parameters. It can be represented as a trajectory in the M-dimensional phase space $\mathbb{R}^M$. For example, in Figure 13.1 we have a two-dimensional trajectory in the pressure–temperature (P–T) space.

A (fuzzy) state S in a phase space $\mathbb{R}^M$ is a fuzzy set, which can be described by fuzzy logic expressions composed of predicates describing in numerical or linguistic terms the parameter values in each of M dimensions. For example, the state S_1 corresponding to the red (top right) region in Figure 13.1 can be described by the fuzzy expression

$$S_1 = (\text{VeryLarge } P) \text{ and } (\text{VeryLarge } T),$$

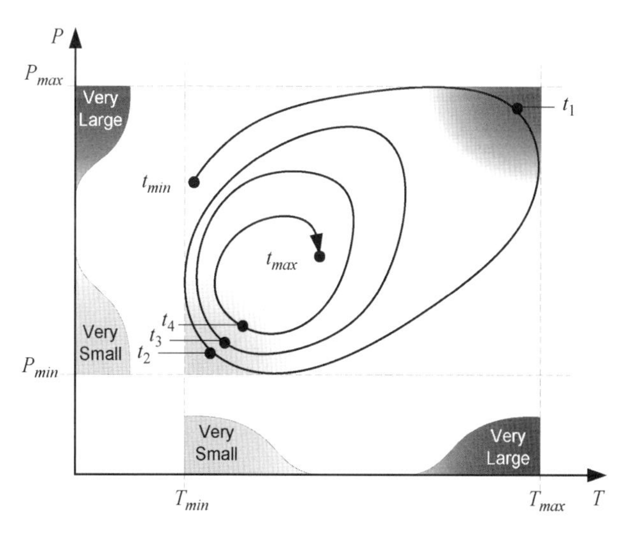

Figure 13.1 Time series as a trajectory in the two-dimensional phase space (P–pressure, T–temperature)

where the linguistic term 'VeryLarge ' is a predicate, and the operator ' and ' stands for the fuzzy logic conjunction. In the same way, the state S_2 corresponding to the cyan (bottom left) region is

$$S_2 = (\text{VerySmall } P) \text{ and } (\text{VerySmall } T).$$

Now, combining the descriptions of the states with the time shift operator shift_{dt} to describe transitions between the states, we can write the following symbolic expression for the environmental scenario 'very low temperature and pressure after very high temperature and pressure':

$$(\text{shift}_{dt=1} S_1) \text{ and } S_2.$$

The only pair of observations in Figure 13.1 that fits the above scenario is the pair (t_1, t_2). Our environmental scenario search engine, ESSE, is designed to mine for phase space transitions like that in very large scientific databases.

In the following subsections we describe the mathematics behind the definition of a fuzzy event scenario and point out hints that can significantly speed up computations.

A classical set A in a space of objects $\mathbb{U}$ can be defined by its indicator function $I_A(u) : \mathbb{U} \to \{0, 1\}$, which is equal to 1 for all elements u from the set A and to 0 otherwise. Figure 13.2 shows the plot of an indicator function of the segment $A = [5, 8]$ as a subset of all real numbers $\mathbb{R}$.

A fuzzy set expresses the degree to which an element belongs to a set. Hence, the indicator function of the fuzzy set is allowed to have values between 0 and 1, which denotes degree of membership of an element in a given set. A fuzzy set A in $\mathbb{U}$ is defined by its *membership function* $\mu_A(u) : \mathbb{U} \to [0, 1]$, which maps each element of $\mathbb{U}$ to its membership grade between 0 and 1. Compare graphs of a membership function for the fuzzy interval [5, 8] and the indicator function for the classical segment [5, 8] (Figure 13.2).

13.2.1 Operators of fuzzy logic

Basic operations of classical set theory (union, intersection, complement) and corresponding operations of mathematical logic (or, and, not) can be generalized for fuzzy sets and fuzzy logic in many different ways. The fuzzy generalization of intersection (logical 'and') is usually called a T-norm operator, a generalization of union (logical 'or') is called a T-conorm or S-norm operator, and the generalization of logical 'not' is called a fuzzy complement operator.

One of the simplest generalizations from classical to fuzzy set theory for two membership functions $\mu_A(u)$, $\mu_B(u)$ is to use the minimum of membership functions for the intersection

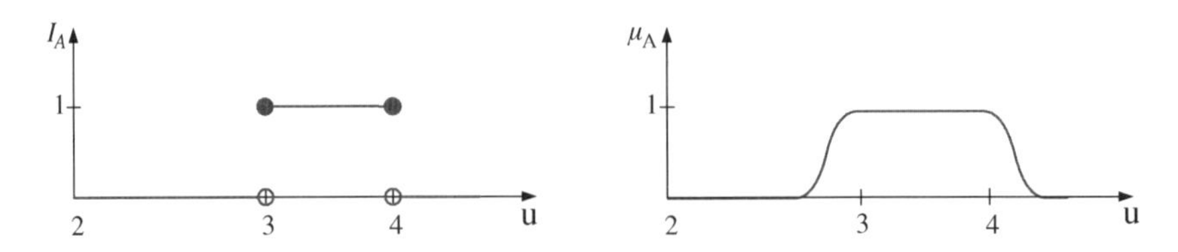

Figure 13.2 Indicator function $I_A(u)$ for the classical set $A = \{x | 5 \leq x \leq 8\}$. Fuzzy membership function $\mu_A(u)$ for the same interval

of the fuzzy sets (fuzzy logic 'and')

$$\mu_{A \cap B} = \min(\mu_A, \mu_B),$$

maximum of membership functions for fuzzy set union

$$\mu_{A \cup B} = \max(\mu_A, \mu_B),$$

and one complement for fuzzy set complement

$$\mu_{\overline{A}} = 1 - \mu_A.$$

In 1980 R. Yager introduced a parametric family of T-norms, T-conorms (Yager, 1980) and fuzzy complements (Yager, 1979). Parameterized by $q \geq 1s$ a family of fuzzy 'and' aggregations for two membership functions is defined by Yager's T-norm operator:

$$T_Y(\mu_A(x), \mu_B(x), q) = 1 - \min\left\{1, \left[(1 - \mu_A(x))^q + (1 - \mu_B(x))^q\right]^{\frac{1}{q}}\right\}.$$

A more general formula for the parametric Yager T-norm operator for fuzzy 'and' aggregation of any $M > 1$ membership functions $\mu_m(x), m = 1, \ldots M$ is

$$T_Y(\mu_m(x), q) = 1 - \min\left\{1, \left[\sum_{m=1}^{M} (1 - \mu_m(x))^q\right]^{\frac{1}{q}}\right\}. \tag{13.1}$$

The resulting surface of values for the multi-dimensional membership function is smoother than using a simple minimum of the aggregating membership functions, which is the limit case of Yager's T-norm for $q = 1$. A parametric family of fuzzy 'or' aggregations for two membership functions is described by Yager's T-conorm operator

$$S_Y(\mu_A(x), \mu_B(x), q) = \min\left\{1, \left[(\mu_A(x))^q + (\mu_B(x))^q\right]^{\frac{1}{q}}\right\}.$$

A more general formula for Yager's 'or' aggregation of any $M > 1$ membership functions is

$$S_Y(\mu_m(x), q) = \min\left\{1, \left[\sum_{m=1}^{M} (\mu_m(x))^q\right]^{\frac{1}{q}}\right\}. \tag{13.2}$$

Yager's fuzzy complements are defined by the formula

$$N_Y(\mu(x), q) = \left(1 - (\mu(x))^q\right)^{\frac{1}{q}}$$

The ESSE search engine is designed to support different libraries of T-norm, T-conorm and complement operators. The results below are obtained using the Yager formulas with the order $q = 5$.

13.2.2 Fuzzy logic predicates

People often use qualitative notions to describe such variables as temperature, pressure and wind speed. In reality, it is difficult to put a single threshold between what is called 'warm' and 'hot'. Fuzzy set theory serves as a translator from vague linguistic terms into strict mathematical objects.

The scenario editor from the ESSE user interface (more details in Section 13.4.3) is used to formulate a set of conditions to be satisfied by the candidate events. The search conditions may be specified in a number of ways, depending on the user's familiarity with the region/data of interest. An expert user can specify exact thresholds as well as limitations that must be maintained on certain parameters. Conditions can also be specified via abstract natural language definitions for each parameter. For instance, temperature limitations can be specified as 'hot', 'cold', or'typical'. The default ESSE library of membership functions formed for each variable (phase space dimension) uses the generic bell 'mother' function (Jang, Sun and Mizutani, 1997):

$$\mu_{gbell}(\tilde{x}; a, b, c) = \frac{1}{1 + \left|\frac{\tilde{x}-c}{a}\right|^{2b}}$$

Here, $\tilde{x}$ stands for the scalar data variable normalized to the range [0,1] , c stands for the centre of the symmetrical 'bell', a for its half-width and $b/2a$ controls its slope. We use here simple range normalization for the variable x:

$$\tilde{x} = \frac{x - x_{\min}}{x_{\max} - x_{\min}},$$

where $x_{\min}$ and $x_{\max}$ stand for the minimal and maximal observable values of x, respectively.

Five membership functions for a linguistic term set {'very small', 'small', 'average', 'large', 'very large'} are plotted in Figure 13.3. Center, slope, and half-width of the bell functions for these linguistic terms are listed in Table 13.1.

In the next plot, Figure 13.4 we present examples of four membership functions from the ESSE numerical fuzzy term set {'less than', 'about', 'between', 'greater than'}. For the normalized variable $\tilde{x}$ the centre, slope and half-width of the bell functions for the numerical terms are listed in Table 13.2.

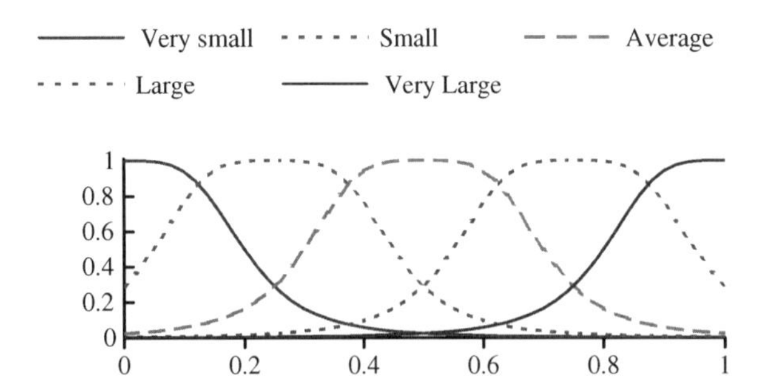

Figure 13.3 Membership functions of the ESSE 'linguistic terms'

Table 13.1 Parameters of membership functions for linguistic terms

Linguistic term	Centre	Slope	Halfwidth
Very small	0	5	0.2
Small	0.25	5	0.2
Average	0.5	5	0.2
Large	0.75	5	0.2
Very large	1	5	0.2

13.2.3 Fuzzy states in time

In the ESSE applications we are searching for events in the environment where the environmental parameters depend on time, as well as membership functions – fuzzy logic predicates $\mu_m(\mathbf{x}(t))$ and fuzzy expressions $E_Y(\mu_m(\mathbf{x}(t)), q)$, which are composed of T-norms, T-conorms and complements over the predicates. We consider the values of the resulting time series $E_Y(\mu_m(\mathbf{x}(t)), q)$ as the 'likeliness' of the environmental state to occur at the time moment t, or, in other terms, to visit a sub-region of the phase space described by the fuzzy expression (see Figure 13.1). We search for the highest values of the membership functions and consider these to be the most likely candidates for the environmental event.

We use a simple climatology analysis to obtain the normalization limits x_{min}, x_{max} used in the calculations of linguistic predicates such as 'very large' from Table 13.1. The limits are set to the minimum and maximum parameter values observed within the continuous or seasonal intervals given by the time constraints of the fuzzy search.

To be able to search for events such as a 'cold day' or a 'cold week' we introduce the concept of an event duration, which may be any multiple of the time step Δt of the input. For example, the time step in the NCEP/NCAR reanalysis is $\Delta t = 6$ hours, so the minimum event duration is also 6 hours, but the event duration may also be 1 day ($4\Delta t$), 1 week ($28\Delta t$) etc. We take a moving average of the input parameters with the time window of the event duration before calculation of membership functions in the fuzzy expression

$$\overline{x}(t_i) = \frac{1}{k} \sum_{j=i}^{i+k-1} x(t_j), \quad t_i = t_0 + i\Delta t.$$

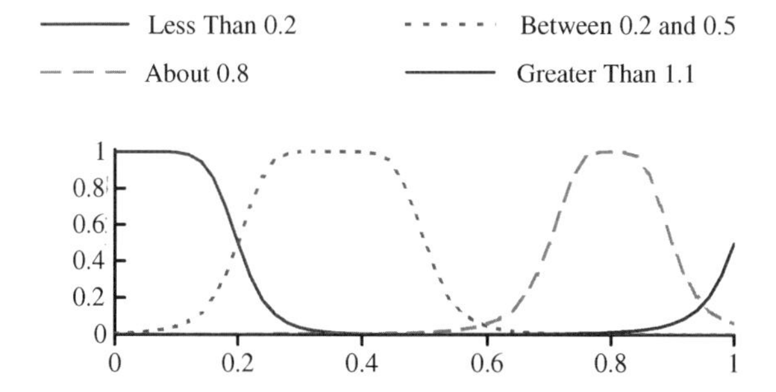

Figure 13.4 Fuzzy membership functions for numerical terms

Table 13.2 Parameters of membership functions for numeric terms

Numerical term	Centre	Slope	Halfwidth
Less than $\tilde{v}$, $\tilde{v} < 0$	$\tilde{v}$	10	0.1
Less than $\tilde{v}$, $\tilde{v} \geq 0$	0	10	$\tilde{v}$
About $\tilde{v}$	$\tilde{v}$	10	0.1
Between $\tilde{v}$ and $\tilde{w}$	$(\tilde{v} + \tilde{w})/2$	10	$\lvert\tilde{w} - \tilde{v}\rvert$
Greater than $\tilde{v}$, $\tilde{v} < 1$	1	10	$\tilde{v}$
Greater than $\tilde{v}$, $\tilde{v} \geq 1$	$\tilde{v}$	10	0.1

For example, when searching for a 'cold day' in the NCEP/NCAR reanalysis, first we have to smooth the air temperature with the time window of 1 day ($k = 4$), then calculate the linguistic predicate 'low' $\mu_{\text{low}}\left(\overline{T}(t_i)\right)$, sort the fuzzy scores in descending order and finally take the times with the highest scores as the candidate events.

The important difference of the averaging operator is the dependence of the input of the neighbour observations in time. Another operator of this class is the time shift, defined by the formula

$$\text{shift}_k \mu(t_i) = \mu\left(t_{i-p}\right).$$

The difference between the averaging and the shift operators is that we average input values, but we shift in time the values of the fuzzy membership function. Thus we have to investigate the properties of the time shift in relation to the fuzzy logical operators T-norm, T-conorm and complement. For any fuzzy logic expression E we have

$$\text{shift}_k \text{shift}_l = \text{shift}_{k+l},$$

$$\text{shift}_k E(\mu_1, \mu_2, \ldots) = E(\text{shift}_k \mu_1, \text{shift}_k \mu_2, \ldots).$$

We need the time shift operator to define a multiple-state scenario. For example, to find an abrupt air pressure drop, we can use a two-state scenario with fuzzy ' and ' of the 'very large' and time-shifted 'very small' predicates for pressure $P(t)$:

$$S(t) = T_Y\left(\mu_{\text{VeryLarge}}(P(t)), \text{shift}_{dt=1\,\text{day}}\, \mu_{\text{VerySmall}}(P(t)), q\right).$$

Following this example, a two-state scenario with the fuzzy expressions for states E_1 and E_2 with the time delay between the steps $k\Delta t$ can be defined as a Yager T-norm conjunction of the time-shifted expressions

$$S(t) = T_Y(E_1, \text{shift}_k E2, q).$$

Generalization of the formula for more than two states is straightforward.

To have the result of the fuzzy search in the form of a ranked list of the k most likely dates (times) of the events, we sort the scenario membership function $S(t)$ and select the times of the maximal values separated in time by more than the event duration $k\Delta t$.

13.2.4 Relative importance of parameters

A fuzzy search request may contain conditions which never or very rarely take place at the same time at the specified location, although they can be observed there separately at different times. For example, very high precipitation rate and very high air pressure are unlikely to occur simultaneously. The fuzzy search for such a combination of conditions may return an empty set of candidate dates and times. We decrease the probability of empty fuzzy search results by introducing the concept of importance of the input parameters. The importance ω_n is a constant weight of a given parameter in the range between 0 and 1. More important parameters are given higher weight, with the condition that the highest priority is then normalized to one. Then instead of membership functions $\mu_n(x(t_i))$ in the fuzzy expressions we use 'optimistic' values $\max(\mu_n(x(t_i)), 1 - \omega_n)$. For parameters with unit importance we use the original membership functions as before, and the parameters with zero importance are not used in the search at all.

13.2.5 Fuzzy search optimization

For performance analysis we have composed a test request with a fuzzy search query for an atmospheric front over the last 50 years for nine different geographic points:

$$(\text{Small } P) \text{ and } (\text{Large shift}_{1d}\ P) \text{ and } (\text{Large } Wind_v) \text{ and } (\text{Small shift}_{1d}\ Wind_v)$$
$$\text{and } (\text{Small shift}_{1d} Wind_u)$$

Execution profiles of this request have shown that computation of Yager operators (Equations (13.1) and (13.2)) consumes most of the CPU time: more than seven out of 12 seconds. The most time consuming operations in these formulas are calculations of powers q and $\frac{1}{q}$, where q is a parameter of the Yager operator family, which we consider to be fixed after initialization of the function library.

We know that values of the fuzzy operands are in the range of [0, 1]. Thus we can use a lookup table to accelerate the calculation of the powers in the Yager functions. We have divided the unit segment [0, 1] into $2^k - 1$ equal steps. At each step we pre-calculate the values of x^q and $x^{\frac{1}{q}}$, and store them in a table. Later, when we need to calculate the T-norm or T-conorm functions, we simply select them as the nearest tabular value. The question is how to choose the number of quantization steps k to have both a good approximation and a short calculation time.

To see how the approximation error depends on the quantization step, we need to define how to estimate the discrepancy between the fuzzy search results obtained with and without quantization of the Yager functions. As a quantitative measure of the difference between the time series of the fuzzy search scores we can use a *root mean square error* (*RMSE*). To measure the change (overlap) in a ranked list of the k most likely date–times of the events found by the fuzzy search, we have introduced a special norm DIFF.

To take in account that only the fist n candidates are considered from a potentially very long fuzzy search score list, we had to modify the well known formula for RMSE as follows:

$$RMSE_n = \frac{RMSE_n(x, y) + RMSE_n(y, x)}{2},$$

where

$$RMSE_n(x, y) = \sqrt{\frac{\sum_{t \in times_x} (x(t) - y(t))^2}{n}}$$

and

$$RMSE_n(y, x) = \sqrt{\frac{\sum_{t \in times_y} (y(t) - x(t))^2}{n}}.$$

Here $times_x$ is the set of the first n most likely candidates for the search result x, and $times_y$ is the same size subset for the search result y. We have to symmetrise $RMSE_n$ because in general $times_x \neq times_y$.

The above norm measures the difference between score values of two result sets. But in many search cases the order of the selected candidates is more important than their scores. To measure the stability of the order in the candidate list, we introduce another norm $DIFF$:

$$DIFF = DIFF_{times_x \cap times_y} + DIFF_{times_x \setminus times_y} + DIFF_{times_y \setminus times_y},$$

where

$$DIFF_{times_x \cap times_y} = \frac{1}{n} \sum_{t \in times_x \cap times_y} \left(\arctan r_x(t) - \arctan r_y(t)\right)^2,$$

$$DIFF_{times_x \setminus times_y} = \frac{1}{n} \sum_{t \in times_x \setminus times_y} \left(\arctan r_x(t) - \arctan(||times_y|| + 1)\right)^2,$$

$$DIFF_{times_y \setminus times_x} = \frac{1}{n} \sum_{t \in times_y \setminus times_x} \left(\arctan r_y(t) - \arctan(||times_x|| + 1)\right)^2.$$

Here $r_x(t)$ and $r_y(t)$ are the ranks of the same search result t in $times_x$ and $times_y$, respectively, and the "logistic" function arctan is used to enhance the importance of the rank difference for the high-ranked candidates.

In Figures 13.5 and 13.6 we compare norms $RMSE$ and $DIFF$ respectively for the search results with and without quantization for different numbers of quantization steps k and the size of the result set $n = 100$.

As expected, the search time difference vanishes when the number of quantization steps k tends to infinity. When the search result difference was compared to the time required for calculations, we chose the "optimal" number of steps as $k = 10$, where the computation time of the test search request was reduced from more than 7 to 1.4 seconds.

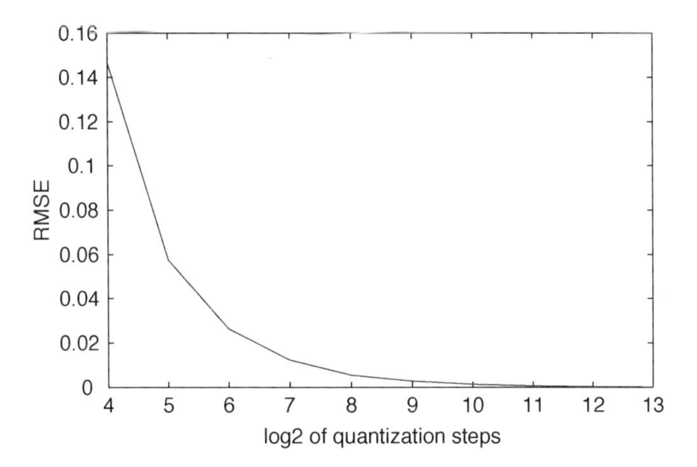

Figure 13.5 *RMSE* of the test search results as a function of the number of quantization steps for the first 100 events list

13.3 Software architecture

13.3.1 Database schema optimization

To optimize the database for retrieving long time series at several geographic grid points we stored one-dimensional binary arrays of floats representing time series as compressed *binary large objects* (*BLOBs*). Transfer overheads are considerable for large data sets, because disk transfer rates are relatively small (20–100 Mb/s). Compression of the data saves time on disk reads. Due to the nature of the data value distribution, the GZIP compression algorithm (Deutsch, 1996) managed to pack the data significantly. The total data size was reduced from 279 to 117 GB.

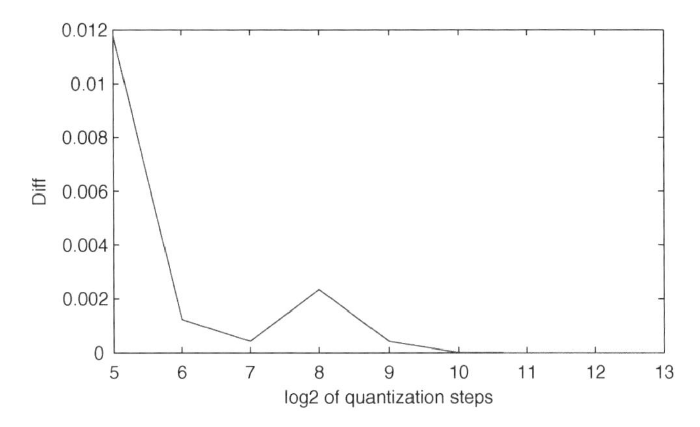

Figure 13.6 *DIFF* of the test search results as a function of the number of quantization steps for the first 100 events subset

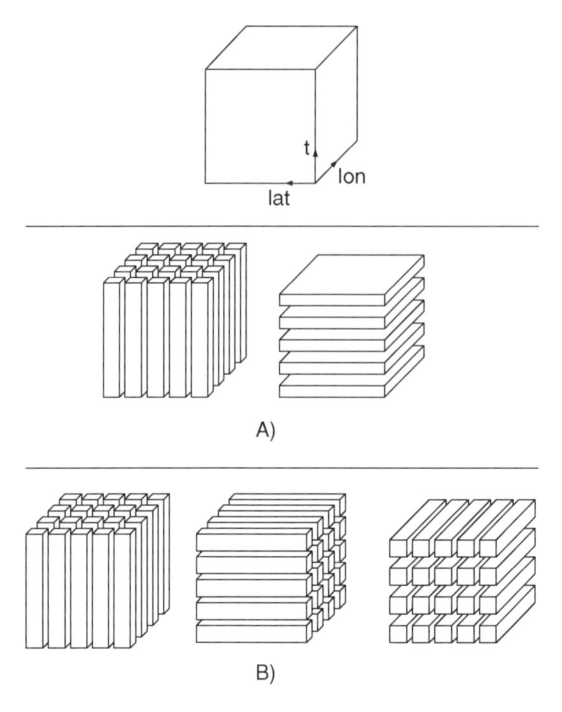

Figure 13.7 Data cube slicing schemes (a contiguous block corresponds to a binary data field)

An obvious shortcoming of the proposed database structure is that it is optimized for time series retrieval. For a query that requests environmental parameter values for a given time across a considerable geographic region one needs to retrieve the whole time series BLOB for each geographic point, decompress and retrieve a single data value from it. Thus, the number of data read from the database is orders of magnitude larger than the final result produced by the query.

To be able to efficiently run both time series and geographic distribution queries we have decided to have multiple data sets. This approach is widely used in OLAP databases. We have already sliced the 'lat–lon–time' data cube along the time dimension to provide good performance for time series retrieval. In addition, we made slices along the lat–lon hyperplane inside the cube (Figure 13.7(a)). Looking for a more general solution, we can make slices along each dimension (Figure 13.7(b)). Additional slices will reduce the amount of retrieved 'non-relevant' data for different types of data query, thus improving data retrieval and post-processing time at the expense of some additional storage space.

The data access component described in Section 13.3.3 automatically decides which data set to address depending on the incoming query properties. For a given query it computes the number of slices n to be read from the grid sliced database and corresponding number of slices m to be read from the time sliced database. Considering that most of the time is spent on disk reads, it selects the slicing schema based on the number of data to be read. Thus query execution times are approximated by the following expression:

$$t_{\text{grid}} = t^1_{\text{grid}} \cdot n$$
$$t_{\text{time}} = t^1_{\text{time}} \cdot m$$

where t^1_{grid} is the estimated time to retrieve one data row from the lat–lon sliced database and, t^1_{time} is the estimated time to retrieve one data row from the time sliced database. The smaller value defines the database to use.

13.3.2 Data grid layer

We have chosen the OGSA-DAI software package as the foundation for our service interface that will give access to environmental databases to grid users. The aim of the OGSA-DAI project is to develop middleware to assist with access and integration of data from separate sources via the grid (Karasavvas *et al.*, 2004; Antonioletti *et al.*, 2005). This is achieved through conformance with emerging Web service standards and a flexible internal architecture.

With OGSA-DAI different types of data resource – including relational, XML and files – can be exposed via Web services. Requests to OGSA-DAI Web services have a uniform format irrespective of the data resource exposed by the service. This does not imply the use of a uniform query language. Actions specified within each request may be specific to the given data resources. For example, a request to a relational database may contain an SQL SELECT statement, while an XML storage request could be specified using XPath syntax. Data produced as a result of the request can be delivered to clients, other OGSA-DAI Web services, URLs, FTP servers, GridFTP servers or files.

Data resource abstraction is at the base of the OGSA-DAI architecture. A data resource is a named entity that represents some storage engine and a set of data behind the engine. Each data resource has a location represented by a URL. A client can easily determine the type of a data resource by querying the list of activities, i.e. operations supported by the data resource.

Besides activities associated with a data resource query and manipulation, the OGSA-DAI package contains general-purpose activities for data transformation (e.g. data compression and decompression) and for data delivery, including delivery to/from other OGSA-DAI servers over the network. Multiple activities may be linked together within a single request so that output from one activity flows into the input of another one. Consequently, each OGSA-DAI request supplied by a client is actually a data-flow graph represented in the form of a 'perform' XML document. An OGSA-DAI client is also able to create a distributed data-flow by submitting several coordinated perform documents to separate OGSA-DAI containers.

Environmental data access scenarios include mostly the retrieval of one-dimensional time series or two-dimensional geographic grids from three- or four-dimensional arrays of data. To support these kinds of operation we have developed a separate OGSA-DAI data resource component, the ESSE Data Resource, and a set of corresponding activity components (Table 13.3).

We have added a data mining activity `fuzzySearch` to the OGSA-DAI service as an implementation of the Environmental Scenario Search Engine (see Section 13.2). This moves data mining processing from the client to a server and significantly reduces the number of data to be transferred over the network. The activity element in the perform document carries a description of an environmental scenario in the form of a combination of fuzzy conditions on the parameters obtained by `getData` activities. The activity output is a time series of fuzzy likeliness of the occurrence of the scenario at each moment in time.

Figure 13.8 gives an example of typical distributed data processing in the context of an OGSA-DAI system. It shows two grid servers each having an extended OGSA-DAI package installed.

A client submits a perform document to the lower container. This document contains fuzzy search data mining activity that scans three time series for a specific environmental event.

Table 13.3 ESSE components added to the OGSA-DAI

Component	Description
`EsseDataResource`	Represents environmental databases (Section 13.3.3).
`GetMetadataActivity`	Query activity. Returns the description of the data maintained by the `EsseDataResource`.
`GetDataActivity`	Query activity. Returns one or several time series from the `EsseDataResource` (Section 13.3.3).
`GetNetCdfActivity`	Query activity. Serializes a data subset into a NetCDF file and returns a URL to that file.
`FuzzySearchActivity`	Transformation activity. Receives one or more time series from `getData` and returns fuzzy membership function values (Section 13.2).
`DataProcessActivity`	Transformation activity. Receives one or more time series from `getData` and returns a computed time series (Section 13.3.4).

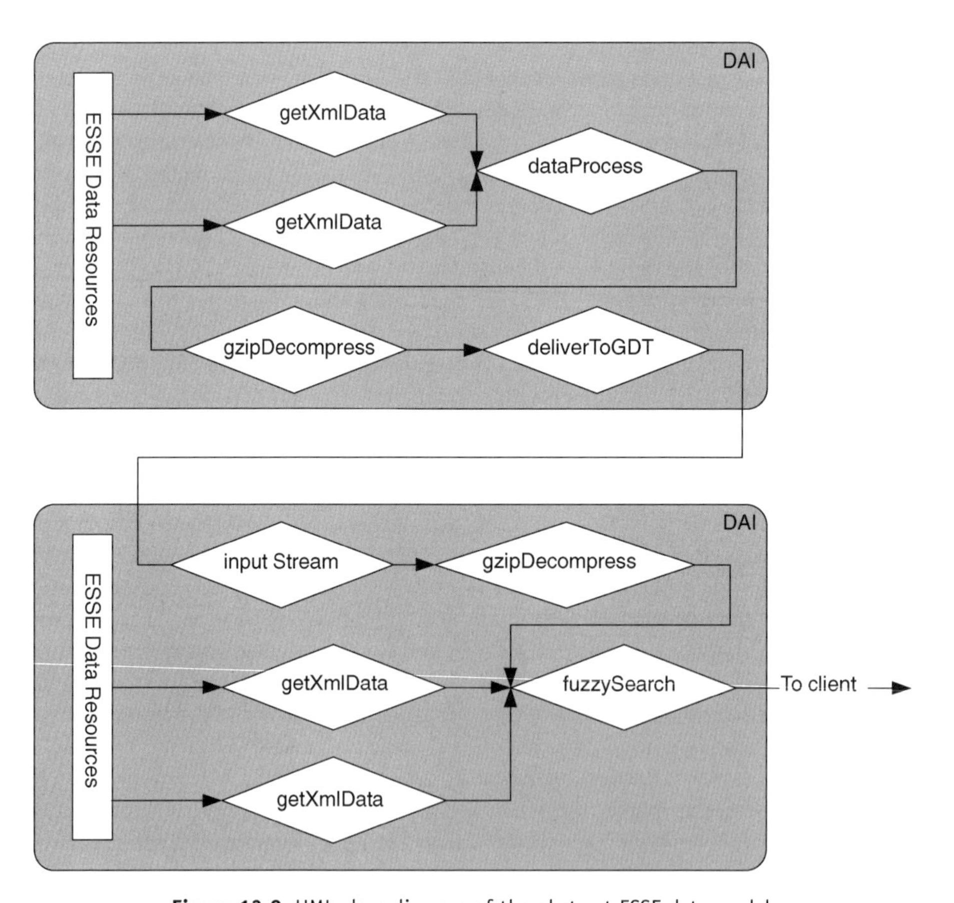

Figure 13.8 UML class diagram of the abstract ESSE data model

Two of the time series are requested from the local database with the help of two `getData` activities. The third time series must come from a remote site. Next, the client submits the second perform document to the upper container. This makes the OGSA-DAI engine in this container request two separate time series available at the site, perform computation with the `dataProcess` activity, compress the serialized result using the `gzipCompress` activity and deliver it to the lower site with the `deliverToGDT` activity.

13.3.3 ESSE data resource

Processing of an actual perform document by the extended OGSA-DAI engine involves the coordinated work of one or more `GetData` activities and the ESSE Data Resource. Each of the `GetData` activities provides the name of a data source, the name of an environmental parameter, the geographic area and the time interval required. The ESSE Data Resource passes this query data to appropriate database engines and returns the results to the activities.

The routing of the query is the responsibility of the Data Source API level. The database structures may differ in schemas, i.e. representation of surface coordinates, time and parameters, and BLOB packing methods, but once we have fixed the request parameters and the result structure we are able to implement a data access object that has the specified interface and is able to access specific databases.

We call the structure returned by the Data Source API and subsequently transferred between OGSA-DAI activities an ESSE data model (Figure 13.9). It is compatible with the *Common Data Model* (*CDM*)[6], which is a result of convergence of the NetCDF and HDF[7] data models. The ESSE Data Model may be thought of as a set of interrelated data arrays (*variables*) and metadata elements (*attributes*). Some of the arrays characterize grid axes along specific *dimensions*. ESSE data model elements can readily be serialized and deserialized in binary and XML forms to be transferred over the network. The OGSA-DAI engine does not perform serialization while passing ESSE data model elements between activities.

It is common that a complicated perform document comprising both data processing and fuzzy search uses several `GetData` activities that request similar data, e.g. the data for the same environmental parameter for a slightly different time interval. The ESSE Data Resource widens several similar `GetData` requests into a single database query. This kind of optimization improves the system performance significantly. For example, when requesting data on two slightly shifted time intervals the performance gain is about 50 per cent.

13.3.4 ESSE data processor

In many cases data coming directly from an ESSE data source need to be modified before use in a fuzzy search or in the visualization engine. For example, the NCEP Reanalysis database has East and North wind speeds (U- and V-wind environmental parameters), but the absolute wind speed and wind direction may be of interest instead. Another frequent example is a temperature unit conversion between Kelvin, Celsius or Fahrenheit. Applications may need diurnal temperature variance, minimum and maximum temperatures over some time period, and so on. Thus we need a special component that can be programmed to calculate a new time series from several inputs.

[6]http://www.unidata.ucar.edu/software/netcdf/CDM/
[7]http://www.hdfgroup.org/

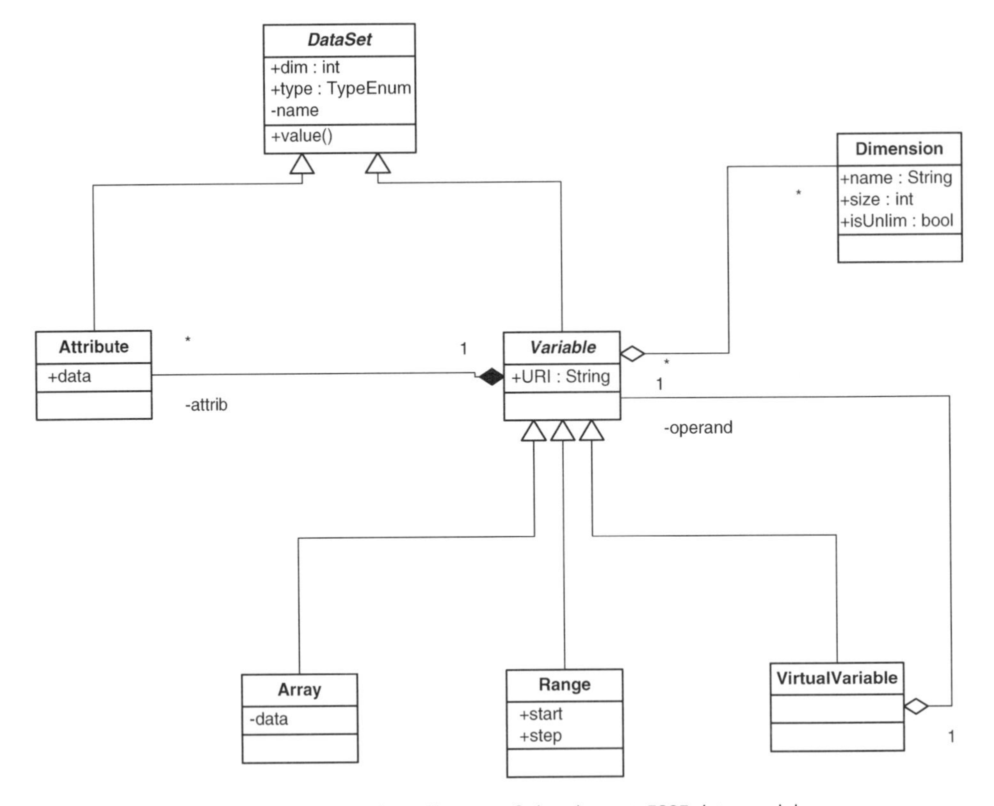

Figure 13.9 UML class diagram of the abstract ESSE data model.

We call this component the ESSE data processor. For the moment our data processor is able to combine several inputs along the time axis. A more general component with spatial functions and more complicated data reorganizations will be considered for future work.

The ESSE data processor receives one or several time series as input and a set of instructions (the program) to process the data. The set of instructions can include arithmetic expressions, elementary functions, moving average, seasonal variations and time shifts. The result is another time series that may contain fewer points than the original data.

The data processor is exposed to OGSA-DAI users as the `dataProcess` activity with multiple inputs and a single output. The data processing expression is represented as an XML-formatted calculation tree in the body of the perform document, which can be parsed into a DOM tree by any standard XML parser.

Suppose we have to calculate $\sqrt{U_c^2 + V_c^2}$, where U_c and V_c are the daily averages of the U- and V-wind components respectively. Then the definition for the `dataProcess` activity in a perform document may look like this:

```
<dataProcess name="dataProcess">
  <attribute name="seasonal" value="false"/>
  <function name="sqrt">
    <function name="sum">
```

```
        <function name="power">
          <function name="average">
            <function name="createInterval">
              <attribute name="length" value="86400"/>
              <attribute name="boundaryTime" value=""/>
              <input name="x"/>
            </function>
          </function>
          <attribute name="additionalArgument" value="2"/>
        </function>
        <function name="power">
          <function name="average">
            <function name="createInterval">
              <attribute name="length" value="86400"/>
              <attribute name="boundaryTime" value=""/>
              <input name="y"/>
            </function>
          </function>
          <attribute name="additionalArgument" value="2"/>
        </function>
      </function>
    </function>
    <outputFormat name="xml"/>
    <output name="dataProcessingOutput"/>
  </dataProcess>
```

Here x and y are the names of the outputs of `getData`, which return from the ESSE data source raw data for the U- and V-wind components. However, if the U- and V-wind components were measured at the same time, then the daily average wind speed can be calculated by the formula $(1/N)\sum_{i=1}^{N}\sqrt{U_i^2 + V_i^2}$, where N is the number of observations per day. In this case we have to move the `createInterval` and `average` elements to the top of the calculation tree.

13.4 Applications

In February 2007 the major results from the new Fourth Assessment Report of the Intergovernmental Panel on Climate Change were released. The Policymaker's Summary of the Fourth Assessment Report's Physical Science Basis (Intergovernmental Panel on Climate Change, 2007) is a collection of statements regarding climate change observations and forecasts, which are formulated in the following fuzzy terms upon agreement of the board of scientific experts.

- *Very likely.* Less than a 10 per cent chance of being wrong.
- *Likely.* Less than a 33 per cent chance of being incorrect.

- *Very high confidence.* At least a 90 per cent chance of being correct.
- *High confidence.* At least an 80 per cent chance of being correct.

To quote some of the main statements in the summary:

Warming of the climate system is unequivocal, as is now evident from observations of increases in global average air and ocean temperatures, widespread melting of snow and ice, and rising global mean sea level.

At continental, regional, and ocean basin scales, numerous long-term changes in climate have been observed. These include changes in Arctic temperatures and ice, widespread changes in precipitation amounts, ocean salinity, wind patterns and aspects of extreme weather including droughts, heavy precipitation, heat waves and the intensity of tropical cyclones (including hurricanes and typhoons).

The combination of the ESSE weather reanalysis data sources with the ESSE fuzzy search engine and data processor becomes a powerful tool for quantitative formalization and rapid validation of such fuzzy statements. Below we present in detail two use cases for estimation of global temperature trends and distribution of temperature extremes in space and time for the last three decades, 1975–2006.

13.4.1 Global air temperature trends

Analysis of air temperature trends requires the interplay of the weather reanalysis data source and the data processor, but it does not involve the ESSE fuzzy search engine. The computation plan is applied in a loop to select data from the global grid with five degrees step in latitude and longitude. It takes less than six minutes to calculate average annual temperatures for the last 30 years on our computational cluster at the Geophysical Centre of the Russian Academy of Science. The output of the data processor in XML format was imported directly into an MS Excel spreadsheet, where we have calculated the appropriate linear regression trends.

The global map indicating the direction and slope of the linear temperature trend in degrees per decade at each grid point (Figure 13.10) was prepared with the mapping toolbox provided with Matlab[8]. The grid point marker was selected based on the sign and the slope of the trend: empty circle (getting colder) and filled circle (getting warmer). The slope is shown by the marker size; a large mark is used for grid-points with a decadal temperature trend larger than 1 degree Celsius, a medium mark is used for decadal trends above 0.5 and a small mark indicates a sign of trends above 0.1 degree.

It is interesting to compare the annual temperature trends reported by the Intergovernmental Panel on Climate Change Third Assessment Report (Intergovernmental Panel on Climate Change, 2001), p. 27), with the map built using the ESSE data processor. The difference in the trends reported by the Intergovernmental Panel on Climate Change in 2001 and by our data grid computations comes from the fact that the Intergovernmental Panel on Climate Change has used a different data source, the Global Temperature Record from the *British Climate Research Unit* (*CRU*) (Brohan *et al.*, 2006), which has well known long-term deviations from the NCEP/NCAR and ERA-40 reanalyses (Simmons *et al.*, 2004). The CRU temperature record has a monthly time step, which means it cannot be directly used in the search for extreme weather events that we present in the next section.

[8] http://www.mathworks.com/

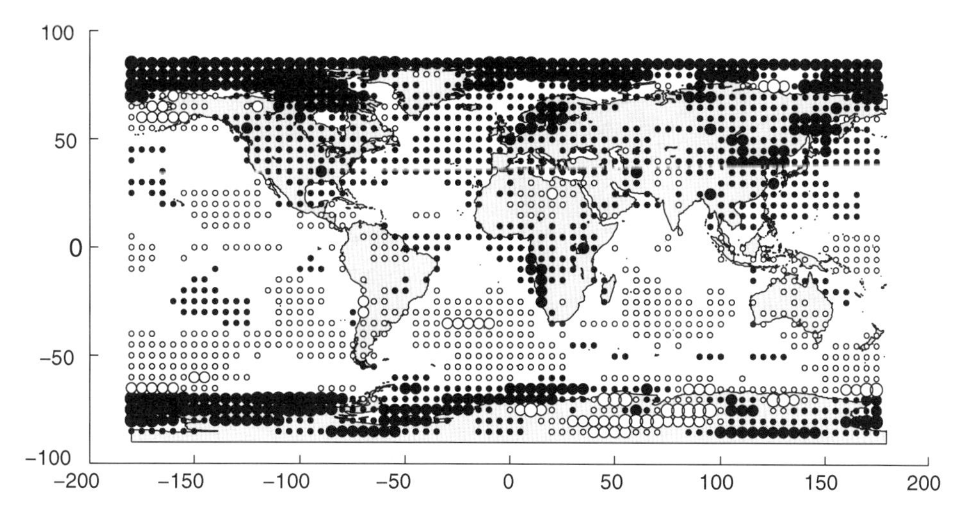

Figure 13.10 Annual temperature trends, 1976–2005, from the NCEP reanalysis data. Empty circles for negative and filled circles for positive temperature trends

13.4.2 Statistics of extreme weather events

Here we illustrate how to apply the ESSE fuzzy search engine and data processor to validate the Fourth Assessment Report (Intergovernmental Panel on Climate Change, 2007) statement that 'it is very likely that hot extremes, heat waves, and heavy precipitation events will continue to become more frequent'.

First, we have to formalize the extreme weather term 'heat wave'. For this demonstration, we used a very simple fuzzy expression, 'very high temperature'. Then we selected from the NCEP/NCAR reanalysis database air temperature time series for the time period 1976–2005 with a six hours time step on the regular grid with five degrees step in latitude and longitude. A fuzzy engine search for 'very high temperatures' over this data set returned 10 extremely hot weather events at each of the five degree grid points, each event having a time of occurrence and fuzzy score. Finally we used the data processor to sort the event occurrence times into three decadal bins: 1976–1985, 1986–1995 and 1996–2005. It took about 1.5 hours to run the search in a loop for each grid point over the five degree grid on our computer cluster.

To visualize the global decadal distribution of 'heat waves', we implemented the following algorithm within the Matlab mapping toolbox. The grid point marker was selected based on the decadal bin with the maximum number of weather extremes: empty circle (getting cold) for the decade 1976–1985 and filled circle (getting hot) for the decade 1996–2005. The size of the grid point depends on the 'certainty' of the weather extreme distribution: the maximum size grid point is plotted for the locations where 9–10 out of 10 events have occurred in the same decade and the minimum size grid point is used to indicate 4–6 event occurrences (Figure 13.11).

13.4.3 Atmospheric fronts

The following example involves a search for a E–W atmospheric front near Moscow described by three parameters, 'air pressure', 'E–W wind speed' (U-wind) and 'N–S wind speed'

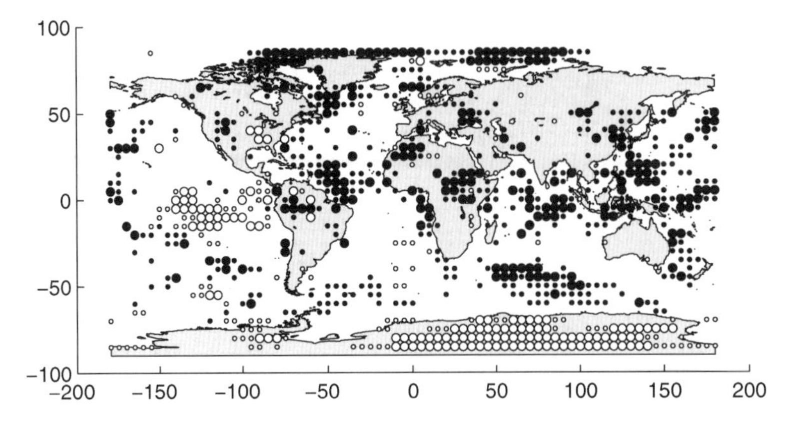

Figure 13.11 Distribution of 10 hottest days by decade. Empty circles for hot decade 1976–1985 and filled circles for 1996–2005

(V-wind) with subsequent fuzzy states

(a) (Small pressure) and (Large V-wind-speed)
(b) (Large pressure) and (Small U-wind speed) and (Small V-wind-speed).

Here we demonstrate an interactive data mining session using our *Investigation of Distributed Environmental Archives System* (*IDEAS*)[9] portal. A typical interactive data mining use case involves three actors, namely a user, the IDEAS portal Web application and the ESSE fuzzy search engine Web service, and consists of the following steps.

(a) The user logs in to the IDEAS portal and receives a list of the currently available (distributed) data sources. For each data source the list has abridged metadata such as name, short description, spatial and temporal coverage, parameter list and link to full metadata description.

(b) The user selects the environmental data source based on the short description or by metadata keyword search (e.g. NCEP/NCAR Reanalysis). The portal stores the data source selection on the server side in the persistent 'data basket' and presents a GIS map with the spatial coverage of the data source.

(c) The user selects a set of 'probes' (representing spatial locations of interest, e.g. Moscow) for the the searching event. IDEAS stores the selected set of 'probes' and presents a list of all the environmental parameters available from the selected data source and a fuzzy constraints editor on the parameters values which represent the event (Figure 13.12).

(d) The user selects some of the environmental parameters and sets the fuzzy constraints on them for the searching event (e.g. low pressure, high V-wind speed).

[9]http://teos1.wdcb.ru/esse

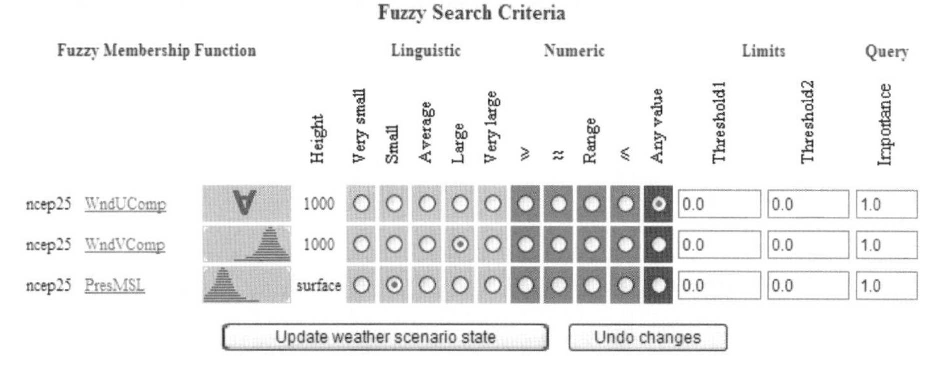

Figure 13.12 ESSE fuzzy state editor

(e) Multiple subsequent environment states can be grouped to form the actual environmental scenario. For example, we need to define the two different states mentioned above. Adding and removing fuzzy states is done via a Web-form shown in Figure 13.13.

(f) The IDEAS portal stores the searching environment states and sends them in a request to the ESSE fuzzy search Web service in the XML format.

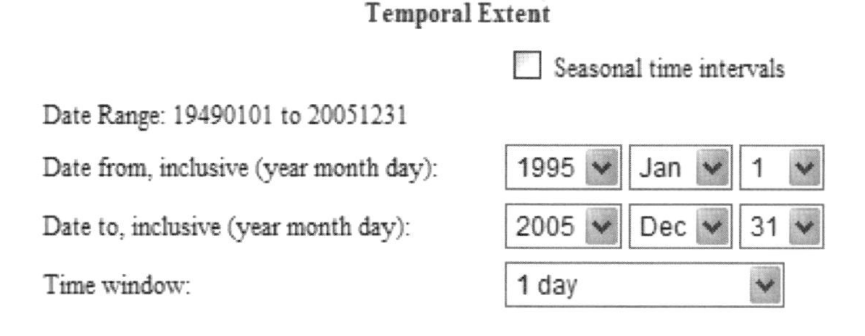

Weather Scenario Fuzzy States

Database	Parameter	Height	Units
ncep25	WndUComp	1000	m/s
ncep25	WndVComp	1000	m/s
ncep25	PresMSL	surface	Pa

S_0 S_1

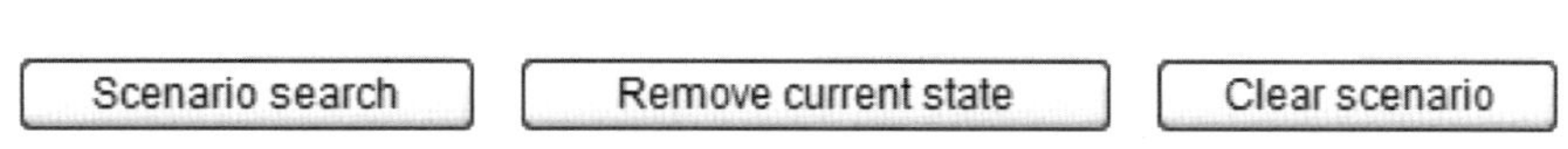

Figure 13.13 Environmental scenario state form

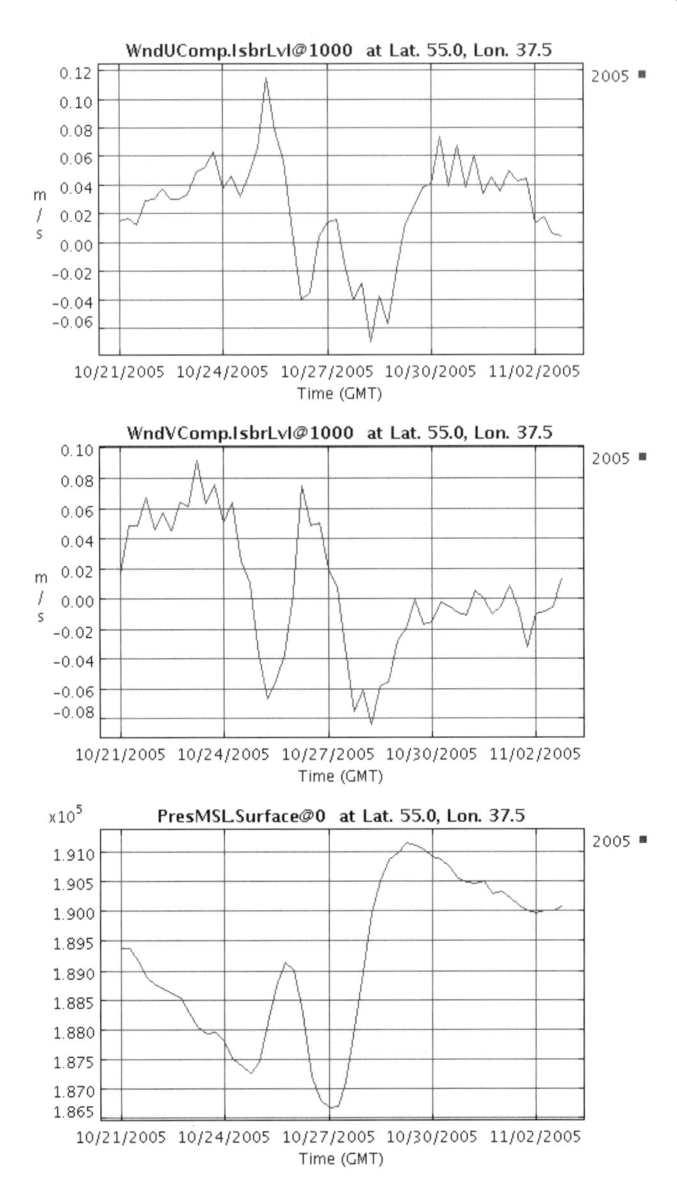

Figure 13.14 Air temperature (top) and pressure (bottom) in Moscow for the event of 13–14 June 2002

(g) The fuzzy search Web service collects data from the data source(s) for the selected parameters and time interval, performs the data mining and returns to the IDEAS portal a ranked list of candidate events with links to the event visualization and data export pages.

(h) The user visualizes interesting events and requests the event-related subset of the data for download from the data source in the preferred scientific format (XML,

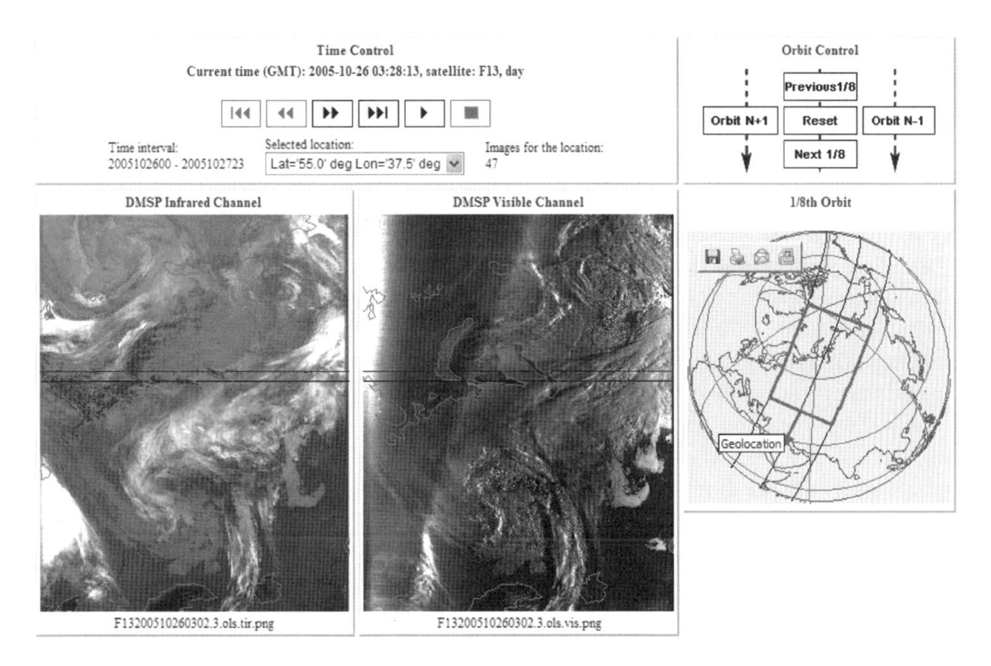

Figure 13.15 DMSP weather images

NetCDF, CSV table). Currently there are three visualization types available: time series (Figure 13.14), animated volume rendering using Vis5D[10] viewer and DMSP satellite images (Figure 13.15).

In Figure 13.14, rectangles show the two environmental states within the discovered event on 26 October 2005: (low pressure–high wind) and (high pressure–low wind).

In the associated day-time *Defense Meteorological Satellite Program* (*DMSP*) satellite images [11] (Figure 13.15), we can clearly see the E–W front passing above Moscow. There are two types of DMSP image available: visible and infrared images. The image are shown together with the actual satellite orbit and visibility sector. Satellite images along with other kinds of visualization may serve as an additional means of verification for the environmental event data mining results.

The IDEAS portal use case variations include data mining for the environmental events, described by parameters from multiple data sources, which may consist of multiple states (e.g. extremely cold day followed by magnetic storm).

13.5 Conclusions

The changing data environment in today's environmental science domain requires the use of new tools, techniques and data sources to meet the key environmental challenges. We

[10]http://www.ssec.wisc.edu/ billh/vis5d.html

[11]http://www.ngdc.noaa.gov/dmsp/

see the grid and grid related tools as being a key part of facing the challenge provided by greater data volumes, distribution and diversity. Specifically, virtualization of data resources as we described using the OGSA-DAI container provides the best hope of integrating data across domains and services. Currently, researchers must face data in different models for visualization, computation and transport and a significant effort is expended just to access the bits. Grids provide a standard way of accessing the data via common service and of integrating services to meet research needs. We have also shown that using standard grid tools provides for the reuse of common data utilities, the possibility for the development of common tools like data mining and the reuse of key components such as those included in the Globus toolkit. Another key feature of grid versus alternative access based on Web services is that a grid is fundamentally asynchronous. In environmental science, archives are now approaching petabyte size and the operations that reveal new discoveries are in general non-trivial computationally. Using grid services and grid tools gets around the problem of the asynchronous nature of such queries. One area where the grid shows great promise as we have demonstrated in our application is in work-flow management. Directing data to available grid services gives the researcher the opportunity to make use of community tools and techniques that have never been available before. Our data mining tool set (described above), for example, is a good use of a grid data service. As the community grows more familiar with the topics, we expect the emergence of more tools and techniques exposed as grid services, such as quality control, forecasting and analysis suites.

Our application of the grid stands as an example of the coming systems which will make use of the grid in support of environmental scientific research. This paradigm should open an unprecedented age of discovery and access, which will have significant real-world impact in the environment.

References

Akima, H. (1978), 'A method of bivariate interpolation and smooth surface fitting for irregularly distributed data points', *ACM Transactions on Mathematical Software* **4**, 148–159.

Antonioletti, M., Atkinson, M. P., Baxter, R., Borley, A., Hong, N. P. C., Collins, B., Hardman, N., Hume, A. C., Knox, A., Jackson, M., Krause, A., Laws, S., Magowan, J., Paton, N. W., Pearson, D., Sugden, T., Watson, P. and Westhead, M. (2005), 'The design and implementation of grid database services in OGSA-DAI', *Concurrency and Computation: Practice and Experience* **17**(2–4), 357–376.

Brohan, P., Kennedy, J., Harris, I., Tett, S. and Jones, P. (2006), 'Uncertainty estimates in regional and global observed temperature changes: a new dataset from 1850', *Journal of Geophysical Research* **111**, D12106.

Deutsch, P. (1996), GZIP file format specification version 4.3, Request for Comments 1952, IETF.

Intergovernmental Panel on Climate Change (2001), *Climate Change 2001: Synthesis Report*, Cambridge University Press. Contribution of Working Groups I, II, and III to the Third Assessment Report to the Third Assessment Report of the Intergovernmental Panel on Climate Change.

Intergovernmental Panel on Climate Change (2007), Summary for policy makers, *in* S. Solomon, D. Q. M. Manning, Z. Chen, M. Marquis, K. Averyt, M. Tignor and H. Miller, eds, 'Climate Change 2007: The Physical Science Basis. Contribution of Working Group I to the Fourth Assessment Report of the Intergovernmental Panel on Climate Change', Cambridge University Press.

Jang, J.-S. R., Sun, C.-T. and Mizutani, E. (1997), *Neuro-Fuzzy and Soft Computing*, Prentice-Hall.

Kalnay, E., Kanamitsu, M., Kistler, R., Collins, W., Deaven, D., Gandin, L., Iredell, M., Saha, S., White, G., Woollen, J., Zhu, Y., Chelliah, M., Ebisuzaki, W., Higgins, W., Janowiak, J., Mo, K., Ropelewski, C., Wang, J., Leetmaa, A., Reynolds, R., Jenne, R. and Joseph, D. (1996), 'The NCEP/NCAR 40-year reanalysis project', *Bulletin of American Meteorological Society* **77**(3), 437–471.

Karasavvas, K., Antonioletti, M., Atkinson, M. P., Hong, N. P. C., Sugden, T., Hume, A. C., Jackson, M., Krause, A. and Palansuriya, C. (2004), Introduction to OGSA-DAI services, *in* P. Herrero, M. S. Pérez and V. Robles, eds, 'SAG', Vol. 3458 of *Lecture Notes in Computer Science*, Springer, pp. 1–12.

Mesinger, F., Dimego, G., Kalnay, E., Mitchell, K., Shafran, P., Ebisuzaki, W., Jovi, D., Wollen, J., Rogers, E., Berbery, E., Ek, M., Fan, Y., Grumbine, R., Higgins, W., Li, H., Lin, Y., Manikin, G., Parrish, D., and Shi, W. (2006), 'North American Regional Reanalysis', *Bulletin of American Meteorological Society* **87**, 343–360.

Simmons, A., Jones, P., da Costa Bechtold, V., Beljaars, A., Kallberg, P., Saarinen, S., Uppala, S., Viterbo, P. and Wedi, N. (2004), ERA40 project report series, *in* 'Comparison of trends and variability in CRU, ERA-40 and NCEP/NCAR analyses of monthly-mean surface air temperature', number 18, European Centre for Medium Range Weather Forecasts.

Uppala, S., Kallberg, P., Simmons, A., Andrae, U., da Costa Bechtold, V., Fiorino, M., Gibson, J., Haseler, J., Hernandez, A., Kelly, G., Li, X., Onogi, K., Saarinen, S., Sokka, N., Allan, R., Andersson, E., Arpe, K., Balmaseda, M., Beljaars, A., van de Berg, L., Bidlot, J., Bormann, N., Caires, S., Chevallier, F., Dethof, A., Dragosavac, M., Fisher, M., Fuentes, M., Hagemann, S., Holm, E., Hoskins, B., Isaksen, L., Janssen, P., Jenne, R., McNally, A., Mahfouf, J.-F., Morcrette, J.-J., Rayner, N., Saunders, R., Simon, P., Sterl, A., Trenberth, K., Untch, A., Vasiljevic, D., Viterbo, P. and Woollen, J. (2005), 'The ERA-40 re-analysis', *Quarterly Journal of the Royal Meteorological Society* **131**, 2961–3012.

Yager, R. (1979), 'On the measure of fuzziness and negation, part I: membership in the unit interval', *International Journal of Man–Machine Studies* **5**, 221–229.

Yager, R. (1980), 'On a general class of fuzzy connectives', *Fuzzy Sets and Systems* **4**, 235–242.

14 Data pre-processing using OGSA-DAI

Martin Swain and **Neil P. Chue Hong**

ABSTRACT

Data pre-processing and data management are challenging and essential features of grid-enabled data mining applications. Powerful data management grid middleware is available, but it has yet to be fully exploited by grid-based data mining systems. Here, the Open Grid Services Architecture – Data Access and Integration (OGSA-DAI) software is explored as a uniform framework for providing data services to support the data mining process. It is shown how the OGSA-DAI activity framework already provides powerful functionality to support data mining, and that this can be readily extended to provide new operations for specific data mining applications. This functionality is demonstrated by two application scenarios, which use complex workflows to access, integrate and preprocess distributed data sets. Finally, OGSA-DAI is compared with other available data handling solutions, and future issues in the field are discussed.

14.1 Introduction

Data management grid middleware has evolved to address the issues of distributed, heterogeneous data collections held across dynamic virtual organizations (Finkelstein, Gryce and Lewis-Bowen, 2004; Laure, Stockinger and Stockinger, 2005). Many of the principles and technologies developed can also be used to assist in the manipulation of data for the purposes of data mining. The *Open Grid Services Architecture – Data Access and Integration* (*OGSA-DAI*) software (Antonioletti *et al.*, 2005), provides a uniform framework to access and integrate data on grid systems. It is based around the concept of distinct data operations being implemented as pluggable *activities*, which can be chained together to create data-centric workflows. In addition, an important feature of OGSA-DAI is its application programming interface, which allows developers to quickly implement and deploy data pre-processing operations that are specific to grid-based data mining applications. This chapter introduces OGSA-DAI and demonstrates how it can be used to support the pre-processing and management of data for grid-based data mining.

Data Mining Techniques in Grid Computing Environments Edited by Werner Dubitzky

14.2 Data pre-processing for grid-enabled data mining

From the perspective of data mining, data management functionality is required to process data until it is in a suitable format for the data mining algorithms. This can be seen as a pipeline of operations that deliver, filter, transform and integrate the data – a data-centric workflow. In grid environments such a pipeline must be able to handle the inherently distributed and heterogeneous nature of grid data resources.

The data management and manipulation tasks required to support the data mining process include the following.

(a) *Data cleaning.* May involve dealing with missing values, incomplete data, erroneous values, massive data and specialized cleaning methods for particular application domains. Various metrics and methodologies exist to determine data quality and include, for example, calculating statistical summaries of the data to provide indications of the data quality and other important characteristics.

(b) *Data integration.* Is concerned, particularly in a grid environment, with the federation or combination of distributed data sets. Distributed data sets can be heterogeneous, and may belong to more than organization.

(c) *Data selection.* Involves the filtering of data to display only a subset of the entire data set.

(d) *Data transformation.* Involves a variety of tasks related to redefining data. For example, continuous numeric values may be transformed into discrete character values, or mathematical functions may be applied to combine or otherwise manipulate data values. Data sets may be divided into training and testing data sets for cross-validation, and formatted according to the requirements of different data mining algorithms.

(e) *Data mining.* Generates data models, such as association rules, clusters or decision trees, which must be stored, annotated and exported for future use.

In this chapter relational databases are mainly used in the examples to demonstrate OGSA-DAI's functionality, because relational databases are widely used in data mining applications. A typical application scenario is as follows: assuming that high-quality data reside in a number of tables in distributed databases with different schemas, data integration can be performed by initially selecting relevant data from the tables, transferring them and combining them into a new, single table. The data are then filtered, for example by using SQL to select subsets of the data, and then further processed by aggregating them and performing value mappings. At this stage the data may still be stored in a database, but, in most grid systems, they must be converted into one or more files before being scheduled for analysis with a data mining algorithm.

In this scenario, throughout the data preparation stages the data are kept in the relational database representation. This allows the powerful functionality of grid data management middleware such as OGSA-DAI to support the process of data selection, transformation and manipulation.

14.3 Using OGSA-DAI to support data mining applications

While early grid applications focused principally on the storage, replication and movement of file-based data, many applications now require full integration of database technologies and other structured forms of data through grid middleware. In particular, many data mining applications benefit from a consistent interface to data resources, and OGSA-DAI provides a framework which allows these interfaces to be brought together and extended easily.

OGSA-DAI provides services to interact with data resources and perform data manipulation operations via a document-oriented interface. This process is described by Antonioletti *et al.* (2005). In summary, clients interact with remote data resources by specifying a set of *activities* in a document called the *request document*. This document is sent to the data service, the entire set of operations is executed by the service and a single *response document* is returned to the client. The response document informs clients of their request document's execution status, and it can also be used to return data to a client once all activities have been executed.

Using OGSA-DAI, complex workflows can be created that perform powerful transformations on data in preparation for data mining. Workflows can also be used to bring together many disparate data resources. Standard data resources supported by OGSA-DAI include relational databases, XML databases and some types of file. Additional data sources can be utilized by OGSA-DAI through the development of data resource implementation classes to manage access to and use of the data resource. This process is described in more detail in the OGSA-DAI documentation (OGSA-DAI Team, 2007). Data resources need to support three distinct functions: getting the resource ID, destroying the resource and creating resource accessors.

A resource accessor enables activities to interact with the data resource in the presence of a specified security context and ensures that the resource is accessed correctly given the nature of the client. As a result of this, a cohesive security policy can be applied across multiple activities and resources. The resource accessor thus provides the first level of mediation between different formats, presenting activities with a well defined way of accessing the information held in different types of data source.

In the following sections OGSA-DAI's activity framework is explained in detail, and it is shown how it can be used to access, integrate, preprocess and transform data for grid-enabled data mining applications.

14.3.1 OGSA-DAI's activity framework

The activity model in OGSA-DAI allows new functionality to be created, deployed and used in OGSA-DAI services in a similar way to plug-ins in many other pieces of software. This provides two important benefits to developers: a well defined, composable, workflow unit, and an easy way to extend functionality without needing to understand the underlying infrastructure.

Before we describe activities in detail, let us consider a simple example, as shown in Figure 14.1. This demonstrates many of the different aspects of OGSA-DAI workflows.

- A workflow consists of a number of units called activities. An activity is an individual unit of work in the workflow and performs a well defined data related task such as running an SQL query, performing a data transformation or delivering data.
- Activities are connected. The outputs of activities connect to the inputs of other activities.
- Data flow from activities to other activities and this is in one direction only.
- Different activities may output data in different formats and may expect their input data in different formats. Transformation activities can transform data between these formats.

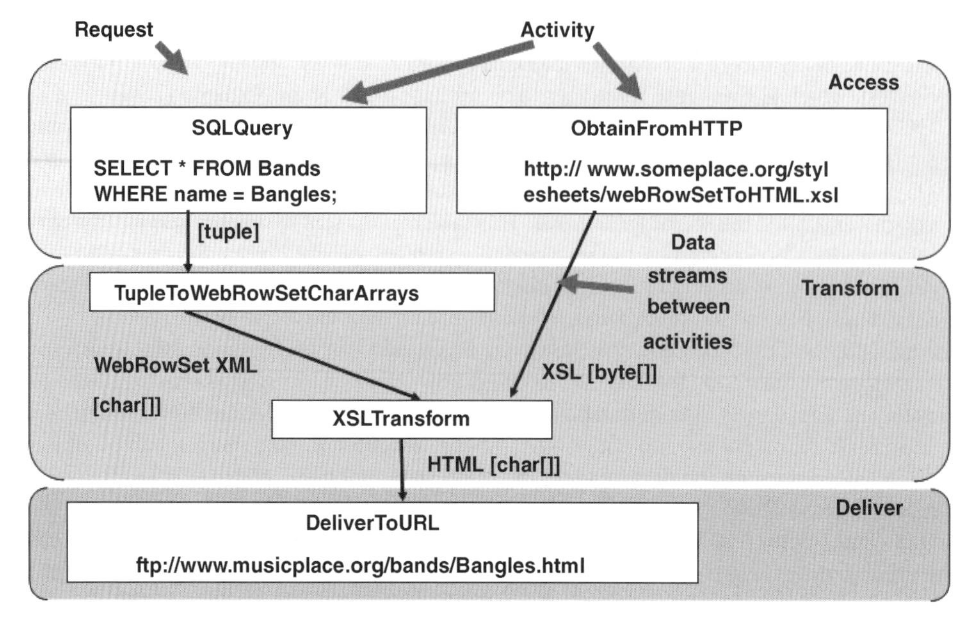

Figure 14.1 A simple example of an OGSA-DAI workflow in a request document

- Workflows, and therefore OGSA-DAI itself, do not just carry out data access, but also data updates, transformations and delivery.

It should also be clear from this example that OGSA-DAI's utility for data mining comes from its ability to act as a general data processing framework, optimized for integration tasks. It is easy to add different types of data source; indeed, it is possible to work with different data models. For instance, although the example above concentrates on tuple-processing based on the relational model, projects such as SEEGEO have used OGSA-DAI to carry out feature-based processing and integration (which is prevalent in the GIS community) with data from the MOSES project (Townend *et al.*, 2007).

The anatomy of an activity An activity can have zero or more named inputs and zero or more named outputs. Blocks of data flow from an activity's output into another activity's input. Activity inputs may be optional or required. OGSA-DAI does not attempt to validate the connection between activities to ensure that the objects written to the output are compatible with those expected at the connected input. Errors like this will be detected by the activities themselves when the workflow is executed.

In Figure 14.2, the SQLQuery activity is passed an SQL expression as an input and the results of the query are passed to the TupleToCSV activity, which transforms this data into *comma-separated value* (*CSV*) format, showing a simple use of OGSA-DAI to convert data retrieved from a database.

The standard activity library A key feature of OGSA-DAI which is of benefit to data mining is the availability of a consistent, complete set of basic data access and integration activities. These activities must be well defined to allow composability.

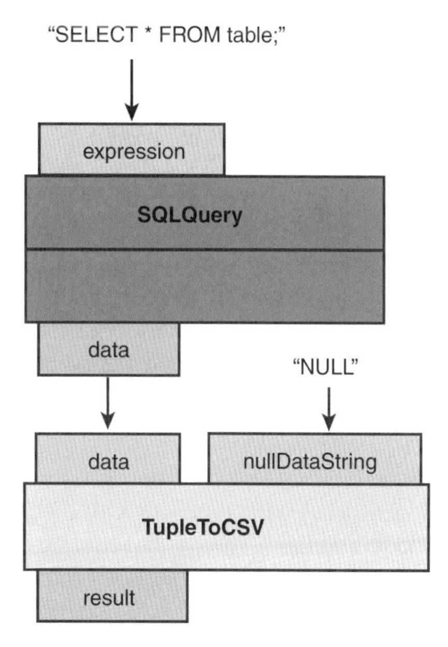

Figure 14.2 Activity inputs and outputs

A generic activity representation is used where inputs and parameters have been homogenized. A special kind of input, called an input literal, is used to represent parameters in requests. This can be seen in the previous example, Figure 14.2, where the SQL expression `SELECT * FROM table;` is an example of an input literal. Since we have a common passing mechanism for all inputs, clients are free to choose whether an input value is specified in the request or is obtained from the output of another activity. This second method can be clearly seen in Figure 14.1; whilst it might be expected that the XSL transform document provided to the XSLTransform activity is a 'real' document, it is just the output of a previous activity in the workflow and could have just as easily been generated as the result of a query on a database.

The set of recommended inputs and outputs defined in the activity library set includes Java's basic types (`Object`, `String`, `Integer`, `Long`, `Double`, `Number`, `Boolean`) as well as some other types such as `char[]`, `byte[]`, `Clob`, `Blob`. We have also defined two types – `tuple` and `MetadataWrapper` – that are of particular importance.

(a) A `tuple` is used to represent a row of relational data. Each element of a tuple represents a column. Relational resources are commonly used in many OGSA-DAI scenarios and using a preferred (and recommended) format has many benefits, one of which is that we need only one set of transformation activities, e.g. for projection, sorting etc. Data conversion activities are used to convert tuples to CSV or WebRowSet if required.

(b) A `MetadataWrapper` can wrap any object that the users want to be treated as metadata by activities. It is left to individual activities to handle metadata blocks as they see fit. OGSA-DAI is agnostic of metadata representations and allows users to use any kind of domain-specific metadata they might need.

The set of all designed and implemented activities currently supported (Karasavvas, Atkinson and Hume, 2007) includes different means to transform data between formats. The main difference is between activities that transform between two specific formats, and activities that apply a transformation based on inputs provided to the activity.

An example of the former is the TupleToCSV activity, which takes a list of tuples and converts it into a CSV-formatted output suitable for many clients. Additional activities of this type can be easily implemented provided that the input and output formats, along with the transformation, can be specified in Java code.

The more generic transforms may be used to support a particular type of transformation framework, or to allow more dynamic transformations to be carried out during a more complex workflow. An example is the XSLTransform activity, which can use an XSL transform document, which may have been produced as the result of an earlier stage in the workflow.

The activity API for developers The design of the activity inputs and outputs yields a simple API for activity developers. The interaction with XML, which is used by the document-orientated interface, are kept to a minimum. There is no requirement to parse XML fragments to elicit activity input and output names and configuration settings. The activity framework provides this information in a standard set of Java objects. Activity developers also do not need to write XML Schema documents describing their activities.

Activities can implement 'extension' interfaces to specify what they require from the activity framework, to be able to execute, for example data resources, spawning of other activities, production of monitoring events or definition of a security context. This provides a method for activity developers to specify to the activity framework their initialization and configuration requirements and removes some of the overhead of developing this themselves.

Processing OGSA-DAI lists The OGSA-DAI activity framework and engine, which execute the datacentric workflows from clients, support a number of constructs that benefit the developers of data mining applications.

An important construct, particularly when processing large numbers of streaming data, is the list. Lists are implemented in OGSA-DAI by using blocks that are inserted into a stream of data to mark the beginning and end of a group of data blocks. A list groups related data together as one unit. For an example, consider the SQLQuery activity again, in Figure 14.2. SQLQuery can dynamically take any number of SQL expressions as input. The output for each expression is a number of rows represented as tuples. Without a way to group the output tuples we would have no way to differentiate between the results of queries on table1 and table2. We can use lists (illustrated with the square brackets in Figure 14.3) to group the output of each execution of the query. This results in two lists, one with the output of the first query and one with that of the second query.

Lists play an important role in the streaming of data through activities. They allow activities to be described in terms of different granularities instead of different input and output types. For example, an activity could take a simple string per execution or a list of strings per execution and then iterate over the strings internally. In our standard activity library we have defined activities in their lowest sensible granularity to enhance composability. To homogenize granularity, OGSA-DAI provides utility activities that allow lists to be flattened, ListRemover, or flattened and then reconstructed, ListProcessingWrapper. An example of using lists for data filtering is given in Section 14.3.2.

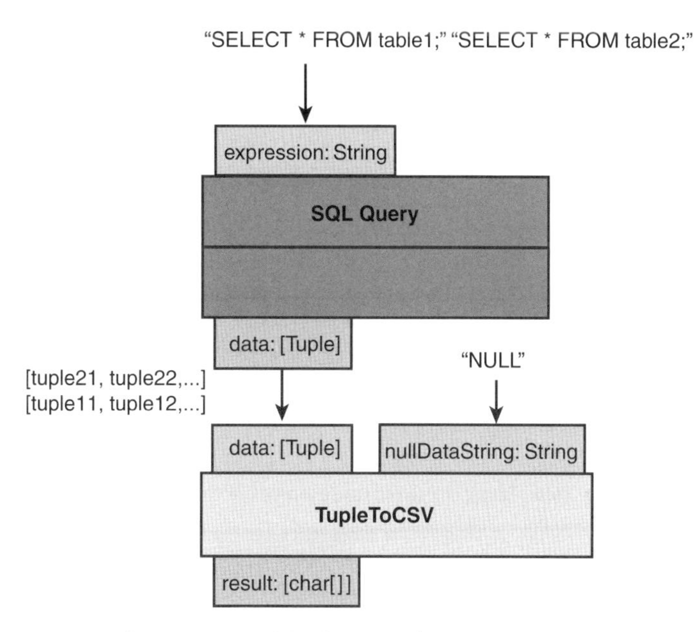

Figure 14.3 Activity inputs and outputs using lists

14.3.2 OGSA-DAI workflows for data management and pre-processing

The overall approach of the OGSA-DAI activity framework allows data in different formats and different models to be handled efficiently.

Data federation

OGSA-DAI is able to federate a number of data resources through the same data service. This can be configured by the service administrator or, alternatively, by clients. This flexibility means that clients can set up and use their own federations according to their application-specific requirements. In this scenario the client is aware of the individual data resources that make up the group.

Workflows for multiple data resources

OGSA-DAI also supports workflows that target multiple data resources, which means that different activities within the same workflow can be targeted at different data resources, as shown in Figure 14.4. This allows OGSA-DAI to execute workflows where a transfer of data from data resource R1 to data resource R2 does not include transferring data external to the data service – a prime requirement for efficient data mining. It also enables data resources to be configured and maintained independently, but effectively treated as a single collection for the purposes of the OGSA-DAI workflow, making it easier to develop and share workflows.

Filtering and integrating data using activities

Given the power of lists in OGSA-DAI and the ability of all activities to iteratively process lists, it is possible to use activities to filter data as part of the workflow. Again, like the transformation

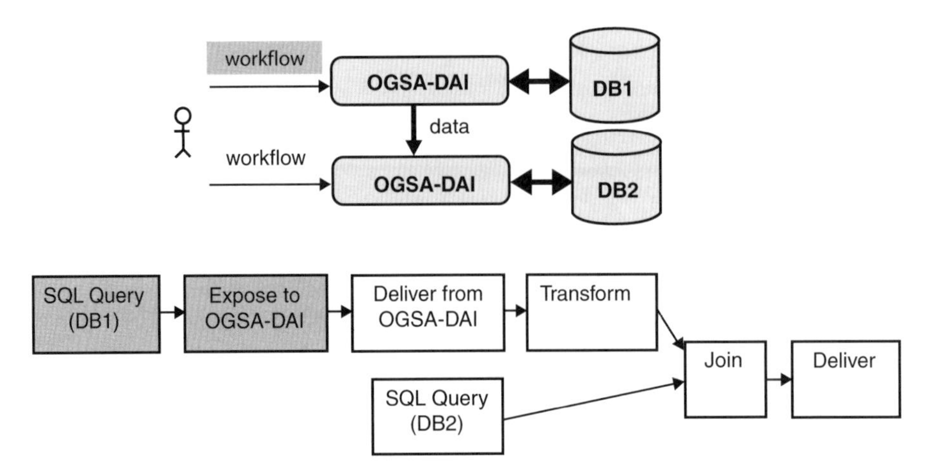

Figure 14.4 Targeting two data resources with one workflow

activities described earlier, filtering can either be done by specific activities or by activities that take input to define the filter. Combinations can also be used to achieve different effects: the supplied `TupleSplit`, `TupleToWebRowSetCharArrays` and `XSLTransform` activities would allow simple filtering to remove first names from a complex name field.

Simple data integration and reformatting can be achieved easily using iteration, for instance, a 'Stitch' activity which takes two inputs and concatenates then into a single output, e.g. First Names and Surname. In its simplest form, without using lists, it will receive one entry on each input at each iteration. This means one First Name is stitched to one Surname. In the presence of lists, the behaviour is more complex. We can group sets of First Names in a list, which counts as a single object for each iteration. Thus we can stitch multiple First Names to each Surname. Depending on the output format chosen, this could be to represent multiple first names for a single person, or multiple family members sharing the same surname. It is also possible to use a `ControlledRepeat` activity before another activity to repeat a particular value until the other data stream finishes.

Asynchronous delivery

OGSA-DAI implements the concept of asynchronous delivery through data sink and data source resources. A request can be submitted that creates a data source or a data sink resource. Clients can then interact with these resources using dedicated data source or data sink services, which only support interaction with data source and sink resources, rather than via a single overloaded data service. Furthermore operations can be used to query the state of these resources, via their dedicated services, and to terminate them if required. This leads to a more consistent handling of data sources and sinks – they are simply resources, and like data resources or resource groups they can be created and utilized via activities or dedicated services.

Asynchronous delivery is also the recommended way of interacting with data resources as it means that the workflow does not block waiting for the response from the OGSA-DAI service. This allows multiple workflows to be targeted at a service. In many cases, workflows associated with pre-processing for data mining will end with the results being inserted into a database again. In these cases, asynchronous delivery methods are not used to deliver data, but may be used to provide status responses.

14.4 Data pre-processing scenarios in data mining applications

In this section it is shown how OGSA-DAI version WSRF-2.2 has been used in data mining applications from the DataMiningGrid project (Stankovski *et al.*, 2008). In the DataMiningGrid OGSA-DAI clients and services were used in a number of scenarios. One approach was to calculate a summary of the data sets accessed, consisting of global properties and statistical values. This data summary was subsequently used by different activities for data preprocessing.

14.4.1 Calculating a data summary

The data summary was developed using the OGSA-DAI APIs as follows, and is shown in Figure 14.5.

Initially input to the OGSA-DAI client is defined by the user, who selects a data service and defines an SQL query to be executed against one of the service's data resources. Then:

(a) The client constructs a request document with this query, sends it to the data service for execution and retrieves the metadata associated with the query.

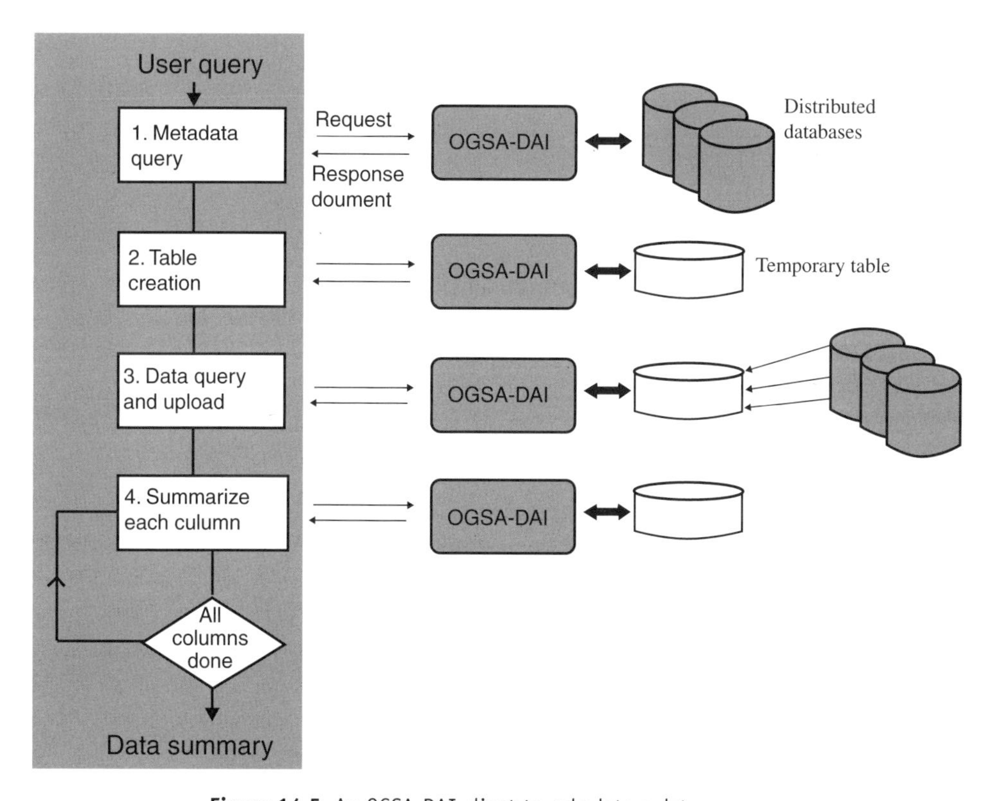

Figure 14.5 An OGSA-DAI client to calculate a data summary

(b) The client processes the metadata returned by the query and a request is made by the client to a data service, which creates a temporary table according to the query metadata.

(c) Having created this temporary table, a request is made to retrieve the query data and load them into the temporary table.

(d) It is now possible to calculate statistical properties of this data set. This is performed by the client issuing a series of requests, consisting of SQL queries, to summarize the values in each column. The values calculated may depend on the application. Typically, for numeric data the average, standard deviation, maximum, minimum and variance are calculated, while for character data this involves finding all the distinct values in the column. This data set summary is stored at the client and used in subsequent requests or routines as required.

It is also possible to develop specific activities to perform this data summary instead of using SQL. However, SQL was used for a number of reasons: first, it was much quicker to develop the client using SQL queries rather than developing new activities; and second, because database systems have been optimized to efficiently execute queries and may contain sophisticated technology for data processing and knowledge discovery, it was desirable to push processing onto the database server rather than overloading the OGSA-DAI server.

In the following sections we show how this data summary was used in two data mining applications.

14.4.2 Discovering association rules in protein unfolding simulations

This application was developed in cooperation with a storage repository and data warehouse for protein folding and unfolding simulations called P-found (Silva *et al.*, 2006). The aim of the P-found system is to enable researchers to perform data mining on protein folding data sets that reside in scientific laboratories, distributed over a number of different geographic areas. This is facilitated by performing common analyses on the simulation data, and storing the results of these analyses in data warehouses located at each laboratory. The data set comprising a single simulation can be hundreds of megabytes, or even a few gigabytes in size, while archived collections of simulation data can easily reach hundreds of gigabytes. A typical data mining exercise that might be performed on this repository is described by Azevedo *et al.* (2005).

OGSA-DAI clients may be used to integrate distributed P-found databases: the clients first access data from a number of repositories, store the data in a temporary table and then calculate a data summary. In the DataMiningGrid these data were streamed from the temporary table, and a number of data filtering operations and data transformations were applied before the data were delivered to a file. The process is shown in Figure 14.6

(a) Data from multiple databases have been uploaded to a temporary table, a data summary calculated and now all data are selected from the table.

(b) Data are processed column by column by using the OGSA-DAI projection activities to extract each column from the data stream. Transformation activities are applied to the data. These activities were developed by the DataMiningGrid project, and they include; (*a*) windowing or aggregation; here the window size is set to 50, so that every 50 values are averaged to give one value; (*b*) normalization; the normalization range was defined to be between 0 and 1; and (*c*) discretization; the continuous numeric values are mapped onto a set of ordinal values. The columns of preprocessed data are loaded into another temporary table.

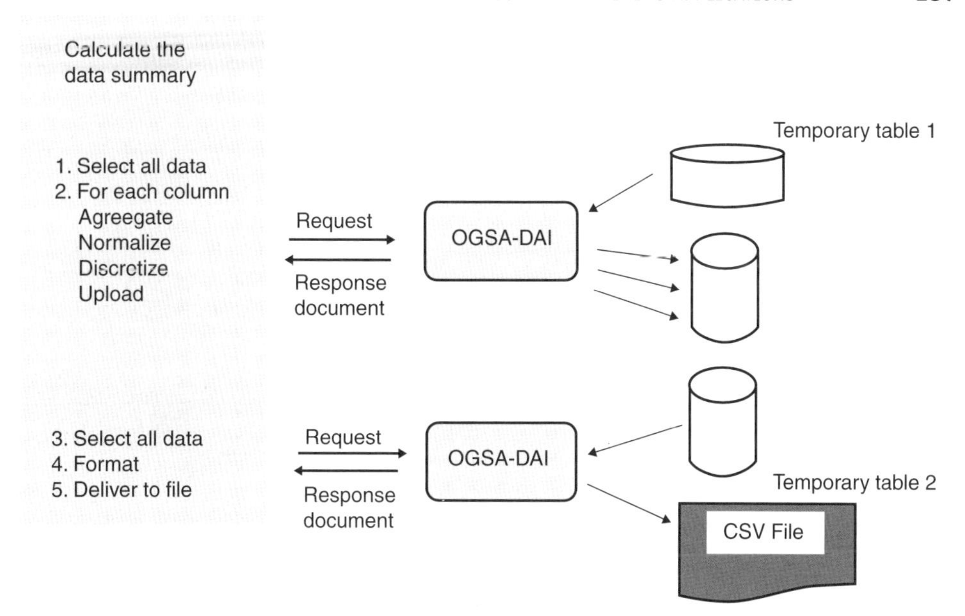

Figure 14.6 An OGSA-DAI client to perform data transformation based on values generated by a data summary

(c) The data are accessed from the table.

(d) At this stage additional functionality can be applied to convert this transformed data set into different data formats. In this case we used the format of the CAREN association algorithm (Azevedo *et al.*, 2005), which is similar to a CSV format except with the variables separated by spaces instead of commas.

(e) Finally, the data are written to a file that is then scheduled for processing.

14.4.3 Mining distributed medical databases

This application classifies data accessed from distributed medical databases for a nutrition-related study performed on Slovenian children. OGSA-DAI clients were developed to query nine databases located in different geographic regions of Slovenia, and to generate training and testing data sets for cross-validation. These data sets were formatted according to the requirements of the Weka Toolkit's decision tree algorithms (Witten and Frank, 2005).

Figure 14.7 shows the data processing workflow: data are extracted from distributed relational databases and integrated into a single database table, and then divided up and formatted for cross-validation.

To perform the data integration a temporary database table was created on one relational data resource, each database was queried in turn and the retrieved data were streamed to the resource hosting the temporary database, where it is loaded into the table. In this application, each database has the same schema and is implemented using MySQL.

Cross-validation was performed using an activity developed by the DataMiningGrid, based on an existing OGSA-DAI activity that returns a fraction of a data set, called `randomSample`. The cross-validation activity streams through the tuples, and according to a user-defined probability determines whether to output the tuples to one of two streams leading to delivery in

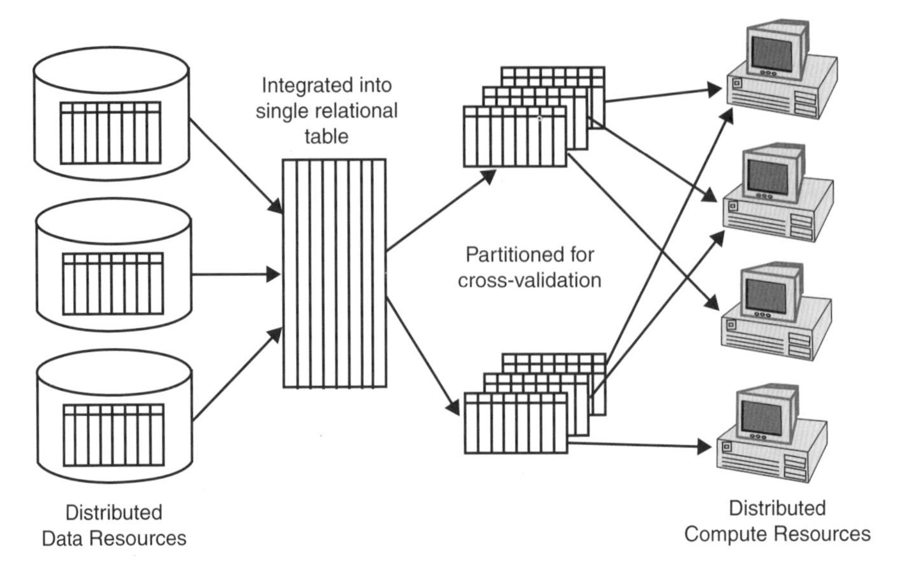

Figure 14.7 Data mining distributed data

files containing either the training or the test data sets. However, before writing the data to the files they are formatted according to Weka's *attribute-relation file format* (*ARFF*) structure (Witten and Frank, 2005). This formatting utilizes the data summary calculated earlier; in particular, one property determined by the data summary is the complete set of values that character attributes may have.

For example, the ARFF 'nominal' attribute, for the well known Iris machine learning data set, would be formatted using the complete set of names for the Iris class, and the header line in the ARFF file would be @ATTRIBUTE class {Iris setosa, Iris versicolor, Iris virginica}.

While the OGSA-DAI client creates the header of the ARFF file using properties from the data summary, the body of the ARFF file is formatted using OGSA-DAI's CSV activity.

14.5 State-of-the-art solutions for grid data management

OGSA-DAI is not the only solution currently available for data in the grid domain. *Storage Resource Broker* (*SRB*) (Rajasekar, Wan and Moore, 2002), developed by the San Diego Supercomputer Center, provides access to collections of data primarily using attributes or logical names rather than using the data's physical names or locations. SRB is primarily file oriented, although it can also work with various other data object types. OGSA-DAI on the other hand takes a database-oriented approach to its access mechanisms. Although SRB is currently not open source, whilst OGSA-DAI is, a new product called iRods (Rajasekar *et al.*, 2006) is being developed, which will be available under an open source license. iRods characterizes the management policies for a community's data collections in terms of rules and state information, which are interpreted to decide how the system is to respond to various requests and conditions.

WebSphere Information Integrator (*WSII*), a commercial product from IBM, provides data searching capabilities spanning organizational boundaries and a means for federating and replicating data, as well as allowing for data transformations and data event publishing to take place. A detailed comparison between previous versions of OGSA-DAI and WSII was performed by Lee *et al.* (2005).

Mobius (Hastings *et al.*, 2005), developed at Ohio State University, provides a set of tools and services to facilitate the management and sharing of data and metadata in a grid environment. To expose XML data in Mobius, the data must be described using an XML Schema, which is then shared via their Global Model Exchange. Data can then be accessed by querying the Schema using, for example, XPath. OGSA-DAI, in contrast, does not require an XML Schema to be created for each piece of data; rather, it directly exposes this information (data and metadata/schema) and relies on the resource's intrinsic querying mechanisms to query its data.

These three projects all provide mechanisms to share data across organizational boundaries, however they complement the functionality provided by OGSA-DAI.

14.6 Discussion

OGSA-DAI and related grid data management tools are good for managing high quality, well structured data that can be processed in a flexible way, often with fairly standardised approaches. OGSA-DAI is a solution for data integration problems, which are an inherent aspect of data mining in grid environments where data resources are characterized by heterogeneity, and distributed over many independent administrative domains.

While OGSA-DAI is a solution for data management tasks that need to be closely integrated with data pre-processing and transformation, previous versions have not been specifically designed for data processing. As grid-enabled data processing tasks grow in complexity and intensity, it is necessary to use purpose-built data processing middleware and scheduling systems. Nonetheless, for many complex data processing tasks, it is essential to store intermediate data products along the data processing pipeline, and it is clear that sophisticated data management software is, and will continue to be, an attractive partner for more compute intensive applications.

With the release of OGSA-DAI 3.0, many of the challenges of incorporating data processing into data-centric workflows have been addressed. The revised workflow model allows data processing services to be orchestrated, and workflows to be presented as services, with their own interfaces, in their own right. OGSA-DAI 3.0 also addresses some of the basic challenges of data integration and allows more types of data source to be presented through a uniform interface. Nevertheless, the challenges of dealing with data evolution and complex query execution remain important for grid data solutions to tackle.

14.7 Open Issues

Grid data management solutions are now able to provide uniform interfaces to interact with increasingly diverse data resources, including the database systems often used by data mining applications. As more applications are deployed on grids, users increasingly demand more sophisticated interfaces, fueling the development of flexible architectures than can be easily extended to meet these diverse needs.

Perhaps the most pressing need is concerned with data exchange. Grid systems need to support a wide range of data formats for different algorithms and storage devices. Frequent data conversions can be a significant burden, not only on the system's efficient operation, but also on developers' time. Data management systems and data processing algorithms need to converge on a standard data representation that would streamline the data pre-processing pipeline.

In the longer term, grid data solutions will integrate data virtualization technologies with data pre-processing operations. Data virtualization technologies are concerned with creating a logical structure for data resources, with the eventual aim of totally insulating users from details concerning the physical locations of data. Currently, data virtualization solutions provide users with transparent views of file collections. However, they should ultimately integrate distributed heterogeneous resources, including databases, files, and other storage devices. Integrating data pre-processing with these data management solutions will enable virtual data warehouses to be realized, allowing arbitrary data sets to be instantiated as and when required.

14.8 Conclusions

Effective data management functionality is an essential component of grid-enabled data management systems. OGSA-DAI is a sophisticated, extendable solution for developing workflows concerned with all aspects of data pre-processing for data mining. One of the most appealing aspects of OGSA-DAI is its extendable activity framework: this offers a relatively easy way for developers to provide grid-enabled data pre-processing operations that are an integral component of state-of-the-art data management functionality.

Research on grid data mining systems usually focuses much more on the actual data mining rather than the data handling middleware, which is, arguably, the foundation of the whole process. By demonstrating the use of standard data management middleware in typical grid-enabled data mining scenarios, it is hoped that this balance will be corrected, and that future systems will be more willing to fully utilize the powerful solutions that are already available.

References

Antonioletti, M., Atkinson, M., Baxter, R., Borley, A., Hong, N. P. C., Collins, B., Hardman, N., Hume, A. C., Knox, A., Jackson, M., Krause, A., Laws, S., Magowan, J., Paton, N. W., Pearson, D., Sugden, T., Watson, P. and Westhead, M. (2005), 'The design and implementation of grid database services in OGSA-DAI', *Concurrency and Computation: Practice and Experience* **17** (2–4), 357–376.

Azevedo, P., Silva, C., Rodrigues, J., Loureiro-Ferreira, N. and Brito, R. (2005), Detection of hydrophobic clusters in molecular dynamics protein unfolding simulations using association rules, *in* J. L. Oliveira, V. Maojo, F. Martín-Sánchez and A. S. Pereira, eds, 'ISBMDA', Vol. 3745 of *Lecture Notes in Computer Science*, Springer, pp. 329–337.

Finkelstein, A., Gryce, C. and Lewis-Bowen, J. (2004), 'Relating requirements and architectures: a study of data grids', *Journal of Grid Computing* **2** (3), 207–222.

Hastings, S., Langella, S., Oster, S., Kurc, T., Pan, T., Catalyurek, U., Janies, D. and Saltz, J. (2005), Grid-based management of biomedical data using an XML-based distributed data management system, *in* 'Proceedings of the 2005 ACM Symposium on Applied Computing (SAC'05)', ACM Press, New York, pp. 105–109.

Karasavvas, K., Atkinson, M. and Hume, A. (2007), 'Redesigned and new activities', OGSA-DAI Design Document Version 1.7, http://www.ogsadai.org.uk

Laure, E., Stockinger, H. and Stockinger, K. (2005), 'Performance engineering in data grids', *Concurrency and Computation: Practice and Experience* **17**(2–4), 171–191.

Lee, A., Magowan, J., Dantressangle, P. and Bannwart, F. (2005), 'Bridging the integration gap, Part 1: Federating grid data', IBM Developer Works, http://www.ibm.com/developerworks/grid/library/gr-feddata/index.html

OGSA-DAI Team (2007), 'OGSA-DAI 3.0 documentation', http://www.ogsadai.org.uk/documentation/.

Rajasekar, A., Wan, M. and Moore, R. (2002), MySRB and SRB – components of a data grid, *in* 'The 11th International Symposium on High Performance Distributed Computing', pp. 301–310.

Rajasekar, A., Wan, M., Moore, R. and Schroeder, W. (2006), A prototype rule-based distributed data management system, *in* 'High Performance Distributed Computing (HPDC) workshop on Next Generation Distributed Data Management'.

Silva, C. G., Ostropytsky, V., Loureiro-Ferreira, N., Berrar, D., Swain, M., Dubitzky, W. and Brito, R. M. M. (2006), P-found: the protein folding and unfolding simulation repository, *in* 'Proceedings of the 2006 IEEE Symposium on Computational Intelligence in Bioinformatics and Computational Biology', pp. 101–108.

Stankovski, V., Swain, M., Kravtsov, V., Niessen, T., Wegener, D., Kindermann, J. and Dubitzky, W. (2008), 'Grid-enabling data mining applications with DataMiningGrid: an architectural perspective', *Future Generation Computer Systems* **24**, 259–279.

Townend, P., Xu, J., Birkin, M., Turner, A. and Wu, B. (2007), Modelling and simulation for e-social science: current progress, *in* 'Proceedings of the 2007 UK e-Science All Hands Meeting'.

Witten, I. and Frank, E. (2005), *Practical Machine Learning Tools and Techniques. 2nd edn*, Morgan Kaufmann.

Index